Advances in Machine Fault Diagnosis

Advances in Machine Fault Diagnosis

Editors

Toomas Vaimann
Grover Zurita Villarroel
Anders Brandt

MDPI • Basel • Beijing • Wuhan • Barcelona • Belgrade • Manchester • Tokyo • Cluj • Tianjin

MDPI

Editors

Toomas Vaimann
Tallinn University of Technology
Estonia

Grover Zurita Villarroel
Universidad Privada Boliviana
Bolivia

Anders Brandt
University of Southern Denmark
Denmark

Editorial Office
MDPI
St. Alban-Anlage 66
4052 Basel, Switzerland

This is a reprint of articles from the Special Issue published online in the open access journal *Applied Sciences* (ISSN 2076-3417) (available at: https://www.mdpi.com/journal/applsci/special_issues/Advances_Machine_Fault_Diagnosis).

For citation purposes, cite each article independently as indicated on the article page online and as indicated below:

LastName, A.A.; LastName, B.B.; LastName, C.C. Article Title. *Journal Name* **Year**, *Volume Number*, Page Range.

ISBN 978-3-0365-5109-8 (Hbk)
ISBN 978-3-0365-5110-4 (PDF)

Cover image courtesy of Toomas Vaimann

Contents

About the Editors

Toomas Vaimann

Dr. Toomas Vaimann received the B.Sc., M.Sc., and Ph.D. degrees in electrical engineering from the Tallinn University of Technology, Estonia, in 2007, 2009, and 2014, respectively. He is currently a Senior Researcher with the Department of Electrical Power Engineering and Mechatronics, Tallinn University of Technology. He has been working in several companies as an electrical engineer. His main research interest includes diagnostics of electrical machines. He is a senior member of IEEE, member of the Estonian Society of Moritz Hermann Jacobi and the Estonian Society for Electrical Power Engineering.

Grover Zurita Villarroel

Prof. Dr. Grover Zurita Villarroel has postdoctoral studies at the University of New South Wales (UNSW), Sydney Australia (2002). His M.Sc. and Ph.D degrees in Mechanical engineering at the Luleå University of Technology, Sweden, in 1993 and 2001, respectively. Dr. Zurita, currently, works as director of the laboratory of Industrial Innovation Technology at the Universidad Privada Boliviana (UPB). He worked at "Universidad Polytechnic Salesiana de Cuenca (Ecuador)", as Prometeo Senior Researcher between 2014 to 2015. He was a visiting Professor at "Chongqing Technology and Business University, (China)", September–October 2015. He was Head of Division of Electromecanica Engineering at UPB", between 2003 and 2014. Researcher at the University of "Luleå University of Technology, Sweden", with research projects with Swedish company, Scania CV. Expert in vibrational analysis and digital signal processing, applied to machine diagnosis.

Anders Brandt

Anders Brandt is a Professor and Head of Department of Mechanical and Production Engineering at Aarhus University in Denmark. Anders is a well-known educator, researcher, and consultant within the fields of applied signal analysis, condition monitoring, and structural health monitoring, with over 30 years of experience from industry and academia. He is the author of "Noise and Vibration Analysis—Signal Analysis and Experimental Procedures", Wiley 2011, and has conducted over 250 short courses on various topics of vibration analysis.

applied sciences

MDPI

Editorial

Advances in Machine Fault Diagnosis

Toomas Vaimann

Department of Electrical Power Engineering and Mechatronics, Tallinn University of Technology, 12616 Tallinn, Estonia; toomas.vaimann@taltech.ee

1. Introduction

The growing need for intelligent machines, the outreach for more efficient use of the machines in industry, and the development of Industry 4.0 and 5.0 ideologies have pushed forward the field of machine fault diagnosis. Research in the field is receiving significant attention in academia and industry due to the importance of identifying underlying causes of machine faults. The overall objective of different machine fault diagnosis methods is to develop an effective diagnosis procedure. Recent methodological advances permit compressive machine fault diagnosis, providing detailed information, essential for the prevention of future machine failures.

Some of the most promising approaches for the continuous advancement of fault detection and diagnosis technologies are advanced digital signal processing, vibration-based condition monitoring, modal and operational mode analysis, neural network analysis, and machine learning. Moreover, there is often a combination and cross-usage of the methods. With the development of different IT solutions, the number and usability of methods is rapidly growing. Even some of the older and known methods, proposed years ago but discarded due to problems with automation or the need for additional computational resources, have found new life and promising results due to advances in measurement technologies and computational power. The opportunities opened by Internet of things (IoT) and cloud computing services are breaking new ground and shifting paradigms in the field.

Artificial intelligence (AI) has become one of the most transformative technological revolutions since, e.g., the invention of the steam or electric engines. Robustness, precision automated (online) learning, and the capacity to handle complex data are some of the AI attributes that hold significant potential for machine fault diagnosis. In hand with IoT and cloud computing, the emerging AI-based diagnostic methods are proving themselves to be powerful tools for the future.

The main objective of this Special Issue (SI) was to gather state-of-the-art research contributing recent advances in machine fault diagnosis and, hopefully, to outline future research directions in the field.

2. Review of Issue Contents

This SI compiles 11 papers from authors and research groups active in the field of machine fault diagnosis. As the topic itself is interdisciplinary, the papers presented in this SI have different viewpoints. Some look at the electrical side of machines, while others concentrate on the mechanical issues. However, most of the papers cover the integral part of measurement technology and related signal processing, essential for successful prognostic and diagnostic procedures. One of the papers is a comprehensive review paper, while the other ten focus on research aspects of machine fault diagnosis.

The only review paper of the SI [1] provides an overview of the trends and challenges in intelligent condition monitoring of electrical machines using machine learning. As the world is moving toward Industry 4.0 standards, the problems of limited computational power and available memory are decreasing day by day. A significant amount of data with

Citation: Vaimann, T. Advances in Machine Fault Diagnosis. *Appl. Sci.* **2021**, *11*, 7348. https://doi.org/10.3390/app11167348

Received: 7 July 2021
Accepted: 3 August 2021
Published: 10 August 2021

a variety of faulty conditions of electrical machines working under different environments can be handled remotely using cloud computation. Moreover, the mathematical models of electrical machines can be utilized for the training of AI algorithms. Nevertheless, the collection of big data is a challenging task for the industry and laboratory because of limited resources. Some promising machine learning-based diagnostic techniques are presented in terms of their attributes.

The authors of [2] analyzed the frequency interaction of the turbine block in a stand with respect to the magnitude of the error in measuring the turbine's power. They proposed an algorithm for constructing a dynamic analysis during the formation of a wave field of a stand for testing turbines. The research algorithm involved the use of theoretical solutions of nonlinear wave processes using linear oscillations, refined by experiments. The diagnostic model can determine the technical condition of the stand's elements and determine the causes of the discrepancies between the calculated and measured turbine power values. To clarify the stiffness coefficients between the stand's elements, a modal analysis was used to obtain the range of their changes depending on the external dynamic load, which made it possible to assess the impact of changes in the frequency interaction conditions on the turbine power measurement for different test modes. The conditions for amplifying the amplitude of oscillations at their eigenfrequencies were obtained, and the value of the possible deviation of the expected power value at its measurement for specific modes of the turbine was calculated.

Condition monitoring of induction motors using transient modeling and recovery of a nonstationary fault signature was analyzed in [3]. The authors presented the modeling and the broken rotor bar fault diagnostics using time–frequency analysis of the motor current under an extended startup transient time. The transient current-based nonstationary signal was retrieved and investigated for its time–frequency response to segregate the rotor faults and spatial harmonics. For studying the effect of reduced voltage on various parameters and the theoretical definition of the fault patterns, a winding function analysis-based model is presented. Moreover, an algorithm to improve the spectrum legibility was proposed. It was shown that, through efficient utilization of the attenuation filter and consideration of the area containing the maximum power spectral density, the diagnostic algorithm gave promising results.

Machine learning advances and its utilization in machine fault diagnosis were outlined in [4], where the authors dealt with transfer learning-based fault diagnosis under data deficiency. In fault diagnosis studies, data deficiency, meaning that the fault data for the training are scarce, is often encountered, which may greatly deteriorate the performance of the fault diagnosis. To solve this issue, the transfer learning approach was employed to exploit the neural network trained in another (source) domain where enough fault data were available in order to improve the neural network performance of the real (target) domain. While there have been similar attempts of transfer learning in the literature to solve the imbalance issue, they were related to the sample imbalance between the source and target domain, whereas this study considered the imbalance between the normal and fault data. To illustrate this, normal and fault datasets were acquired from the linear motion guide, in which the data at high and low speeds represented the real operation (target) and maintenance inspection (source), respectively. The effect of data deficiency was studied by reducing the number of fault data in the target domain, and comparing the performance of transfer learning, thereby exploiting the knowledge of the source domain and the ordinary machine learning approach without it. By examining the accuracy of the fault diagnosis as a function of the imbalance ratio, it was found that the lower bound and interquartile range of the accuracy were improved greatly by employing the transfer learning approach.

Feature extraction for bearing fault detection using wavelet packet energy and fast kurtogram analysis was presented in [5]. In this paper, an integrated method for fault detection of bearings using wavelet packet energy and a fast kurtogram was proposed. The method consisted of three stages. Firstly, several commonly used wavelet functions were compared to select the appropriate wavelet function for the application of wavelet packet energy. Then, the analyzed signal was decomposed using wavelet packet energy, and the energy of each decomposed signal was calculated and selected for signal reconstruction. Secondly, the reconstructed signal was analyzed by fast kurtogram to select the best central frequency and bandwidth for the band-pass filter. Lastly, the filtered signal was processed using the squared envelope frequency spectrum and compared with the theoretical fault characteristic frequency for fault feature extraction. The procedure and performance of the proposed approach were illustrated and estimated using simulation analysis, proving that the proposed method can effectively extract the weak transients. Moreover, the analysis results of gearbox bearing and rolling bearing cases showed that the proposed method can provide more accurate fault features compared with the individual fast kurtogram method.

Fault diagnosis of industrial robot applications using machine learning was the main topic of [6], which dealt with exploiting generative adversarial networks as an oversampling method for fault diagnosis of an industrial robotic manipulator. Data-driven machine learning techniques play an important role in fault diagnosis, safety, and maintenance of the industrial robotic manipulator. However, these methods require data that are hard to obtain, especially data collected from fault condition states; furthermore, without enough and appropriated (balanced) data, acceptable performance should not be expected. Generative adversarial networks (GANs) are receiving significant interest, especially in the image analysis field due to their outstanding generative capabilities. This paper investigated whether or not GAN can be used as an oversampling tool to compensate for an unbalanced dataset in an industrial manipulator fault diagnosis task. A comprehensive empirical analysis was performed by taking into account six different scenarios for mitigating the unbalanced data, including classical under- and oversampling methods. In all cases, a wavelet packet transform was used for feature generation while a random forest was used for fault classification. Aspects such as loss functions, learning curves, random input distributions, data shuffling, and initial conditions were also considered. A nonparametric statistical test of hypotheses revealed that the GAN-based fault diagnosis outperformed both under- and oversampling classical methods.

Novel models for induction machine diagnosis were described in [7], where the authors presented a hybrid finite element method/analytical model of a three-phase squirrel cage induction motor solved using parallel processing for reducing the simulation time. The growing development of AI techniques can lead toward more reliable diagnostic algorithms. The biggest challenge for AI techniques is that they need a big amount of data under various conditions to train them. These data are difficult to obtain from the industries because they contain low numbers of possible faulty cases, as well as from laboratories, because a limited number of motors can be broken for testing purposes. The only feasible solution is mathematical models, which in the end can become part of advanced diagnostic techniques. The benefits of analytical and numerical models for their speed and accuracy, respectively, can be exploited by making a hybrid model. Moreover, the concept of cloud computing can be utilized to reduce the simulation time of the finite element model. In this paper, a hybrid model solved on multiple processors in a parallel fashion was presented. The results depicted that, by dividing the rotor steps among several processors working in parallel, the simulation time was considerably reduced.

Under different degradation conditions, the complexity of natural oscillation of the piston pump will change, as stated by the authors of [8]. Given the difference in the characteristic values of the vibration signal under different degradation states, this paper presented a degradation state recognition method based on improved complete ensemble empirical mode decomposition with adaptive noise (ICEEMDAN) and extreme gradient boosting (XGBoost) to improve the accuracy of state recognition. Firstly, ICEEMDAN was proposed to alleviate the mode mixing phenomenon, which was used to decompose the vibration signal and obtain the intrinsic mode functions (IMFs) with less noise and more physical meaning; subsequently, the optimal IMF was found using the correlation coefficient method. Then, the time domain, frequency domain, and entropy of the effective IMF were calculated, and the new characteristic values which can represent the degradation state were selected by principal component analysis (PCA) to realize dimension reduction. Lastly, the abovementioned characteristic indices were used as the input of the XGBoost algorithm to achieve the recognition of the degradation state. In this paper, the vibration signals of four different degradation states were generated and analyzed through the piston pump slipper degradation experiment. By comparing the proposed method with different state recognition algorithms, it was found that the method based on ICEEMDAN and XGBoost was accurate and efficient, with an average accuracy rate of more than 99%.

Planetary gearboxes were the research object in [9]. The movement of the gear is a typical complex motion and is often under variable conditions in real environments, which may render vibration signals of planetary gearboxes nonlinear and nonstationary. It is more difficult and complex to achieve fault diagnosis than to effectively fix the axis gearboxes. A fault diagnosis method for planetary gearboxes based on an improved complementary ensemble empirical mode decomposition (ICEEMD)/time–frequency information entropy and variable predictive model-based class discriminate (VPMCD) was proposed in this paper. First, the vibration signal of planetary gearboxes was decomposed into several intrinsic mode functions (IMFs) using the ICEEMD algorithm, which was used to determine the noise component as a function of the magnitude of the entropy, as well as to remove the noise components. Then, the time–frequency information entropy of the intrinsic modal function under the new decomposition was calculated and regarded as the characteristic matrix. Lastly, the fault mode was classified using the VPMCD method. The experimental results demonstrated that the proposed method can not only solve the fault diagnosis of planetary gearboxes under different operation conditions, but also be used for fault diagnosis under variable operation conditions.

The authors of [10] dealt with the condition monitoring of rotating machinery, in an effort to avoid disastrous failure and guarantee safe operation. Vibration-based fault diagnosis showed the most attractive properties for fault diagnosis of rotating machinery (FDRM). Recently, Lempel–Ziv complexity (LZC) was investigated as an effective tool for FDRM. However, LZC only performs single-scale analysis, which is not suitable for extracting the fault features embedded in a vibrational signal over multiple scales. In this paper, a novel complexity analysis algorithm, called hierarchical Lempel–Ziv complexity (HLZC), was developed to extract the fault characteristics of rotating machinery. The proposed HLZC method considers the fault information hidden in both low-frequency and high-frequency components, resulting in a more accurate fault feature extraction. The superiority of the proposed HLZC method in detecting the periodical impulses was validated by using simulated signals. Meanwhile, two experimental signals were utilized to prove the effectiveness of the proposed HLZC method in extracting fault information. Results showed that the proposed HLZC method had the best diagnosing performance compared with LZC and multiscale Lempel–Ziv complexity methods.

Last, but not least, the authors of [11] researched problems arising in the bearings and shafts of machines. Bearing preload significantly affects the running performance of a shaft-bearing system including the fatigue life, wear, and stiffness. Due to the mounting error, the bearing rings are often angularly misaligned. The effects of the combined bearing preload and angular misalignment on the fatigue life of ball bearings and a shaft-bearing system were analyzed in this paper. The contact force distribution of the angular contact ball bearings in the shaft-bearing system was investigated using the system model. The system model included the bearing model, whereas the shaft model was verified by comparing it with the manufacturer's manual and the results from other theoretical models, revealing a difference within 3% between the results from the present bearing model and the manufacturer manual. The global optimization method was used to solve the ball element displacements and friction coefficients, which improved the computation efficiency of the model. The fatigue life results showed that the system life at the optimal angular misalignment was more than 1.5 times that without angular misalignment at the low preload value, and this ratio decreased as the preload value increased.

3. Conclusions

This SI presents new and innovative research addressing some of the many scientific and technological challenges associated with the quickly evolving field of machine fault diagnosis. It must be emphasized that, with the growing efficiency needs and higher degrees of automation, in hand with novel opportunities emerging as a result of IT solutions and AI possibilities, the necessity of machine fault diagnosis is growing and the advances in the field are accelerating. The advanced techniques used for diagnostic and prognostic processes rely on a deeper understanding of automated and numeric methods and models, in order to ensure a safer and more innovative future for all fields of industry. The studies contained within this volume provide a valuable basis for the research and engineering community dealing with machine diagnosis, highlighting open trends for future development.

Funding: This research received no external funding.

Conflicts of Interest: The authors declare no conflict of interest.

References

1. Kudelina, K.; Vaimann, T.; Asad, B.; Rassõlkin, A.; Kallaste, A.; Demidova, G. Trends and Challenges in Intelligent Condition Monitoring of Electrical Machines Using Machine Learning. *Appl. Sci.* **2021**, *11*, 2761. [CrossRef]
2. Petrochenkov, A.; Sal'nikov, A.; Bochkarev, S.; Ilyushin, P. Analysis of the Frequency Interaction of the Turbine Block in the Stand for the Magnitude of the Error in Measuring the Turbine's Power. *Appl. Sci.* **2021**, *11*, 4149. [CrossRef]
3. Asad, B.; Vaimann, T.; Belahcen, A.; Kallaste, A.; Rassõlkin, A.; Ghafarokhi, P.; Kudelina, K. Transient Modeling and Recovery of Non-Stationary Fault Signature for Condition Monitoring of Induction Motors. *Appl. Sci.* **2021**, *11*, 2806. [CrossRef]
4. Cho, S.; Kim, S.; Choi, J. Transfer Learning-Based Fault Diagnosis under Data Deficiency. *Appl. Sci.* **2020**, *10*, 7768. [CrossRef]
5. Zhang, X.; Zhu, J.; Wu, Y.; Zhen, D.; Zhang, M. Feature Extraction for Bearing Fault Detection Using Wavelet Packet Energy and Fast Kurtogram Analysis. *Appl. Sci.* **2020**, *10*, 7715. [CrossRef]
6. Pu, Z.; Cabrera, D.; Sánchez, R.; Cerrada, M.; Li, C.; Valente de Oliveira, J. Exploiting Generative Adversarial Networks as an Oversampling Method for Fault Diagnosis of an Industrial Robotic Manipulator. *Appl. Sci.* **2020**, *10*, 7712. [CrossRef]
7. Asad, B.; Vaimann, T.; Belahcen, A.; Kallaste, A.; Rassõlkin, A.; Iqbal, M. The Cluster Computation-Based Hybrid FEM–Analytical Model of Induction Motor for Fault Diagnostics. *Appl. Sci.* **2020**, *10*, 7572. [CrossRef]
8. Guo, R.; Zhao, Z.; Wang, T.; Liu, G.; Zhao, J.; Gao, D. Degradation State Recognition of Piston Pump Based on ICEEMDAN and XGBoost. *Appl. Sci.* **2020**, *10*, 6593. [CrossRef]
9. Wang, Y.; Fan, Z.; Liu, H.; Gao, X. Planetary Gearbox Fault Diagnosis Based on ICEEMD-Time-Frequency Information Entropy and VPMCD. *Appl. Sci.* **2020**, *10*, 6376. [CrossRef]
10. Han, B.; Wang, S.; Zhu, Q.; Yang, X.; Li, Y. Intelligent Fault Diagnosis of Rotating Machinery Using Hierarchical Lempel-Ziv Complexity. *Appl. Sci.* **2020**, *10*, 4221. [CrossRef]
11. Zhang, Y.; Zhang, M.; Wang, Y.; Xie, L. Fatigue Life Analysis of Ball Bearings and a Shaft System Considering the Combined Bearing Preload and Angular Misalignment. *Appl. Sci.* **2020**, *10*, 2750. [CrossRef]

applied sciences

Review

Trends and Challenges in Intelligent Condition Monitoring of Electrical Machines Using Machine Learning

Karolina Kudelina *, Toomas Vaimann, Bilal Asad, Anton Rassõlkin, Ants Kallaste and Galina Demidova

Department of Electrical Power Engineering and Mechatronics, Tallinn University of Technology, 19086 Tallinn, Estonia; toomas.vaimann@taltech.ee (T.V.); bilal.asad@taltech.ee (B.A.); anton.rassolkin@taltech.ee (A.R.); ants.kallaste@taltech.ee (A.K.); galina.demidova@taltech.ee (G.D.)
* Correspondence: karolina.kudelina@taltech.ee

Abstract: A review of the fault diagnostic techniques based on machine is presented in this paper. As the world is moving towards industry 4.0 standards, the problems of limited computational power and available memory are decreasing day by day. A significant amount of data with a variety of faulty conditions of electrical machines working under different environments can be handled remotely using cloud computation. Moreover, the mathematical models of electrical machines can be utilized for the training of AI algorithms. This is true because the collection of big data is a challenging task for the industry and laboratory because of related limited resources. In this paper, some promising machine learning-based diagnostic techniques are presented in the perspective of their attributes.

Keywords: fault diagnostics; machine learning; artificial intelligence; pattern recognition; neural networks

Citation: Kudelina, K.; Vaimann, T.; Asad, B.; Rassõlkin, A.; Kallaste, A.; Demidova, G. Trends and Challenges in Intelligent Condition Monitoring of Electrical Machines Using Machine Learning. *Appl. Sci.* **2021**, *11*, 2761. https://doi.org/10.3390/app11062761

Academic Editor: Cecilia Surace

Received: 22 February 2021
Accepted: 17 March 2021
Published: 19 March 2021

1. Introduction

Nowadays, electrical machines and drive systems are being used in many applications and play a significant role in industries. As electrical machines are used in different applications, the maintenance question is of great importance. Today, there are plenty of condition monitoring methods to detect failures in electrical equipment. In general, diagnostic techniques can be divided into the following groups [1–5]:

- Noise and vibration monitoring,
- Motor-current signature analysis (MCSA),
- Temperature measurement,
- Electromagnetic field monitoring,
- Chemical analysis,
- Radio-frequency emissions monitoring,
- Acoustic noise measurement,
- Model and artificial intelligence-based techniques.

Generally, stresses that impact electrical machines' operation can be classified into four main categories, also known as TEAM (Thermal, Electric, Ambient, and Mechanical) stresses. Because of these stresses, faults tend to appear in the machine.

Statistically, 36% of all motor failures are related to the stator winding faults [6]. Usually, winding failures develop from a turn-to-turn short circuit [7]. Without timely maintenance, this fault can grow to phase-to-phase or phase-to-ground short circuits [8]. Due to the fact that this inter-turn fault is hardly detectable in the early stages of its development, this topic is mainly challenging in the electrical machine industry [9]. From the point of view of reliability, in this case, one of the most critical points is electrical machines' insulation [10]. Insulation plays a significant role during the design processes [11]. The insulation condition can be defined by chemical, mechanical, or electrical analysis of the insulating materials [12].

Mechanical faults make a significant proportion of overall faults in the form of eccentricity, broken rotor bars, cracked end rings, damaged bearings, etc. [13]. A broken rotor bar is a widespread and frequently occurring fault. In the machine, this fault can be caused by high operating temperature, cracks in the bar, or natural degradation [14]. Some effects can indicate a broken rotor bar: torque oscillations, high radial speed, sparking, rotor asymmetry [15]. This fault is difficult to be exposed at the early stages, but it is equally essential for avoiding negative and catastrophic consequences in production.

Another mechanical fault that occurs in an electrical machine is eccentricity. Eccentricity faults refer to the inconsistent air gap between the rotor and the stator. The air gap eccentricity exceeding 10% is considered a fault [16]. There is a variety of eccentricity types: static eccentricity (SE), dynamic eccentricity (DE), and elliptic eccentricity [17]. Additionally, there are cases when mixed eccentricity occurs in electrical machines. Eccentricity is mainly caused by improper installation, bolts lack or missing, shaft misalignment, or rotor imbalance [18]. Eccentricity faults can cause additional noise and vibration [19]. When the eccentric fault becomes severe, it will cause friction between the stator and the rotor and, as a result, affect the regular operation of the motor.

At the same time, another widely spread mechanical failure is bearing faults. The production of bearings is carried out under stringent requirements for quality. However, the bearing's real lifespan can be significantly decreased due to different ambient and manufacturing factors, such as material fatigue, improper placement, contamination, improper lubrication, and bearing currents [20]. Constant monitoring of the bearing parameters, such as temperature measurement, timely lubricant analysis, noise, and vibration measurement, could significantly decrease the risk of bearing damage [21].

The distribution of all the faults mentioned above depends mainly on the motor's parameters, such as machine type, size, rated voltage, etc. To increase the reliability of the machine, many parameters must be monitored [22]. The main faults and their signatures are shown in Table 1.

Table 1. Signatures of the main faults in electrical machines.

Fault Signatures	Winding Short Circuit [23,24]	Rotor Broken Bar(s) [25]	Eccentricity [26,27]	Bearing Faults [28]
vibration	✔	✔	✔	★
current	★	★	✔	★
temperature	✔	✔	✔	✕
magnetic flux changes	★	★	★	✔
chemical analysis	✔	✕	✕	✕
torque changes	★	★	✔	✕

★—the most preferable parameter for condition monitoring; ✔—parameter can be used for condition monitoring; ✕—parameter cannot be used for condition monitoring.

As shown in Figure 1, three main types of machine maintenance can be expressed to be applied in practice: corrective, preventive, and predictive maintenance [29].

In the case of corrective maintenance, also known as reactive maintenance, all needed repairs are assumed to be done after the failure has already occurred. However, this solution is appropriate only for small and insignificant workstations, where unexpected failure does not lead to economic or catastrophic consequences. Alternatively, many manufacturers assume preventive maintenance to the machine to avoid fatal outcomes. In this case, the electrical equipment needs to be regularly checked by the manufacturers through scheduled and specified inspections.

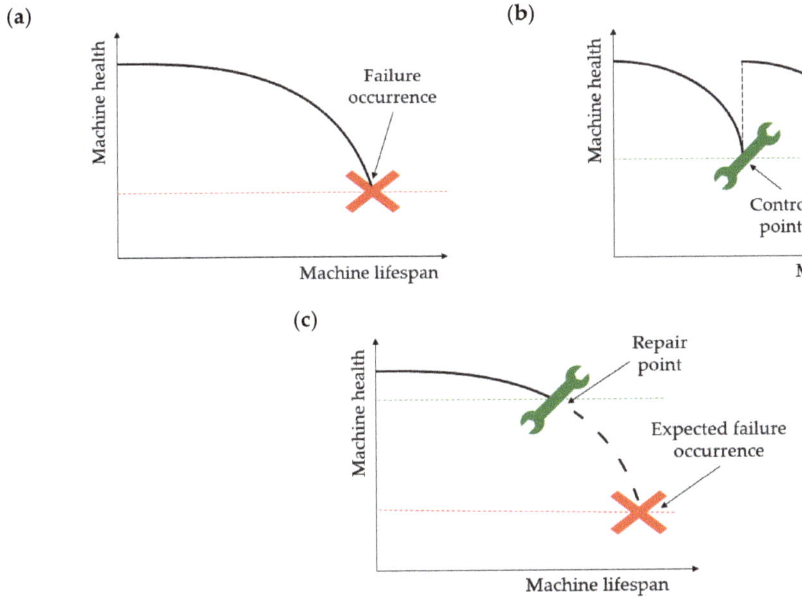

Figure 1. Maintenance types: (**a**) corrective maintenance, (**b**) preventive maintenance, (**c**) predictive maintenance.

Although this solution can prolong machine lifespan, this schedule-based condition monitoring approach provides very little information on the remaining useful lifetime (RUL) of the devices and does not allow for their prognostic and full exploitation [30]. Moreover, because of the scheduled controls in production, it usually means a partial or total shutdown of the manufacturing process, leading to inefficient resource usage and extra operating costs.

To decrease shutdown costs and minimize downtime, manufacturers switch their production over predictive maintenance [31,32]. Condition monitoring is an essential component of predictive maintenance that allows forecasting a further failure based on electrical equipment's working conditions. A schematic illustration of the condition monitoring is shown in Figure 2. As can be seen, condition monitoring consists of several stages. The accuracy of measuring systems largely depends on the sensors used for data acquisition. Signal processing is one of the essential stages in condition monitoring.

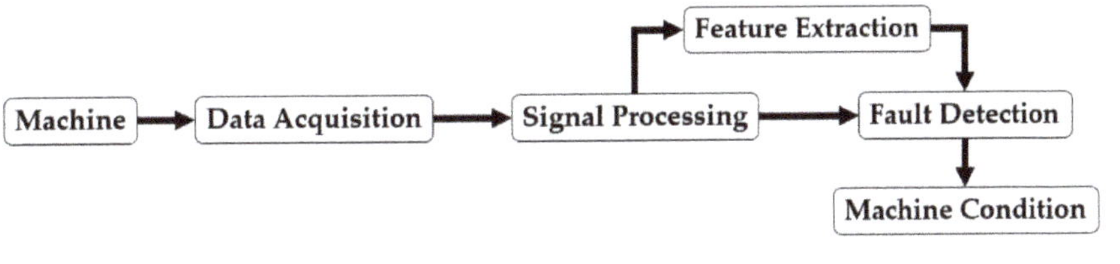

Figure 2. General diagram of decision models.

For feature extraction, to predict and teach the system to detect faults in the future, the system needs a more powerful tool. Moreover, as the data amount is increasing worldwide and computer science is rapidly developing, it is reasonable to remake production under advanced approaches using artificial intelligence (AI). There are widely used thermal imaging in industry to monitor the fault at the early stages of development [33]. In this case, as an example, different variants of machine learning (ML) algorithms can be used for fault detection. These algorithms, as well as their comparison, are described in the following chapters.

2. Diagnostic Possibilities with Machine Learning

Many types of research about intelligent health monitoring refer to machine learning (ML) [34–36]. ML is a study of computer science and artificial intelligence that is not oriented directly to problem solution but rather learning in the process by applying solutions to many similar problems [37]. Typical tasks of ML are classification and regression, learning associations, clustering, and other machine learning tasks, such as reinforcement learning, learning to rank, and structure prediction [38]. ML is closely related to data mining, which can discover new data patterns in large datasets. The main difference is that ML is concentrated on adaptive behavior and operative usage, while data mining focuses on processing extensive amounts of data and discovering unknown patterns. Based on the dataset, so-called training data, ML algorithms can build a model that predicts and makes decisions. There are many types as well as algorithms of ML. These algorithms can be supervised, unsupervised, semi-supervised, and reinforcement [39]. Figure 3 shows the most common methods used in machine learning.

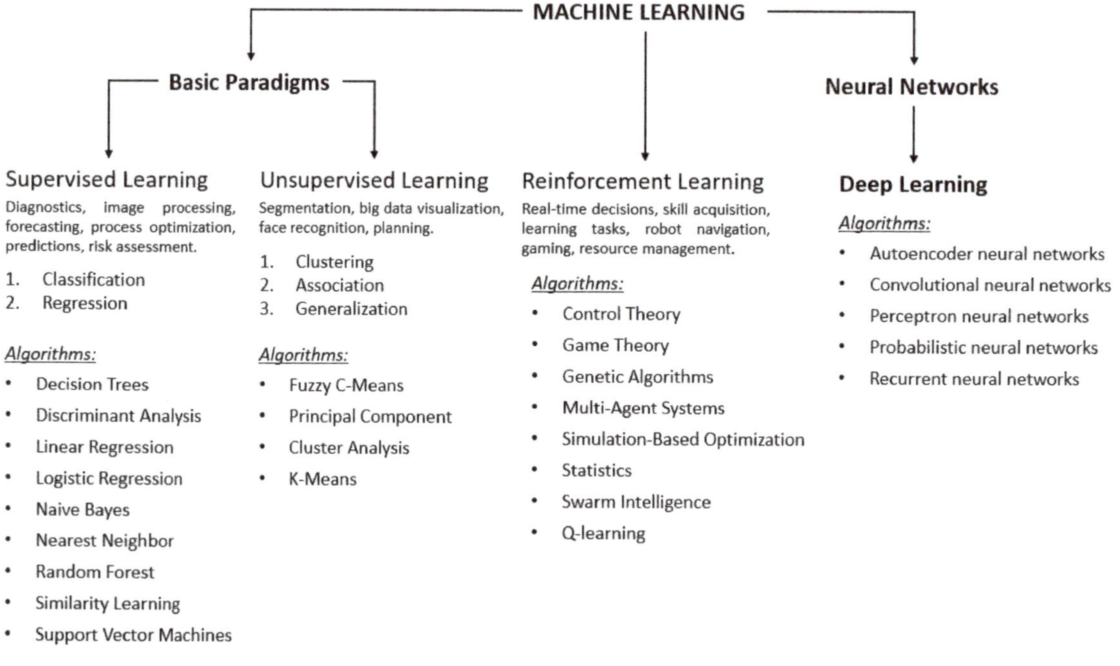

Figure 3. Algorithms of machine learning.

The basic paradigms of ML are supervised and unsupervised algorithms. Supervised ML, also known as "learning with a teacher," is a type of learning from examples, where the training set (situation) and test set (required solution) are set [40,41]. Those training sets are challenging to obtain from industry and laboratories. Because of the limited number of faulty machines working in the industry due to scheduled maintenance (preventive) and in laboratories, a limited number of destructive tests can be performed for training purposes. Moreover, data collection with more than one fault (composite faults) in the same machine is not straightforward in both scenarios. Thanks to the increasing computational power of computers and cloud computation, the mathematical models of electrical machines can train AI algorithms. A comparison of different types of mathematical models of induction motors and their attributes can be found in [42,43].

At the same time, unsupervised ML, also known as "learning without a teacher", is a type of learning where patterns are to be discovered from unknown data [44,45]. In this case, there is only training data, and the aim is to group objects into clusters and/or reduce a large amount of the given data. Sometimes, industrial systems use semi-supervised algorithms in order to get a more precise outcome. In this case, some cases have both training set and test set, while some have only training data.

Differently from basic approaches, reinforcement ML focuses on understanding patterns in repetitive situations and their generalization [46]. The purpose is to minimize errors and increase accuracy; the machine learns to analyze the information before each step. Moreover, the machine aims to get the maximum reward (benefit) from the learning, which is set in advance, such as minimum resource spending, reaching the desired value, minimum analyzing time, etc.

One group of widely used intelligent condition monitoring methods, which can be successfully applied to condition monitoring of many machine parameters, is artificial neural networks (ANNs). ANNs can be supervised, unsupervised, and reinforced. Many studies mistakenly consider NNs as a separate field from machine learning groups. However, NNs and deep learning are related to computer science, artificial intelligence, and machine learning. A diagram of NNs related fields is shown in Figure 4.

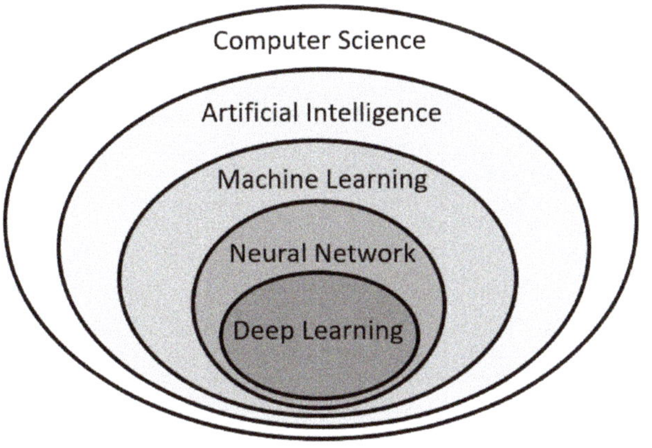

Figure 4. Neural network-related fields.

Machine learning is a powerful tool with a broad set of different algorithms that can be applied for solving many problems. These algorithms, as well as other applications, are described in more detail in the following chapters.

3. Supervised Machine Learning

Supervised ML includes a variety of function algorithms that can map inputs to desired outputs. Usually, supervised learning is used in the classification and regression problems: classifiers map inputs into pre-defined classes, while regression algorithms map inputs into a real-value domain. In other words, classification allows predicting the input category, while regression allows predicting a numerical value based on collected data. The general algorithm of supervised learning is shown in Figure 5.

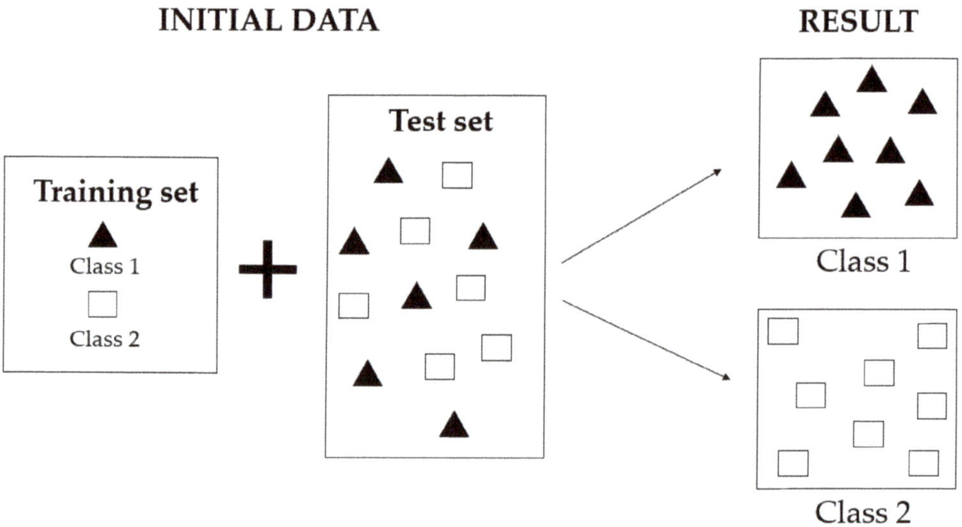

Figure 5. Supervised learning algorithm.

Unsupervised learning aims to discover features from labeled examples so it is possible to analyze unlabeled examples with possibly high accuracy. Basically, the program creates a rule according to what the data are to be processed and classified.

Among supervised algorithms, the most widely used are the following algorithms: linear and logistic regression [47,48], Naive Bayes [49,50], nearest neighbor [51,52], and random forest [53–56]. In condition monitoring and diagnostics of electrical machines, the most suitable supervised algorithms are decision trees [57–59] and support vector machines [60–62].

3.1. Decision Trees

A decision tree (DT) is a decision support tool extensively used in data analysis and statistics. Special attention has been paid to DTs in artificial data mining. DTs' goal is to create a model that predicts the target's value based on multiple inputs. The structure of DTs can be represented by branches and leaves. The branches contain attributes on which the function depends, while leaves contain the values of the function. The other nodes contain attributes by which the decision cases are different. An example of the DT algorithm is shown in Figure 6.

Among other decision models, DTs are the simplest and need a little amount of data to succeed. Moreover, this algorithm can be a hybrid model with another decision model in achieving a more accurate outcome. However, these models are unstable. A little amount of input data can lead to a significant change in the decision tree structure, leading to inaccurate results. Additionally, regression algorithms can fail in the case of decision trees.

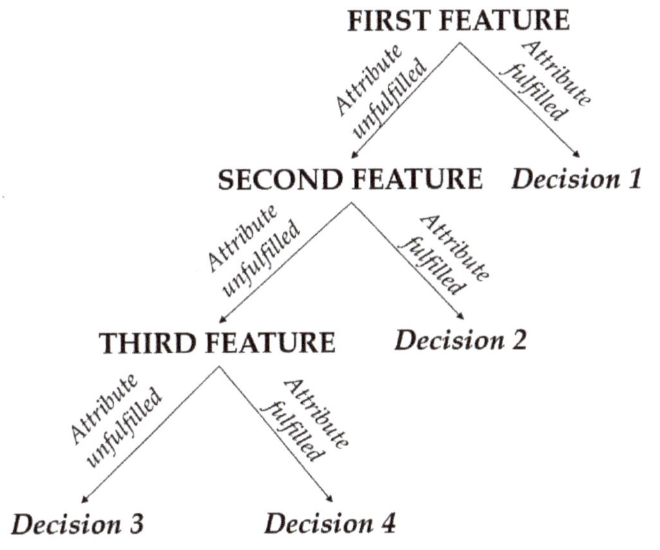

Figure 6. Decision tree diagram.

3.2. Support Vector Machines

Another widely used condition monitoring set of ML algorithms are the support vector machines (SVM). This is a set of supervised models used for regression, novelty detection tasks, feature reduction, and SVM, which is preferable in classification objectives [63]. In linear classification, each datapoint is represented as a vector in n-dimensional space (n—the number of features). Each of these points belongs to only one of two classes. Figure 7 shows an example of data classification.

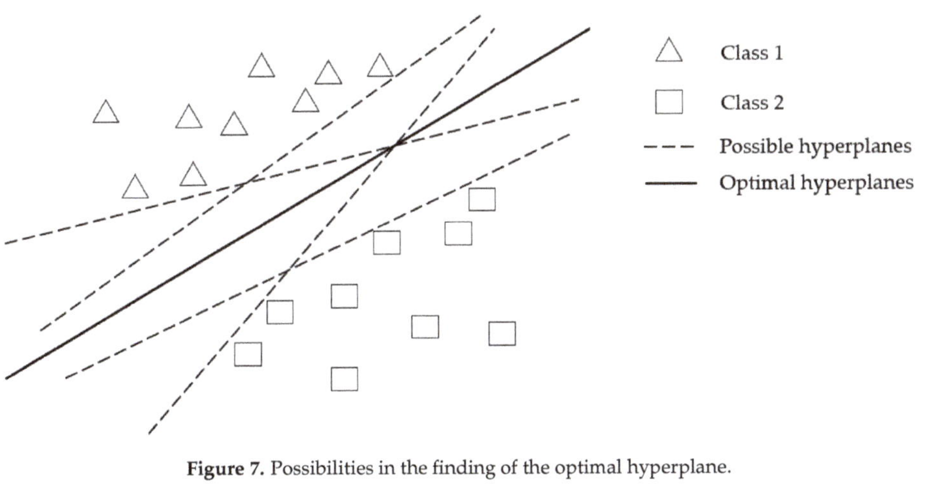

Figure 7. Possibilities in the finding of the optimal hyperplane.

In the picture, two data classes are represented: Class 1 (triangles) and Class 2 (squares). The aim is to separate these points by a hyperplane of dimension (n − 1), ensuring a maximum gap between them. There are many possible hyperplanes. Maximizing the gap between classes contributes to a more confident classification and helps to find an optimal hyperplane. As shown in Figure 8, to detect the optimal hyperplane, it is essential to find support vectors that can be defined in as closer position to the hyperplane as possible.

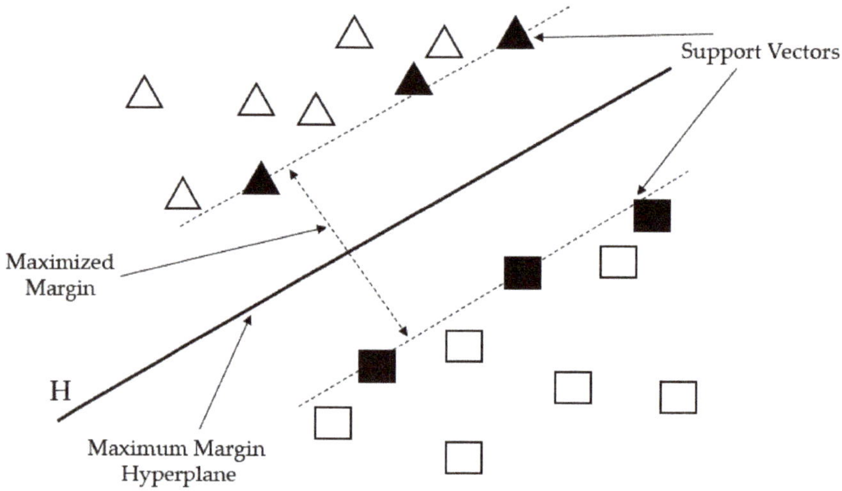

Figure 8. Support vectors and optimal hyperplane in linear classification.

In addition to linear classification, SVMs can deal with non-linear classification using the kernel trick, also known as the kernel machine. As shown in Figure 9, the processing algorithm is similar to the linear one, but the kernel function replaces the datapoints.

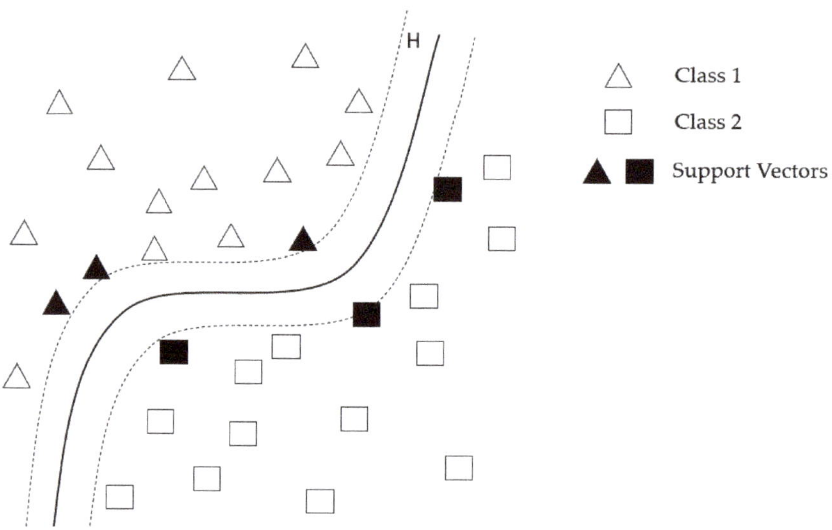

Figure 9. Support vectors and optimal hyperplane in non-linear classification.

SVM is a good solution when there is no initial information about the data. This method is highly preferred because of the little computation power needed to produce results with significant accuracy. Although kernel machine is a great advantage of SVM, its managing is a complicated task. Moreover, it can take a long time to make large amounts of data processed, so SVM is not preferable in large datasets.

Supervised ML approaches are widely applicable for condition monitoring of electrical machines. Many relevant kinds of research can be found in the literature. The authors in [64] proposed a new signal processing method for fault diagnosis of low-speed machinery based on DT approaches. In [65], the authors applied statistical process control and supervised ML techniques to diagnose wind turbine faults and predict maintenance needs. The researchers in [66] presented a semi-supervised ML method that uses the DT algorithm's co-training to handle unlabeled data and applied to fault classification in electric power systems. In [67], the authors proposed a RUL prediction method of lithium-ion batteries using particle filter and support vector regression.

4. Unsupervised Machine Learning

Unsupervised ML includes algorithms that can learn spontaneously to perform a proposed task without intervention from a teacher. Unsupervised learning is often contrasted with supervised learning when an outcome is known, and it is required to find a relationship between system responses. In unsupervised learning, as shown in Figure 10, the program tries to find similarities between objects and divide them into groups if there are similar patterns. These groups are called clusters. Among supervised algorithms, the most widely used are the following algorithms: cluster analysis, fuzzy c-means [68,69], and k-means [70]. In the diagnosis of electrical machines, principal component analysis is the most frequently used algorithm [71–73].

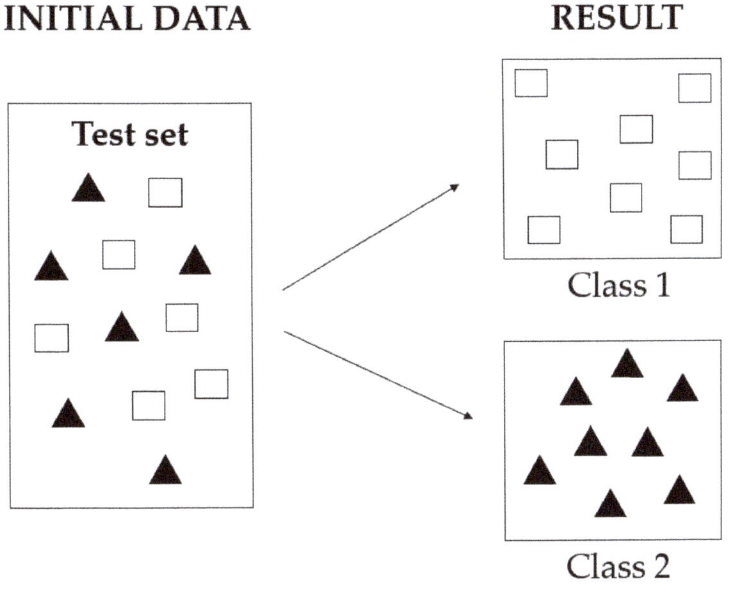

Figure 10. Unsupervised learning algorithm.

More frequently, the dataset is so large that it is difficult to interpret and distinguish the necessary information. Principal component analysis (PCA) is one of the most spread algorithms to reduce the data's dimensions while losing the least amount of information. PCA can be interpreted geometrically, as shown in Figure 11.

The algorithm of SVM is as follows:

(a) Points with specific coordinates are designated on the plane.
(b) The direction of the maximum data change is selected, and a new axis PCA is drawn through the experimental points.
(c) Experimental points are to be projected on the axis PCA.
(d) It is assumed that all the points were initially projected on the axis PCA, and all deviations from this axis can be considered as noise.

If noise is considerable, another axis can be added perpendicular to the first one to describe the data's remaining change. As a result, there is a new representation, which has a smaller number of variables, where all variables are considered, and none of them are deleted. An insignificant part of the data is separated and turns into noise. The main components give the initially hidden variables that control the data device.

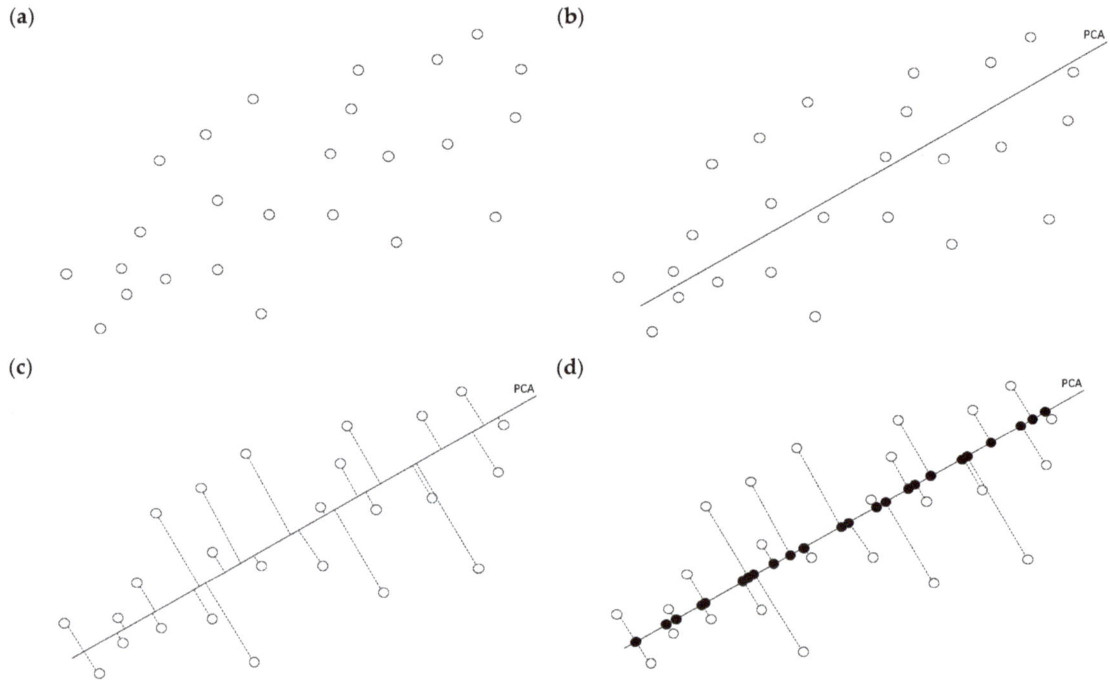

Figure 11. Support vectors and optimal hyperplane in non-linear classification: (**a**) initial dataset, (**b**) optimal vector determination, (**c**) projection of initial dataset on the vector, (**d**) new data parameters definition.

PCA is the most common approach to dimensionality reduction. It is a useful tool for the visualization of large datasets. One of PCA's main advantages is that components are independent of each other, and there is no correlation between them. It can significantly reduce the training time. At the same time, these independent values can become less interpretable. Besides applying PCA, there is still information loss, and the data analysis is relatively less precise than the original values.

Many studies are available in the literature where unsupervised algorithms are used for the analysis of high-dimensional datasets. In [74], the authors applied a new method to the fault diagnosis of rolling bearings in the field of high-dimensional unbalanced fault diagnosis data based on PCA for better classification performance. In [75], researchers used a PCA-based method to monitor non-linear processes. The researchers in [76] proposed a PCA-based hybrid method for monitoring linear and non-linear industrial processes.

5. Reinforcement Learning

Reinforcement learning (RL) is one of the ML methods, where the system (agent) learns by interacting with some environment. Different from supervised algorithms, there is no need for labeled data pairs. RL is mainly focused on finding a balance between an unknown environment and existing knowledge. The general algorithm of reinforcement learning is shown in Figure 12.

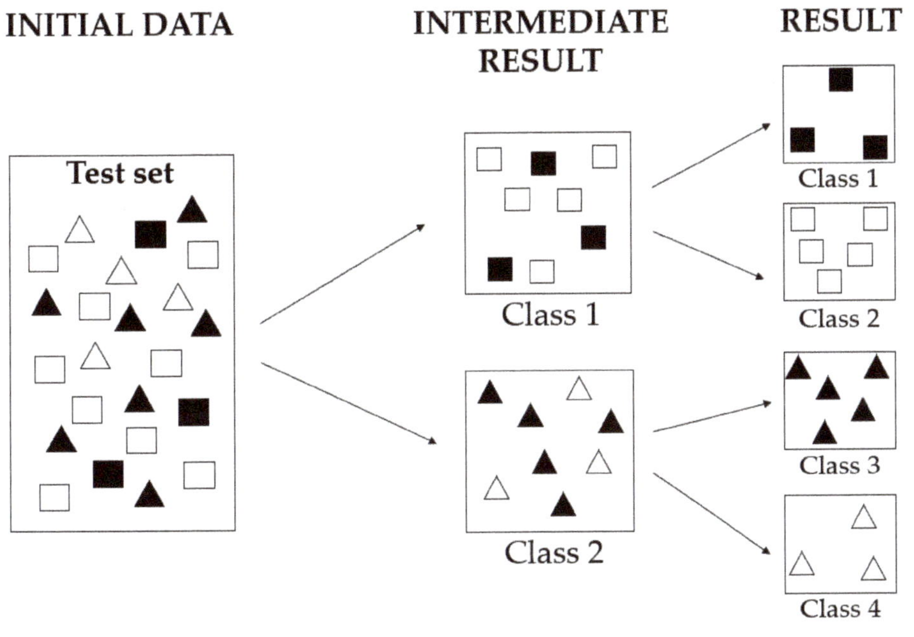

Figure 12. Reinforcement learning algorithm.

One of the algorithms, which can be used in data mining and cluster analysis, is swarm intelligence [77–79]. Swarm intelligence (SI) describes a decentralized and self-organized system's collective behavior, which is considered an optimization method. SI system consists of agents (boids) that interact with each other and the environment. SI should be a multi-agent system with self-organized behavior, which could exhibit a reasonable behavior. This algorithm can adapt to changes and converge fast at some optima. Simultaneously, solutions are dependent sequences of random decisions and can be trapped in local minimum in complex tasks.

At the same time, the more frequently used reinforcement algorithm in condition monitoring is the genetic algorithm [80–82]. A genetic algorithm (GA) is a tool for solving optimization problems and modeling random selection using natural selection mechanisms in the environment. A distinctive feature of the GA is the emphasis on using the "crossing" operator, which uses the instrumental role of crossing in wildlife.

In the case of GA, the problem is formalized so that its solution can be encoded in the form of a vector of genes (genotype), where each gene has some value. In classical implementations of GA, it is assumed that the genotype has a fixed length. However, there are GA variations that are free from this limitation. The general diagram of GA is shown in Figure 13.

Basically, the optimization algorithm with the usage of GA is as follows:

(a) There is a task, and many genotypes of the initial population are to be created.
(b) This initial set of data is to be assessed using the "fitness function," which determines how well each initial population's genotype solves the task.
(c) After this, the best coincidences are to be selected in the population for the next generations.
(d) The best coincidences obtain new solutions. This process repeats until the task is fulfilled and a resultant population is created.

The main benefit of GA is that specified knowledge about the domain is not needed. GA generates a solution through genetic operators. Moreover, a result can contain more than one appropriate solution. However, GA sometimes suffers from degeneracy. The degeneracy can occur if multiple chromosomes represent the same solution. The same shapes of chromosomes occur repeatedly. In this case, the optimal solution is not guaranteed.

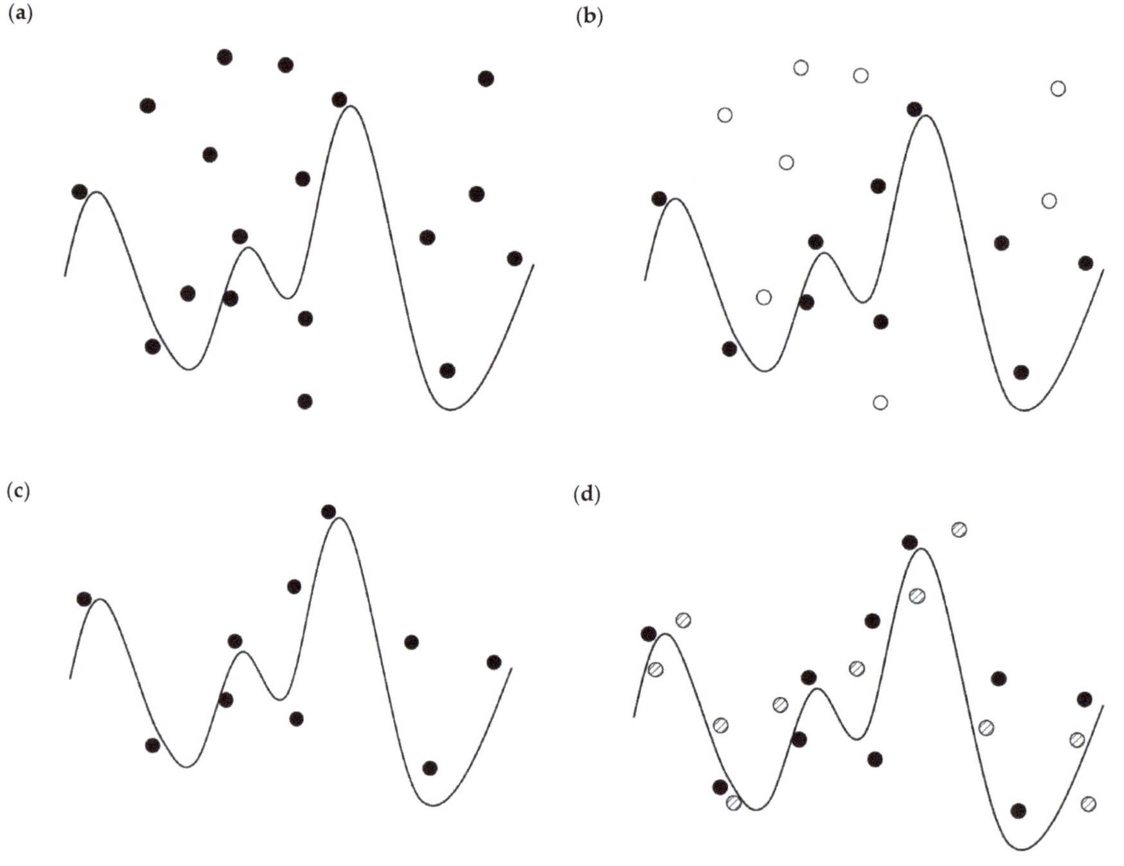

Figure 13. Genetic algorithm diagram: (**a**) creation of initial population, (**b**) application of fitness function, (**c**) selection of the best coincidences, (**d**) creation of resultant population.

Nonetheless, GA is an efficient tool for industrial processes optimization. In [83], researchers proposed a new method based on GAs that can be used for both fault-type classification and RUL prediction. The authors in [84] proposed a method based on genetic mutation particle swarm optimization for gear faults diagnosis. In [85], researchers proposed a GA-based method to optimize and improve the photovoltaic array accuracy.

6. Neural Networks

ANNs have been proved as quite approving tools for condition monitoring and prediction of RUL due to their adaptability, nonlinearity, and arbitrary function approximation ability [86,87]. The main advantage of NNs is that they can outperform nearly every other ML algorithm. This method is supposed to analyze and model processes of damage propagating and predict further failures based on collected data. The main tasks that neuron networks deal with are [88,89]:

1. Classification,
2. Prediction,
3. Recognition.

Artificial neural networks originate from attempts to reproduce biological nervous systems' ability to learn and correct errors by modeling the brain's low-level structure. To create artificial intelligence, you need to build a system with a similar architecture. The architecture of an ANN is shown in Figure 14.

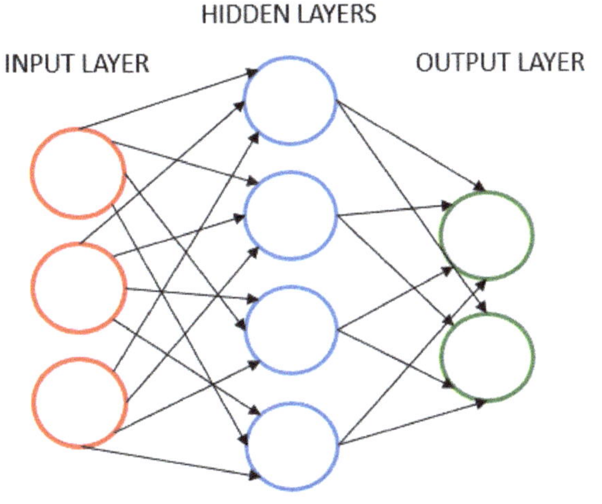

Figure 14. Neural network architecture.

ANNs consist of machine learning algorithms that constitute the human brain with connected signals called neurons. Neurons, both biological as well as artificial, consist of the cell body, dendrite (input), synapse (connection), and axon (output). As seen from the picture, the simplest model of an artificial neural network has three layers of neurons. The first (input) layer is connected to a middle (hidden) layer. The hidden layer is connected to the final (output) layer. In case of the neural networks, to solve a given problem, it is necessary to collect training data. A training dataset is a collection of observations, of which the values of the input and output variables are defined and specified. The neurons transfer a signal from the input layer to the output. The input layer neurons receive data from the outside environment (measuring system, sensors) and, after processing them, transmit signals through the synapses to the neurons of the hidden layer. The neurons of the hidden process receive signals and transmit them to the neurons of the output layer. Basically, the neuron is a computing unit that receives information, performs simple calculations on it, and transfers it further.

Neural networks are not being programmed; they are learning. Learning is one of the main advantages of neural networks over traditional algorithms. Technically, training consists of finding the coefficients of connections between neurons. In the process of training, the neural network can identify complex dependencies between input and output data and perform generalizations. This means that in case of successful training, the network will be able to return the correct result based on data absent in the training sample and incomplete or partially distorted data.

If a neural network consists of more than three layers, which is an increasing tendency nowadays, the algorithm can be considered a deep learning or deep neural network (DNN). Generally, deep learning is one of the ML techniques in ANNs which analyzes big machinery data with more precise results.

NNs have been considered as a universal tool in solving many problems. However, each method has its own limitations, and NNs are no exception. Usually, NNs are used as a hybrid with some other condition monitoring techniques. All the limitations of ANNs and other mentioned ML techniques are given in the following section.

Different types of NN are used for different parameters monitoring. In the literature, a variety of applications can be found. The authors in [90] proposed a novel intelligent fault diagnosis method based on multiscale convolutional NN to identify different failures of wind turbine gearbox. In [91], the authors proposed an intelligent bearing fault diagnosis method combining compressed data acquisition and deep learning, which provides a new strategy to handle the massive data more effectively. The authors in [92] proposed a deep transfer learning (DTL)-based method to predict the remaining useful life in manufacturing. In [93], the author suggested a novel deep convolutional NN cascading architecture for performing localization and detecting defects in power line insulators. Many algorithms have been developed over the years for the automated identification of partial discharges. In [94], an application of a neural network to partial discharge images is presented, which is based on the convolutional neural network architecture, to recognize the aging of high-voltage electrical insulation.

7. Trends in Condition Monitoring and Discussion

The maintenance of the electrical equipment is a very challenging topic at present. Proper, reliable, and efficient fault diagnostic techniques are becoming more and more essential as the world moves towards Industry 4.0 standards [9]. A major issue related to the prediction and condition monitoring is the reliability of the used methods [95,96]. ML algorithms have given a potent tool for classifications. ML methods are not a novelty; thus, researchers meet different limitations. Nowadays, intelligent condition monitoring methods mentioned in previous chapters are mainly used together as a hybrid to get more precise and robust results of fault diagnostics in industrial systems [97].

The main problem of machine learning and neural networks is the training datasets required for system training. To meet precise results and make accurate predictions, the amount and the quality of data play a significant role. Mostly, the dataset shows irrelevant features, requiring a function to build a model. This function will represent how flexible the model is. The main problem with the data is either overfitting or underfitting.

Big data is a trending challenge nowadays. At the same time, high dimensionality and the limited number of training samples lead to overfitting [98]. Frequently, this problem occurs with neural networks [99]. Overfitting means that there is a very qualified training dataset but a very poor test dataset. Simultaneously, the system cannot perform well if the training set is too small or if the data is too noisy and corrupted with irrelevant features. There can be an underfitting phenomenon where the test dataset is good enough, but training data are inferior. All the examples are shown in Figure 15.

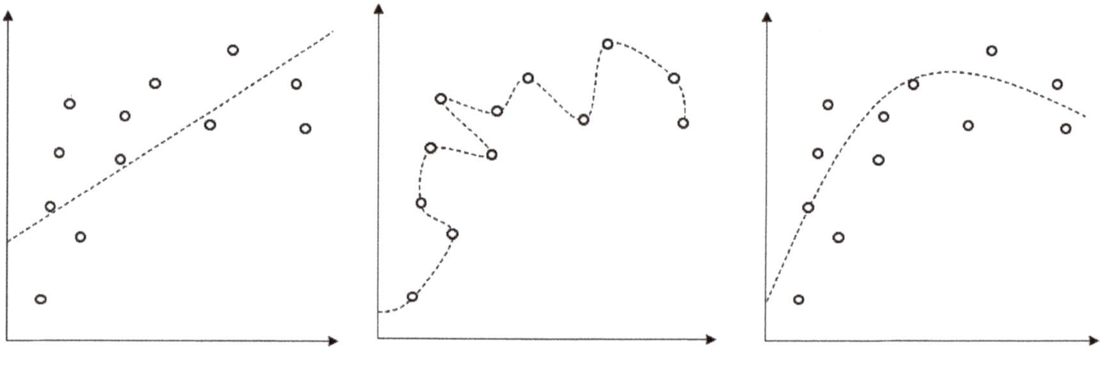

(a) (b) (c)

Figure 15. Data generalization: (**a**) test data is underfitted, (**b**) test data is overfitted, (**c**) test data is balanced.

As shown in Figure 15, both underfitted and overfitted models describe the same dataset. Although the too generalized model does not give the priciest results, at the same time, the overfitted model has a definite idea and is not flexible enough for upcoming new datasets. The challenge is to find a balance between underfitting and overfitting by the usage of different models.

ML is a widespread trend in load forecasting. Many operating decisions, such as reliability analysis or maintenance planning, are based on load forecasts [100]. In this case, artificial neural networks have paid significant attention to proper performance. The main problem overfitted sub-optimization system of ANN that can lead to uncertain forecast results [101]. Working in dynamically changing environments can be a complicated task for NNs. Even if the network has been successfully trained, there is no guarantee that it will work in the future. The market is continually transforming, so today's model can be obsolete tomorrow. In this case, various network architectures must be tested to choose the best one that could follow changes in the environment. Moreover, in the case of NNs, a phenomenon can occur known as catastrophic forgetting. This means that NNs cannot be sequentially trained in several tasks. Each new training set will cause rewriting of all neuron weights, and, as a result, the previously trained data will be forgotten.

Another spread limitation for NNs is the so-called "black box" phenomenon. As was already mentioned, deep learning successfully learns hidden layers of NN architecture mapping inputs and outputs. Approximating the function makes it impossible to study insights into the structure and, as a result, study a cause of a mistake. For this reason, in particular, it is reasonable to choose some other technique or to use NNs in combination with another algorithm.

8. Conclusions

A review of the state of the art, machine learning-based fault diagnostic techniques in the field of electrical machines is presented in this paper. The artificial intelligence-based condition monitoring techniques are becoming more popular as computer power is increasing day by day. Unlike conventional on-board processors responsible for data collection and analysis, the utilization of powerful remote resources using cloud computation gives the freedom of unlimited memory and processing power to handle big data vital for intelligent techniques. Moreover, by effective training of AI algorithms using mathematical models with various faulty conditions, the diagnostic algorithms can be made more reliable.

The collection of these big data is neither possible from industry nor the lab environment. It is not possible from the industry because of the limited number of faulty machines under service. In the lab, a limited number of machines can be broken due to economic constraints. Due to the trend of mounting sensors on the remotely located machines and collecting their data over the cloud, the processing power-related constraints are resolved. Machine learning makes a considerably significant portion of AI techniques. For future work, the studied techniques will be implemented in practice on real industrial objects. Those techniques can use statistical or convention signal processing techniques to detect fault-related patterns and estimate electrical machines' life estimation. Moreover, they give the flexibility to train algorithms under a variety of working conditions. Those conditions may include grid fed, scalar control, low load, and changing load in case of induction machines in particular and for the rest of other machines in general.

Author Contributions: Conceptualization, K.K. and T.V.; methodology, K.K. and B.A.; validation, K.K., B.A., and A.R.; methodology, A.K.; writing—original draft preparation, K.K.; writing—review and editing, T.V. and G.D.; visualization, A.R.; supervision, A.K. All authors have read and agreed to the published version of the manuscript.

Funding: This research has been funded by the Baltic Research Program under Grant "Industrial Internet methods for electrical energy conversion systems monitoring and diagnostics".

Institutional Review Board Statement: Not applicable.

Informed Consent Statement: Not applicable.

Conflicts of Interest: The authors declare no conflict of interest.

References

1. Vaimann, J.T.; Sobra, A.; Belahcen, A.; Rassõlkin, M.; Rolak, A.; Kallaste, A. Induction machine fault detection using smartphone recorded audible noise. *IET Sci. Meas. Technol.* **2018**, *12*, 554–560. [CrossRef]
2. Vaimann, T.; Belahcen, A.; Kallaste, A. Necessity for implementation of inverse problem theory in electric machine fault diagnosis. In Proceedings of the 2015 IEEE 10th International Symposium on Diagnostics for Electrical Machines, Power Electronics and Drives (SDEMPED), Guarda, Portugal, 1–4 September 2015.
3. Nandi, S.; Toliyat, H.A.; Li, X. Condition monitoring and fault diagnosis of electrical motors—A review. *IEEE Trans. Energy Convers.* **2005**, *20*, 719–729. [CrossRef]
4. Wrobel, R.; Mecrow, B.C. A Comprehensive review of additive manufacturing in construction of electrical machines. *IEEE Trans. Energy Convers.* **2020**, *35*, 1054–1064. [CrossRef]
5. Iakovleva, M.E.; Belova, M.; Soares, A. Specific features of mapping large discontinuous faults by the method of electro-magnetic emission. *Resources* **2020**, *9*, 135. [CrossRef]
6. Sarkhanloo, M.S.; Ghalledar, D.; Azizian, M.R. Diagnosis of stator winding turn to turn fault of induction motor using space vector pattern based on neural network. In Proceedings of the 3rd Conference Thermal Power Plants, Tehran, Iran, 18–19 October 2011; pp. 1–6.
7. Muljadi, E.; Samaan, N.; Gevorgian, V.; Li, J.; Pasupulati, S. Circuit current contribution for different wind turbine generator types. In Proceedings of the IEEE PES General Meeting PES 2010, Detroit, MI, USA, 25–29 July 2010.
8. Kudelina, K.; Asad, B.; Vaimann, T.; Rassõlkin, A.; Kallaste, A. Production quality related propagating faults of induction machines. In Proceedings of the 2020 XI International Conference on Electrical Power Drive Systems (ICEPDS), Saint-Petersburg, Russia, 5–6 October 2020.
9. Asad, B.; Vaimann, T.; Rassõlkin, A.; Kallaste, A.; Belahcen, A. Review of Electrical Machine Diagnostic Methods Applicability in the Perspective of Industry 4.0. *Electr. Control Commun.* **2018**, *14*, 108–116. [CrossRef]
10. Stone, G.C.; Boulter, E.A.; Culbert, I.; Dhirani, H. *Electrical Insulation for Rotating Machines: Design, Evaluation, Aging, Testing, and Repair*; John Wiley & Sons: Hoboken, NJ, USA, 2014.
11. Orosz, T. Evolution and modern approaches of the power transformer cost optimization methods. *Period. Polytech. Electr. Eng. Comput. Sci.* **2019**, *63*, 37–50. [CrossRef]
12. Tamus, Z. Ádám Complex diagnostics of insulating materials in industrial electrostatics. *J. Electrost.* **2009**, *67*, 154–157. [CrossRef]
13. Asad, B.; Vaimann, T.; Belahcen, A.; Kallaste, A.; Rassõlkin, A.; Iqbal, M.N. Broken rotor bar fault detection of the grid and inverter-fed induction motor by effective attenuation of the fundamental component. *IET Electr. Power Appl.* **2019**, *13*, 2005–2014. [CrossRef]
14. Asad, B.; Vaimann, T.; Rassõlkin, A.; Kallaste, A.; Belahcen, A. A survey of broken rotor bar fault diagnostic methods of induction motor. *Electr. Control Commun. Eng.* **2019**, *14*, 117–124. [CrossRef]
15. Asad, B.; Vaimann, T.; Belahcen, A.; Kallaste, A.; Rassolkin, A. Rotor fault diagnostic of inverter fed induction motor using frequency Analysis. In Proceedings of the 2019 IEEE 12th International Symposium on Diagnostics for Electrical Machines, Power Electronics and Drives (SDEMPED), Toulouse, France, 27–30 August 2019; pp. 127–133.
16. Rosero, J.A.; Cusido, J.; Garcia, A.; Ortega, J.; Romeral, L. Broken bearings and eccentricity fault detection for a permanent magnet synchronous motor. In Proceedings of the IECON 2006—32nd Annual Conference on IEEE Industrial Electronics, Paris, France, 7–10 November 2006; pp. 964–969.
17. Kallaste, A.; Belahcen, A.; Kilk, A.; Vaimann, T. Analysis of the eccentricity in a low-speed slotless permanent-magnet wind generator. In Proceedings of the 2012 Electric Power Quality and Supply Reliability, Tartu, Estonia, 11–13 June 2012; pp. 1–6.
18. Chen, Y.; Liang, S.; Li, W.; Liang, H.; Wang, C. Faults and diagnosis methods of permanent magnet synchronous motors: A review. *Appl. Sci.* **2019**, *9*, 2116. [CrossRef]
19. Kallaste, A. Low Speed Permanent Magnet Slotless Generator Development and Implementation for Windmills. Ph.D. Thesis, Tallinn University of Technology, Tallinn, Estonia, 2013.
20. Kudelina, K.; Asad, B.; Vaimann, T.; Rassõlkin, A.; Kallaste, A.; Lukichev, D.V. Main faults and diagnostic possibilities of BLDC Motors. In Proceedings of the 2020 27th International Workshop on Electric Drives: MPEI Department of Electric Drives 90th Anniversary (IWED), Moscow, Russia, 27–30 January 2020.
21. Kudelina, K.; Asad, B.; Vaimann, T.; Rassolkin, A.; Kallaste, A. Effect of Bearing Faults on Vibration Spectrum of BLDC Motor. In Proceedings of the 2020 IEEE Open Conference of Electrical, Electronic and Information Sciences (eStream), Vilnius, Lithuania, 30 April 2020.
22. Bin Lee, S.; Stone, G.C.; Antonino-Daviu, J.; Gyftakis, K.N.; Strangas, E.G.; Maussion, P.; Platero, C.A. Condition monitoring of industrial electric machines: State of the art and future challenges. *IEEE Ind. Electron. Mag.* **2020**, *14*, 158–167. [CrossRef]
23. Dos Santos, T.; Ferreira, F.J.; Pires, J.M.; Damásio, C. Stator Winding Short-Circuit Fault Diagnosis in Induction Motors using Random Forest. In Proceedings of the 2017 IEEE International Electric Machines and Drives Conference (IEMDC), Miami, FL, USA, 21–24 May 2017.

24. Ghosh, R.; Seri, P.; Hebner, R.E.; Montanari, G.C. Noise rejection and detection of partial discharges under repetitive impulse supply voltage. *IEEE Trans. Ind. Electron.* **2020**, *67*, 4144–4151. [CrossRef]
25. Wang, Z.; Yang, J.; Li, H.; Zhen, D.; Xu, Y.; Gu, F. Fault identification of broken rotor bars in induction motors using an improved cyclic modulation spectral analysis. *Energies* **2019**, *12*, 3279. [CrossRef]
26. Xu, X.; Han, Q.; Chu, F. Review of electromagnetic vibration in electrical machines. *Energies* **2018**, *11*, 1779. [CrossRef]
27. Sathyan, S.; Aydin, U.; Lehikoinen, A.; Belahcen, A.; Vaimann, T.; Kataja, J. Influence of magnetic forces and magneto-striction on the vibration behavior of an induction motor. *Int. J. Appl. Electromagn. Mech.* **2019**, *59*, 825–834. [CrossRef]
28. Kudelina, K.; Asad, B.; Vaimann, T.; Belahcen, A.; Rassõlkin, A.; Kallaste, A.; Lukichev, D.V. Bearing Fault Analysis of BLDC Motor for Electric Scooter Application. *Designs* **2020**, *4*, 42. [CrossRef]
29. Susto, G.A.; Schirru, A.; Pampuri, S.; McLoone, S.; Beghi, A. Machine Learning for Predictive Maintenance: A Multiple Classifier Approach. *IEEE Trans. Ind. Inform.* **2015**, *11*, 812–820. [CrossRef]
30. Vaimann, T.; Rassõlkin, A.; Kallaste, A.; Pomarnacki, R.; Belahcen, A. Artificial intelligence in monitoring and diagnostics of electrical energy conversion systems. In Proceedings of the 2020 27th International Workshop on Electric Drives: MPEI Department of Electric Drives 90th Anniversary (IWED), Moscow, Russia, 27–30 January 2020.
31. Lei, Y.; Li, N.; Gontarz, S.; Lin, J.; Radkowski, S.; Dybala, J. A Model-Based Method for Remaining Useful Life Prediction of Machinery. *IEEE Trans. Reliab.* **2016**, *65*, 1314–1326. [CrossRef]
32. Bangalore, P.; Tjernberg, L.B. An Artificial Neural Network Approach for Early Fault Detection of Gearbox Bearings. *IEEE Trans. Smart Grid* **2015**, *6*, 980–987. [CrossRef]
33. Glowacz, A. Fault diagnosis of electric impact drills using thermal imaging. *Measurement* **2021**, *171*, 108815. [CrossRef]
34. Leahy, K.; Hu, R.L.; Konstantakopoulos, I.C.; Spanos, C.J.; Agogino, A.M. Diagnosing wind turbine faults using machine learning techniques applied to operational data. In Proceedings of the 2016 IEEE International Conference on Prognostics and Health Management (ICPHM), Ottawa, ON, Canada, 20–22 June 2016.
35. Wu, L.; Kaiser, G.; Solomon, D.; Winter, R.; Boulanger, A.; Anderson, R. Improving efficiency and reliability of building systems using machine learning and automated online evaluation. In Proceedings of the 2012 IEEE Long Island Systems, Applications and Technology Conference (LISAT), Farmingdale, NY, USA, 4 May 2012.
36. Liu, H.; Liu, S.; Liu, Z.; Mrad, N.; Dong, H. Prognostics of damage growth in composite materials using machine learning techniques. In Proceedings of the 2017 IEEE International Conference on Industrial Technology (ICIT), Toronto, OT, Canada, 22–25 May 2017.
37. Helm, J.M.; Swiergosz, A.M.; Haeberle, H.S.; Karnuta, J.M.; Schaffer, J.L.; Krebs, V.E.; Spitzer, A.I.; Ramkumar, P.N. Machine learning and artificial intelligence: Definitions, applications, and future directions. *Curr. Rev. Musculoskelet. Med.* **2020**, *13*, 69–76. [CrossRef] [PubMed]
38. Ławrynowicz, A.; Tresp, V. Introducing machine learning. In *Perspectives on Ontology Learning*; Lehmann, J., Voelker, J., Eds.; IOS Press: Heidelberg, Germany, 2014.
39. Ayodele, T.O. Types of Machine Learning Algorithms. In *New Advances in Machine Learning*; Zhang, Y., Ed.; IntechOpen: London, UK, 2010.
40. Nasteski, V. An overview of the supervised machine learning methods. *Horiz. B* **2017**, *4*, 51–62. [CrossRef]
41. Elforjani, M.; Shanbr, S. Prognosis of bearing acoustic emission signals using supervised machine learning. *IEEE Trans. Ind. Electron.* **2018**, *65*, 5864–5871. [CrossRef]
42. Asad, B.; Vaimann, T.; Belahcen, A.; Kallaste, A.; Rassõlkin, A.; Iqbal, M.N. Cluster computation-based hybrid fem—Analytical model of induction motor for fault diagnostics. *Appl. Sci.* **2020**, *10*, 7572. [CrossRef]
43. Asad, B.; Vaimann, T.; Belahcen, A.; Kallaste, A.; Rassõlkin, A.; Iqbal, M.N. Modified winding function-based model of squirrel cage induction motor for fault diagnostics. *IET Electr. Power Appl.* **2020**, *14*, 1722–1734. [CrossRef]
44. Greene, D.; Cunningham, P.; Mayer, R. Unsupervised Learning and Clustering. In *Machine Learning Techniques for Multimedia: Case Studies on Organization and Retrieval*; Cunningham, P., Cord, M., Eds.; Springer: Berlin/Heidelberg, Germany, 2008; pp. 51–90.
45. Michau, G.; Fink, O. Unsupervised Fault Detection in Varying Operating Conditions. In Proceedings of the 2019 IEEE International Conference on Prognostics and Health Management (ICPHM), San Francisco, CA, USA, 17–20 June 2019; pp. 1–10.
46. Mousavi, S.S.; Schukat, M.; Howley, E. Distributed deep reinforcement learning: An overview. In Proceedings of the SAI Intelligent Systems Conference, London, UK, 21–22 September 2016.
47. Qian, Y.; Ye, M.; Zhou, J. Hyperspectral Image Classification Based on Structured Sparse Logistic Regression and Three-Dimensional Wavelet Texture Features. *IEEE Trans. Geosci. Remote. Sens.* **2013**, *51*, 2276–2291. [CrossRef]
48. Ohsaki, M.; Wang, P.; Matsuda, K.; Katagiri, S.; Watanabe, H.; Ralescu, A. Confusion-matrix-based kernel logistic regression for imbalanced data classification. *IEEE Trans. Knowl. Data Eng.* **2017**, *29*, 1806–1819. [CrossRef]
49. Liu, B.; Blasch, E.; Chen, Y.; Shen, D.; Chen, G. Scalable sentiment classification for Big Data analysis using Naïve Bayes Classifier. In Proceedings of the 2013 IEEE International Conference on Big Data, Santa Clara, CA, USA, 6–9 October 2013; pp. 99–104.
50. Sun, S.; Przystupa, K.; Wei, M.; Yu, H.; Ye, Z.; Kochan, O. Fast bearing fault diagnosis of rolling element using Lévy Moth-Flame optimization algorithm and Naive Bayes. *Ekspolatacja Niezawodn. Maint. Reliab.* **2020**, *22*, 730–740. [CrossRef]
51. Muja, M.; Lowe, D.G. Scalable nearest neighbor algorithms for high dimensional Data. *IEEE Trans. Pattern Anal. Mach. Intell.* **2014**, *36*, 2227–2240. [CrossRef]

52. Tian, J.; Morillo, C.; Azarian, M.H.; Pecht, M. Kurtosis-based feature extraction coupled with k-nearest neighbor distance analysis. *IEEE Trans. Ind. Electron.* **2016**, *63*, 1793–1803. [CrossRef]
53. Kusiak, A.; Verma, A. A Data-Mining Approach to Monitoring Wind Turbines. *IEEE Trans. Sustain. Energy* **2012**, *3*, 150–157. [CrossRef]
54. Ristin, M.; Guillaumin, M.; Gall, J.; Van Gool, L. Learning of random forests for large-scale image classification. *IEEE Trans. Pattern Anal. Mach. Intell.* **2016**, *38*, 490–503. [CrossRef] [PubMed]
55. Waske, B.; Van Der Linden, S.; Benediktsson, J.A.; Rabe, A.; Hostert, P. Sensitivity of support vector machines to random feature selection in classification of hyperspectral Data. *IEEE Trans. Geosci. Remote. Sens.* **2010**, *48*, 2880–2889. [CrossRef]
56. Saberi, A.N.; Sandirasegaram, S.; Belahcen, A.; Vaimann, T.; Sobra, J. Multi-Sensor fault diagnosis of induction motors using random forests and support vector machine. In Proceedings of the 2020 International Conference on Electrical Machines (ICEM), Gothenburg, Germany, 23–26 August 2020.
57. Zhao, Y.; Yang, L.; Lehman, B.; De Palma, J.-F.; Mosesian, J.; Lyons, R. Decision tree-based fault detection and classification in solar photovoltaic arrays. In Proceedings of the 2012 Twenty-Seventh Annual IEEE Applied Power Electronics Conference and Exposition (APEC), Orlando, FL, USA, 5–9 February 2012; pp. 93–99.
58. Kamiński, B.; Jakubczyk, M.; Szufel, P. A framework for sensitivity analysis of decision trees. *Central Eur. J. Oper. Res.* **2018**, *26*, 135–159. [CrossRef] [PubMed]
59. Chen, M.; Zheng, A.; Lloyd, J.; Jordan, M.; Brewer, E. Failure diagnosis using decision trees. In Proceedings of the International Conference on Autonomic Computing, New York, NY, USA, 17–18 May 2004; pp. 36–43.
60. Aydin, I.; Karakose, M.; Akin, E. Artificial immune based support vector machine algorithm for fault diagnosis of induction motors. In Proceedings of the 2007 International Aegean Conference on Electrical Machines and Power Electronics, Bodrum, Turkey, 10–12 September 2007; pp. 217–221.
61. Yi, Z.; Etemadi, A.H. A novel detection algorithm for line-to-line faults in Photovoltaic (PV) arrays based on support vector machine (SVM). In Proceedings of the 2016 IEEE power and energy society general meeting (PESGM), Boston, MA, USA, 17–21 July 2016; pp. 9–12.
62. Soualhi, A.; Medjaher, K.; Zerhouni, N. Bearing health monitoring based on HILBERT–HUANG transform, support vector machine, and regression. *IEEE Trans. Instrum. Meas.* **2015**, *64*, 52–62. [CrossRef]
63. Awad, M.; Khanna, R. Support vector machines for classification. In *Efficient Learning Machines*; Mariette, A., Rahul, K., Eds.; Apress: Berkeley, CA, USA, 2015; pp. 39–66.
64. Song, L.; Wang, H.; Chen, P. Vibration-Based Intelligent Fault Diagnosis for Roller Bearings in Low-Speed Rotating Machinery. *IEEE Trans. Instrum. Meas.* **2018**, *67*, 1887–1899. [CrossRef]
65. Hsu, J.-Y.; Wang, Y.-F.; Lin, K.-C.; Chen, M.-Y.; Hsu, J.H.-Y. Wind turbine fault diagnosis and predictive maintenance through statistical process control and machine learning. *IEEE Access* **2020**, *8*, 23427–23439. [CrossRef]
66. AbdelGayed, T.S.; Morsi, W.G.; Sidhu, T.S. Fault Detection and Classification Based on Co-training of Semisupervised Machine Learning. *IEEE Trans. Ind. Electron.* **2018**, *65*, 1595–1605. [CrossRef]
67. Wei, J.; Dong, G.; Chen, Z. Remaining Useful Life Prediction and State of Health Diagnosis for Lithium-Ion Batteries Using Particle Filter and Support Vector Regression. *IEEE Trans. Ind. Electron.* **2018**, *65*, 5634–5643. [CrossRef]
68. Huang, H.-C.; Chuang, Y.-Y.; Chen, C.-S. Multiple Kernel Fuzzy Clustering. *IEEE Trans. Fuzzy Syst.* **2012**, *20*, 120–134. [CrossRef]
69. Krinidis, S.; Chatzis, V. A robust fuzzy local information c-means clustering algorithm. *IEEE Trans. Image Process.* **2010**, *19*, 1328–1337. [CrossRef] [PubMed]
70. Yu, S.; Tranchevent, L.-C.; Liu, X.; Glanzel, W.; Suykens, J.A.; De Moor, B.; Moreau, Y. Optimized data fusion for kernel k-means clustering. *IEEE Trans. Pattern Anal. Mach. Intell.* **2011**, *34*, 1031–1039. [CrossRef]
71. Hanley, C.; Kelliher, D.; Pakrashi, V. Principal component analysis for condition monitoring of a network of bridge structures. *J. Phys. Conf. Ser.* **2015**, *628*, 012060. [CrossRef]
72. Mazur, K.; Borowa, A.; Brdys, M. Condition monitoring using PCA based method and application to wastewater treatment plant operation. *IFAC Proc. Vol.* **2006**, *39*, 208–213. [CrossRef]
73. He, Q.; Yan, R.; Kong, F.; Du, R. Machine condition monitoring using principal component representations. *Mech. Syst. Signal Process.* **2009**, *23*, 446–466. [CrossRef]
74. Hang, Q.; Yang, J.; Xing, L. Diagnosis of rolling bearing based on classification for high dimensional unbalanced data. *IEEE Access* **2019**, *7*, 79159–79172. [CrossRef]
75. Deng, X.; Tian, X.; Chen, S.; Harris, C.J. Deep principal component analysis based on layer wise feature extraction and its application to nonlinear process monitoring. *IEEE Trans. Control. Syst. Technol.* **2019**, *27*, 2526–2540. [CrossRef]
76. Deng, X.; Tian, X.; Chen, S.; Harris, C.J. Nonlinear process fault diagnosis based on serial principal component analysis. *IEEE Trans. Neural Netw. Learn. Syst.* **2018**, *29*, 560–572. [CrossRef]
77. Abdmouleh, Z.; Gastli, A.; Ben-Brahim, L.; Haouari, M.; Al-Emadi, N.A. Review of optimization techniques applied for the integration of distributed generation from renewable energy sources. *Renew. Energy* **2017**, *113*, 266–280. [CrossRef]
78. Abraham, A.; Guo, H.; Liu, H. Swarm intelligence: Foundations, perspectives and applications. In *Recent Advances in Computational Optimization*; Fidanova, S., Ed.; Springer: Geneva, Switzerland, 2006; pp. 3–25.
79. Xue, B.; Zhang, M.; Browne, W.N. Particle swarm optimization for feature selection in classification: A multi-objective approach. *IEEE Trans. Cybern.* **2012**, *43*, 1656–1671. [CrossRef]

80. Beg, A.H.; Islam, M.Z. Advantages and limitations of genetic algorithms for clustering records. In Proceedings of the 2016 IEEE 11th Conference on Industrial Electronics and Applications, ICIEA 2016, Hefei, China, 5–7 June 2016.
81. Compare, M.; Martini, F.; Zio, E. Genetic algorithms for condition-based maintenance optimization under uncertainty. *Eur. J. Oper. Res.* **2015**, *244*, 611–623. [CrossRef]
82. Baraldi, P.; Canesi, R.; Zio, E.; Seraoui, R.; Chevalier, R. Genetic algorithm-based wrapper approach for grouping condition monitoring signals of nuclear power plant components. *Integr. Comput. Eng.* **2011**, *18*, 221–234. [CrossRef]
83. Trinh, H.C.; Kwon, Y.K. A data-independent genetic algorithm framework for fault-type classification and remaining useful life prediction. *Appl. Sci.* **2020**, *10*, 368. [CrossRef]
84. Ding, J.; Xiao, D.; Li, X. Gear fault diagnosis based on genetic mutation particle swarm optimization VMD and probabilistic neural network algorithm. *IEEE Access* **2020**, *8*, 18456–18474. [CrossRef]
85. Tao, C.; Wang, X.; Gao, F.; Wang, M. Fault Diagnosis of photovoltaic array based on deep belief network optimized by genetic algorithm. *Chin. J. Electr. Eng.* **2020**, *6*, 106–114. [CrossRef]
86. Tian, Z. An artificial neural network method for remaining useful life prediction of equipment subject to condition monitoring. *J. Intell. Manuf.* **2012**, *23*, 227–237. [CrossRef]
87. Saxena, A.; Saad, A. Evolving an artificial neural network classifier for condition monitoring of rotating mechanical systems. *Appl. Soft Comput.* **2007**, *7*, 441–454. [CrossRef]
88. Oong, T.H.; Ashidi, N.; Isa, M. Networks for pattern classification. *Adapt. Evol. Artif. Neural Netw. Pattern Classif.* **2011**, *22*, 1823–1836.
89. Deng, Y.; Ren, Z.; Kong, Y.; Bao, F.; Dai, Q. A Hierarchical fused fuzzy deep neural network for data classification. *IEEE Trans. Fuzzy Syst.* **2014**, *25*, 51–56. [CrossRef]
90. Jiang, G.; He, H.; Yan, J.; Xie, P. Multiscale convolutional neural networks for fault diagnosis of wind turbine gearbox. *IEEE Trans. Ind. Electron.* **2019**, *66*, 3196–3207. [CrossRef]
91. Sun, J.; Yan, C.; Wen, J. Intelligent bearing fault diagnosis method combining compressed data acquisition and deep learning. *IEEE Trans. Instrum. Meas.* **2018**, *67*, 185–195. [CrossRef]
92. Sun, C.; Ma, M.; Zhao, Z.; Tian, S.; Yan, R.; Chen, X. Deep transfer learning based on sparse autoencoder for remaining useful life prediction of tool in manufacturing. *IEEE Trans. Ind. Inform.* **2019**, *15*, 2416–2425. [CrossRef]
93. Tao, X.; Zhang, D.; Wang, Z.; Liu, X.; Zhang, H.; Xu, D. Detection of power line insulator defects using aerial images analyzed with convolutional neural networks. *IEEE Trans. Syst. Man Cybern. Syst.* **2020**, *50*, 1486–1498. [CrossRef]
94. Florkowski, M. Classification of partial discharge images using deep convolutional neural networks. *Energies* **2020**, *13*, 5496. [CrossRef]
95. Belahcen, A.; Gyftakis, K.N.; Martinez, J.; Climente-Alarcon, V.; Vaimann, T. Condition monitoring of electrical machines and its relation to industrial internet. In Proceedings of the 2015 IEEE Workshop on Electrical Machines Design, Control and Diagnosis (WEMDCD), Torino, Italy, 26–27 March 2015.
96. Savard, C.; Iakovleva, E.V. A Suggested improvement for small autonomous energy system reliability by reducing heat and excess charges. *Batteries* **2019**, *5*, 29. [CrossRef]
97. Bicen, Y.; Aras, F. Intelligent condition monitoring platform combined with multi-agent approach for complex systems. In Proceedings of the 2014 IEEE Workshop on Environmental, Energy, and Structural Monitoring Systems Proceedings, Naples, Italy, 17–18 September 2014.
98. Chen, Y.; Jiang, H.; Li, C.; Jia, X.; Ghamisi, P. Deep Feature Extraction and Classification of Hyperspectral Images Based on Convolutional Neural Networks. *IEEE Trans. Geosci. Remote Sens.* **2016**, *54*, 6232–6251. [CrossRef]
99. Bilbao, I.; Bilbao, J. Overfitting problem and the over-training in the era of data: Particularly for Artificial Neural Networks. In Proceedings of the 2017 Eighth International Conference on Intelligent Computing and Information Systems (ICICIS), Cairo, Egypt, 5–7 December 2017.
100. Fan, S.; Hyndman, R.J. Short-term load forecasting based on a semi-parametric additive Model. *IEEE Trans. Power Syst.* **2012**, *27*, 134–141. [CrossRef]
101. Hinojosa, V.H.; Hoese, A. Short-term load forecasting using fuzzy inductive reasoning and evolutionary algorithms. *IEEE Trans. Power Syst.* **2010**, *25*, 565–574. [CrossRef]

applied sciences

MDPI

Article

The Cluster Computation-Based Hybrid FEM–Analytical Model of Induction Motor for Fault Diagnostics

Bilal Asad [1,2,*], Toomas Vaimann [1], Anouar Belahcen [1,2], Ants Kallaste [1], Anton Rassõlkin [1] and M. Naveed Iqbal [1]

1 Department of Electrical Power Engineering and Mechatronics, Tallinn University of Technology, 19086 Tallinn, Estonia; toomas.vaimann@taltech.ee (T.V.); Anouar.Belahcen@aalto.fi (A.B.); ants.kallaste@taltech.ee (A.K.); anton.rassolkin@taltech.ee (A.R.); miqbal@taltech.ee (M.N.I.)

2 Department of Electrical Engineering and Automation, Aalto University, FI-00076 Espoo, Finland

* Correspondence: bilal.asad@taltech.ee

Received: 8 October 2020; Accepted: 26 October 2020; Published: 27 October 2020

Abstract: This paper presents a hybrid finite element method (FEM)–analytical model of a three-phase squirrel cage induction motor solved using parallel processing for reducing the simulation time. The growing development in artificial intelligence (AI) techniques can lead towards more reliable diagnostic algorithms. The biggest challenge for AI techniques is that they need a big amount of data under various conditions to train them. These data are difficult to obtain from the industries because they contain low numbers of possible faulty cases, as well as from laboratories because a limited number of motors can be broken for testing purposes. The only feasible solution is mathematical models, which in the long run can become part of advanced diagnostic techniques. The benefits of analytical and FEM models for their speed and accuracy respectively can be exploited by making a hybrid model. Moreover, the concept of cloud computing can be utilized to reduce the simulation time of the FEM model. In this paper, a hybrid model being solved on multiple processors in a parallel fashion is presented. The results depict that by dividing the rotor steps among several processors working in parallel, the simulation time reduces considerably. The simulation results under healthy and broken rotor bar cases are compared with those taken from a laboratory setup for validation.

Keywords: induction motors; fault diagnosis; modeling; finite element analysis; parallel processing

---

1. Introduction

Electrical machines, particularly induction motors, are indispensable in almost all sectors of our modern-day society. In the form of conveyor belt movers, compressors, electric vehicles, fans, and pumps, etc., they consume more than 50% of the total generated energy worldwide [1]. This fact makes their predictive maintenance very important, to avoid any catastrophic situation. As the world is moving towards industry 4.0, predictive maintenance is becoming more important—as contrasted to preventive or reactive maintenance. Unlike preventive or reactive maintenance, in predictive maintenance we monitor the behavior of an electrical machine and anticipate failures before they occur.

Predictive maintenance allows the servicing of the machine when it needs. By doing so, the system's downtime can exponentially decrease with a resultant decrease in the maintenance cost. A variety of conventional condition-monitoring techniques have been discussed in literature over the past few decades, such as motor current signature analysis (MCSA) [2–5], thermal analysis [6–8], vibration [9], acoustics [10,11], stray flux monitoring [12,13], partial discharges [14], air-gap flux monitoring [15], etc.

Although these techniques are well-established, compatible with a variety of signal-processing techniques, and require fewer computational resources, they possess several drawbacks. The most prominent drawbacks are relevant to expensive sensors—such as in the case of thermal analysis and the poor legibility of fault-based frequency components at the incipient stage, as in motor current signature analysis (MCSA)-based techniques. Moreover, these techniques depend upon various constraints such as machine structure, the industrial environment, external noise, bad load-coupling, poor foundation, and the impact of the drive controller, etc. The segregation of frequency components when there is more than one fault is another challenging task in MCSA-based diagnostic techniques. It becomes worse when the industrial inverters inject several frequency components as well.

The industrial inverters with complex control algorithms are becoming a crucial part of a drive system. In this case, the definition of faults goes beyond the domain of simple machine equations. The use of conventional diagnostic techniques, while neglecting all subsystems of the drive, can increase the false or missed alarm rate.

To avoid all those problems and to make diagnostic algorithms more reliable, advanced model-dependent and artificial intelligence (AI) based techniques can give promising results. The majority of rotating machine faults are degenerative, which makes fault diagnosis a pattern recognition problem. Due to a variety of global signals and different faults, pattern recognition is not a straightforward problem. Therefore, a reliable diagnostic algorithm can be a combination of data processing for feature extraction and recognition through AI techniques. Various AI techniques such as probability-based, classification, statistical learning, mathematical optimization, and convex optimization can be found in the literature [16]. The statistical and classification-based methods are gaining increasing popularity in uses such as support vector machines (SVM) [17–19], artificial neural networks (ANN) [20–22], Bayesian classifiers, Naïve Bayes classifiers [23–25], machine learning [26,27], k-nearest neighbor algorithm [28–30], etc.

Almost all AI-based diagnostic techniques need a large number of data samples under various conditions. Those conditions may include signals under healthy, faulty, loaded, and no-load conditions. Moreover, various kinds of faults with different severity levels under a variety of loading conditions can better train advanced AI-based techniques. The collection of large amounts of data with different constraints is practically impossible both from industry and laboratory environments. Because, first, in industries there are few faulty machines and, secondly, the type and level of faults in industry machines are unknown at first—which is necessary information for training the diagnostic algorithms. In the laboratory, conducting a large number of destructive tests is not economically feasible. The only optimal way is to rely on the accurate mathematical models of the machine. Using mathematical models, almost any kind of fault in any type of machine with different natures of load can be simulated to train the diagnostic algorithm.

A variety of machine modeling techniques are available in the literature, which can be broadly classified into two categories; analytical and numerical. The two-axis theory-based models [31–33] are being effectively utilized for control and analysis. Although those models are simple to understand, comprehensive, and fast, they are not suitable for fault simulations because of various approximations such as sinusoidal stator and rotor windings distribution, uniform air gap, no inter-bar currents, and no material saturation, etc. The multiple coupled circuit theory-based models such as winding function analysis (WFA) [34,35], modified winding function analysis (MWFA) [36,37], and extended MWFA [37] allow the inclusion of practical stator and rotor winding functions, the stator and rotor slots openings, and non-linear functions for material saturation. Those models can be used to simulate the majority of faults with very much less simulation time and computational complexity, but they do not remain straightforward while dealing with different types of machines with complex geometrical features. Similarly, other analytical models such as magnetic equivalent circuit [38], generalized harmonic analysis [39], voltage behind reactance [40], and convolution theorem [41] can be used for the simulation of various faults in induction machines but at the cost of material and geometry-related approximations.

Having the ability to deal with almost all kind of geometries and material properties, and the compatibility to solve various kinds of problems, the finite element method (FEM)-based modeling techniques are gaining heightened popularity. Using FEM, a vast variety of electromagnetic [42,43], thermal [44], fluid dynamics [45], structural [46], and related problems can be solved with incredible accuracy. In FEM, the geometry of the system is divided into a considerable number of mesh elements represented by nodes and the solution of each node leads toward the final solution. Indeed, it requires significantly powerful computational resources and a large memory to save intermediate results. Although modern computers with advanced processors are very strong, they need a long time, from several minutes to days, for the solution of highly unsymmetrical machines. The saving of simulation time for fault diagnostics is very important for the collection of vast amounts of data, which can be used as a benchmark for advanced fault diagnostic techniques.

Many methods to diminish these problems have been presented in the literature such as; the hybrid analytical–FEM model [47,48], the model order reduction [49–51], and sparse subspace learning (SSL) [52], etc. These methods have their own limitations as they rely on statistical and interpolation techniques, which are different for different kinds of machines. Problems such as the reducibility of the model and the precision of the input grid can lead to the increased complexity of the model.

As the world is moving towards industry 4.0 standards and cloud computation, the computational resources are becoming unlimited. These resources can be in the form of software applications, processing power, and data storage. All these resources are very important for big data-based advanced diagnostic techniques such as machine learning [53], deep learning [54], parallel autonomous mining [55], image processing [56], online wireless monitoring through smart sensors [57], and neural networks [58–60], etc. The basic building blocks of the cloud computation are infrastructure as a service (IaaS), platform as a service (PaaS), and software as a service (SaaS). Those building blocks can be utilized for big data storage, custom software development, and computer application utilization respectively.

In order to curtail the complexity related problems of FEM models, the concept of parallel processing by utilizing the cluster of computers is presented in this paper. Unlike most of the papers where the simulation speed of FEM models is increased either by exploiting the symmetry (which is not true in the case of faulty machines) or by data interpolation, in this paper the complete two-dimensional (2D) geometry of a three-phase squirrel cage induction motor is solved on multiple processor cores working in parallel with each other. All inductances are calculated by doing a magneto-static solution of the machine at several rotor positions. The calculated inductances are saved in the three-dimensional (3D) lookup table as a function of the rotor position. The dynamic behavior is then studied in MATLAB/Simulink, and the results are validated by comparing them with the measurements taken from the laboratory test rig.

2. The Motor's Model

The voltage equations of a squirrel-cage induction motor with a stationary stator and short-circuited rotor cage can be described using magnetic coupled circuits theory as follows:

$$\boldsymbol{V_s} = \boldsymbol{I_s R_s} + \frac{d}{dt}(\boldsymbol{L_{ss} I_s} + \boldsymbol{L_{sr} I_r}) \tag{1}$$

$$\boldsymbol{0} = \boldsymbol{I_r R_r} + \frac{d}{dt}(\boldsymbol{L_{rs} I_s} + \boldsymbol{L_{rr} I_r}) \tag{2}$$

where $\boldsymbol{V_s}$, $\boldsymbol{I_s}$, $\boldsymbol{I_r}$, $\boldsymbol{R_s}$, $\boldsymbol{R_r}$, $\boldsymbol{L_{ss}}$, $\boldsymbol{L_{sr}}$, and $\boldsymbol{L_{rr}}$ are the vectors containing the machine's voltage, currents, resistances, and inductances respectively. The stator-stator, stator–rotor, rotor–stator, and rotor—rotor self and mutual inductance matrices ($\boldsymbol{L_{ss}}$, $\boldsymbol{L_{sr}}$, $\boldsymbol{L_{rs}}$, and $\boldsymbol{L_{rr}}$) can be defined as follows;

$$\boldsymbol{L_{ss}} = \begin{bmatrix} L_{aas} & L_{abs} & L_{acs} \\ L_{bas} & L_{bbs} & L_{bcs} \\ L_{cas} & L_{cbs} & L_{ccs} \end{bmatrix}, \tag{3}$$

$$\mathbf{L}_{sr} = \begin{bmatrix} L_{ar1} & L_{ar2} & \cdots & L_{ari} & \cdots & L_{arn} & L_{are} \\ L_{br1} & L_{br2} & \cdots & L_{bri} & \cdots & L_{brn} & L_{bre} \\ L_{cr1} & L_{cr2} & \cdots & L_{cri} & \cdots & L_{crn} & L_{cre} \end{bmatrix}, \tag{4}$$

$$\mathbf{L}_{rr} = \begin{bmatrix} L_{r1r1} & L_{r1r2} & \cdots & L_{r1ri} & \cdots & L_{r1rn} & L_{r1re} \\ L_{r2r1} & L_{r2r2} & \cdots & L_{r2ri} & \cdots & L_{r2rn} & L_{r2re} \\ \vdots & \vdots & \vdots & \vdots & \vdots & \vdots & \vdots \\ L_{rir1} & L_{rir2} & \cdots & L_{riri} & \cdots & L_{rirn} & L_{rire} \\ \vdots & \vdots & \vdots & \vdots & \vdots & \vdots & \vdots \\ L_{rnr1} & L_{rnr2} & \cdots & L_{rnri} & \cdots & L_{rnrn} & L_{rnre} \\ L_{rer1} & L_{rer2} & \cdots & L_{reri} & \cdots & L_{rern} & L_{rere} \end{bmatrix} \tag{5}$$

$$\mathbf{R}_{rr} = \begin{bmatrix} 2(R_b + r_e) & -R_b & 0 & 0 & \cdots & 0 & -R_b & -r_e \\ -R_b & 2(R_b + r_e) & -R_b & 0 & \cdots & 0 & 0 & -r_e \\ 0 & -R_b & 2(R_b + r_e) & -R_b & \cdots & 0 & 0 & -r_e \\ \vdots & \vdots & \vdots & \vdots & \vdots & \vdots & \vdots & \vdots \\ 0 & 0 & 0 & 0 & \cdots & 2(R_b + r_e) & -R_b & -r_e \\ -R_b & 0 & 0 & 0 & \cdots & -R_b & 2(R_b + r_e) & -r_e \\ -r_e & -r_e & -r_e & -r_e & \cdots & -r_e & -r_e & n_b r_e \end{bmatrix} \tag{6}$$

The last rows and columns in $\mathbf{L}_{sr}$, $\mathbf{L}_{rr}$, and $\mathbf{R}_{rr}$ correspond to the end ring values, which can be neglected in case of a perfect symmetrical machine, as the net end ring current is always zero. In unsymmetrical machines, these entries are important to simulate the end ring faults and to avoid the singularity problems while taking the inverse of inductance matrices.

For the ease of implementation, all these matrices can be grouped.

$$\mathbf{V}_s = \begin{bmatrix} v_{as} & v_{bs} & v_{cs} \end{bmatrix}^T \tag{7}$$

$$\mathbf{I}_s = \begin{bmatrix} i_{as} & i_{bs} & i_{cs} \end{bmatrix}^T \tag{8}$$

$$\mathbf{I}_r = \begin{bmatrix} i_{r1} & i_{r2} & \cdots i_{rn} i_{re} \end{bmatrix}^T \tag{9}$$

$$\mathbf{L} = \begin{bmatrix} L_{ss} & L_{sr} \\ L_{rs} & L_{rr} \end{bmatrix} \tag{10}$$

The currents, torque, and speed can be calculated as:

$$\begin{bmatrix} \mathbf{I}_s \\ \mathbf{I}_r \end{bmatrix} = \begin{bmatrix} L_{ss} & L_{sr} \\ L_{rs} & L_{rr} \end{bmatrix}^{-1} \int \left[\begin{bmatrix} Vs \\ 0 \end{bmatrix} - \begin{bmatrix} \mathbf{R}_s & 0 \\ 0 & \mathbf{R}_r \end{bmatrix} \begin{bmatrix} \mathbf{I}_s \\ \mathbf{I}_r \end{bmatrix} \right] dt \tag{11}$$

$$T_e = \mathbf{I}_s^T \left(\frac{d}{d\theta} L_{rs} \right) \mathbf{I}_r \tag{12}$$

In the matrices form:

$$T_e = \frac{1}{2}\left(\frac{p}{2}\right) \begin{bmatrix} \mathbf{I}_s \\ \mathbf{I}_r \end{bmatrix}^T \frac{d}{d\theta} \begin{bmatrix} L_{ss} & L_{sr} \\ L_{rs} & L_{rr} \end{bmatrix} \begin{bmatrix} \mathbf{I}_s \\ \mathbf{I}_r \end{bmatrix} \tag{13}$$

$$J \frac{d}{dt} \omega_m = T_e - T_L - B\omega_m \tag{14}$$

All inductances and resistances need to be calculated with stepping rotor and save them in 3D lookup tables where the third dimension corresponds to the rotor position as shown in Figure 1. All calculations can be done in offline environment using a magneto-static FEM solution and in the online environment the rotor position can be used as an index value to call a corresponding matrix from the lookup table to calculate the performance parameters like speed and torque, etc.

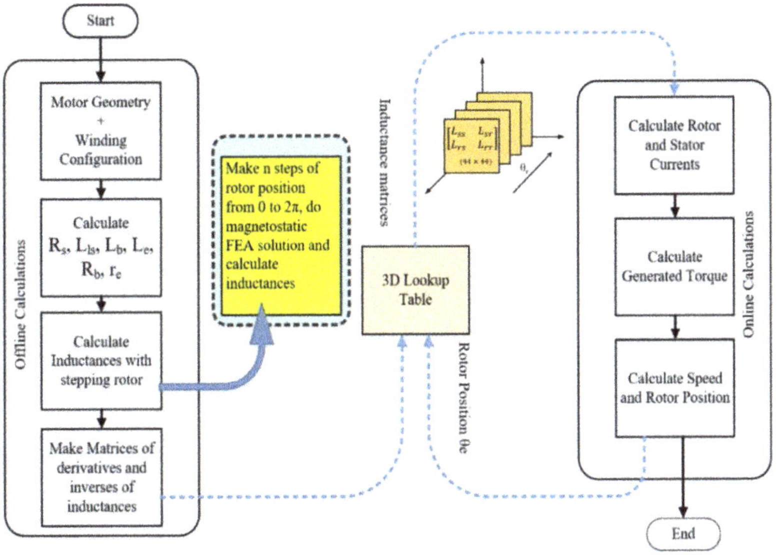

Figure 1. The schematic diagram of inductances calculations and their implementations for dynamic simulation.

3. LAN Network for Cluster Formation

Parallel computing is a form of concurrent computing where several workouts can be performed in the overlapping periods. Generally, any large problem can be divided into n-small problems, which can be solved simultaneously. Unlike traditional serial programs, the divided problem segments should be independent of each other so that they can run on different processors and the solutions can be combined on the client machine at the end. The general schematic diagram of distributed parallel computation is shown in Figure 2. The client machines, job scheduler, WIFI or LAN network, and the worker processors, are the main parts of the distributed cloud computation. The function of the job scheduler is to divide and distribute the segments of the bigger problem into cluster computers. The cluster computers can further divide their portion among their cores in the same manner.

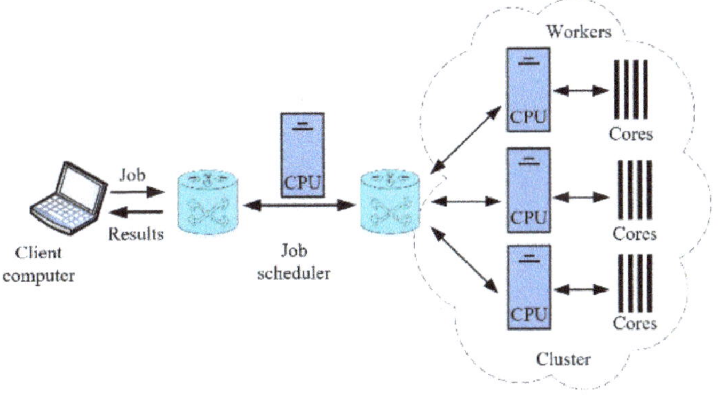

Figure 2. The cluster formation and utilization for parallel processing.

4. Inductances Calculations

In the case of electrical machines, most of the faults such as eccentricity, broken bars, and stator inter-turn short circuits make the machines unsymmetrical. Due to this fact, the complexity reduction techniques such as exploitation of symmetry by considering symmetric and non-symmetric boundary conditions and model order reduction becomes more tedious. The only best optimal and reliable way for fault diagnostics is to simulate the entire machine at various rotor positions. Since the solution at a distinct rotor position is independent of the solution at subsequent rotor positions, the total "n" rotor steps can be divided into various segments. The magneto-static problems of different rotor position sectors such as, $(0 \rightarrow \theta_1)$, $(\theta_2 \rightarrow \theta_3)$, ... , $(\theta_{n\text{-}1} \rightarrow \theta_n)$, $(\theta_{n+1} \rightarrow 2\pi)$ can be divided among the workers for parallel processing.

Figure 3 shows the required steps to calculate the inductances at different rotor positions. The computer cluster consists of four computers making a local area network (LAN). Each computer is Intel(R) Core(TM) i7-7500 CPU @ 3.41 GHz with 8 GB RAM and four cores. The finite element method (FEM) based model is constructed using open-access software FEMM 4.0. For making the model and collecting the results, FEMM is interfaced with MATLAB. After making the machine geometry and winding configuration on FEMM, MATLAB works as a job scheduler. It divides the total number of rotor steps among worker computers and their cores and receives the end results.

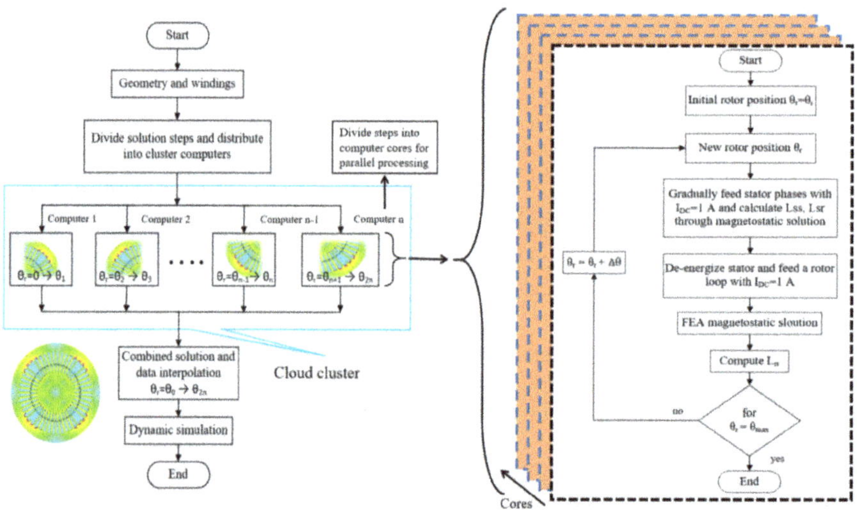

Figure 3. The division of rotor steps among various computers and their cores for parallel computation and the procedure of inductances calculation.

For better accuracy, a considerable number of mesh elements (250,160) with (125,177) nodes are solved having a precision of 1×10^{-8} at each rotor step. The FEMM 4.0 achieves this precision through a conjugate gradient solver with the help of multiple successive approximation iterations. Since the inductance profiles change with the change in the air gap, the selection of an appropriate rotor step size is very important. The changing air gap is the function of stator and rotor slot openings [37], which becomes very prominent when the openings on both sides align with each other and lead to the abruptly changing inductance derivatives. As the rotor and stator, slots are different in number, their least count multiple (LCM) can be a minimum choice regarding the number of rotor steps. By doing so, some phase inductances can be considered as shifted copies of the other phase inductances. In this case, the number of total rotor steps or solution samples will be divisible by the total number of rotor bars and the number of samples corresponding to 120 and 240 degrees for stator phases. For example, if the number of rotor steps is (40 × 48 = 1920), the L_{bb} will be equal to (120 × 1920/360) samples shifted copy of L_{aa}. The same is true in the case of rotor bar inductances, only self and mutual inductances of a single loop need to be calculated, the rest of them are shifted copies. However, special attention is needed if the fault changes all the inductances symmetrically or not.

5. The Simulation Results

At each rotor position, every individual stator phase is energized with a unity DC current, and relevant inductances are calculated by integrating the magnetic vector potential over the coil area as shown by the following equations.

$$L_{self} = \frac{\int_{si} A.J\, da}{i^2} \tag{15}$$

$$L_{mutual} = \frac{n_2}{i_1 a_2}\left(\int_{+si} A_1 da_1 - \int_{-si} A_1 da_2\right) \tag{16}$$

where A is the vector potential, J is the current density, n is the number of turns per phase, i is the phase current, the subscript *"si"* is for surface integral. The first bracket term in Equation (16) is the integration of the vector potential of coils with positive current or the coils pointing out of the page. The second term corresponds to the coils with negative current or pointing into the page, a_2 is the cross-sectional area of the coil, which is approximately equal to the slot area multiplied with the filling factor. The motor's magnetic flux distribution after the energizing phase *"a"* with 1 A DC current is shown in Figure 4. The highlighted slots contain phase *"a"* winding whose surface integral of magnetic potential is equal to self-inductance *Laa* at the specific rotor position as in Equation (15).

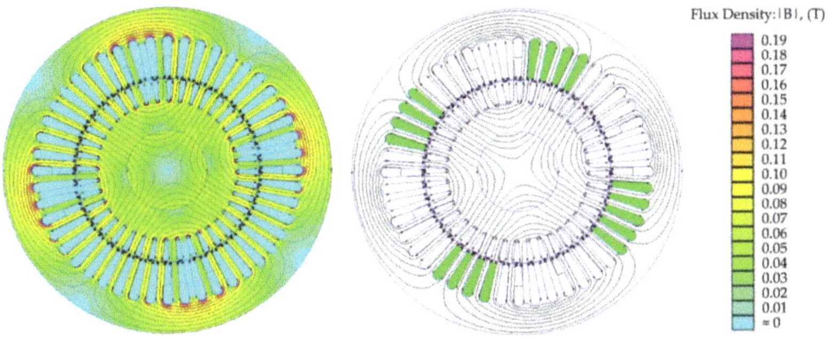

Figure 4. The flux distribution with energized phase *"a"* with I_{DC} = 1 A, and the selection of slots for vector potential integral for calculation of self-inductance L_{aa}.

Similarly, all stator–stator self and mutual and stator–rotor mutual inductances can be calculated. The rotor–rotor self and mutual inductances are calculated by energizing a single rotor loop with unity DC current as shown in Figure 5. Only one rotor loop needs to be energized and solved for its inductances, and the rest of them possess the same solution with a phase shift equal to the angle difference between them.

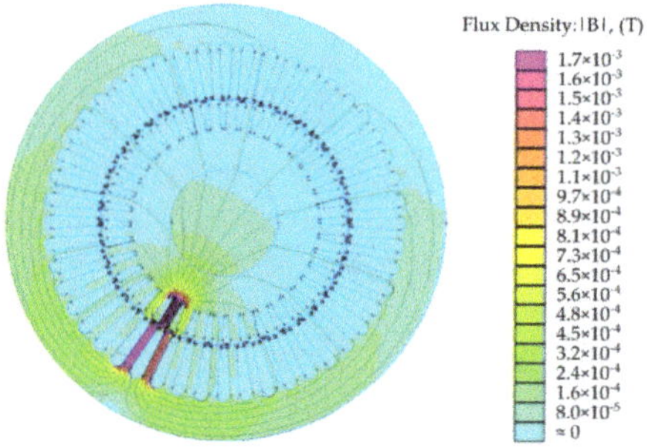

Figure 5. Flux distribution due to energized rotor loop with $I_{DC} = 1$ A for the calculation of rotor inductances.

Since the solution is for the 2D model, the effect of end windings is compensated by using additional end winding leakage inductance and resistance using the following analytical formulas.

$$L_{ew} = \frac{Q_s}{m} q \left(\frac{Z_q}{a} \right)^2 \mu_o l_w \lambda_w \tag{17}$$

where Qs is the number of stator slots, Zq is the number of conductors per slot, a is the number of winding parallel paths per phase, m is the total number of stator phases, and q is the number of slots in a pole captured by an individual phase. lw is the average length of the end winding, λw is the permeance factor which is 0.20 for the motor under investigation. The same formula can be used to calculate the leakage inductance of the rotor end rings. For building the resistance matrix, various stator- and rotor-related resistances are calculated using resistivity formula where the skinning and proximity effects are neglected because of the DC supply current;

$$R = \frac{\rho l}{A} \tag{18}$$

where ρ is the resistivity, A is the cross-sectional area, and l is the length of the conductor. The effective slot area (ESA) is equal to the total area of slot multiplied with the filling factor, which is 0.60 for the machine under investigation. The conductor cross-sectional area can be calculated by dividing the ESA with the total number of conductors in the slot, which are 17 in this case. The resistance of the end windings is included by increasing the length of the per phase conductor corresponding to the length of the end winding. Since the stator and rotor windings are energized with unity DC current, the effects such as proximity, skinning, material saturation, and eddy currents are neglected as the focus is towards the simulation time reduction. However, they can be included in the online section analytically.

Figure 6 shows various inductances as a function of the rotor position. All self and mutual inductances are the functions of the air gap, which changes with the stepping rotor. This phenomenon is evident in the inductance profiles, which are calculated with a rotor step size of 0.1875 degrees. The self- and mutual inductances of stator have five cycles until an angle of 45 degrees, which corresponds to 40 cycles until 360 degrees. This is because the rotor has 40 bars having 40 slot openings. The stator inductances consider the stator air gap as static while rotor associated air gap moves with the moving rotor. The same is true for rotor self- and mutual inductances, which have six cycles per 45 degrees corresponding to 48 cycles till 360 degrees. Where 48 is the number of stator slots and for rotor inductances, the rotor-associated air gap remains static while the stator gap has a relative motion.

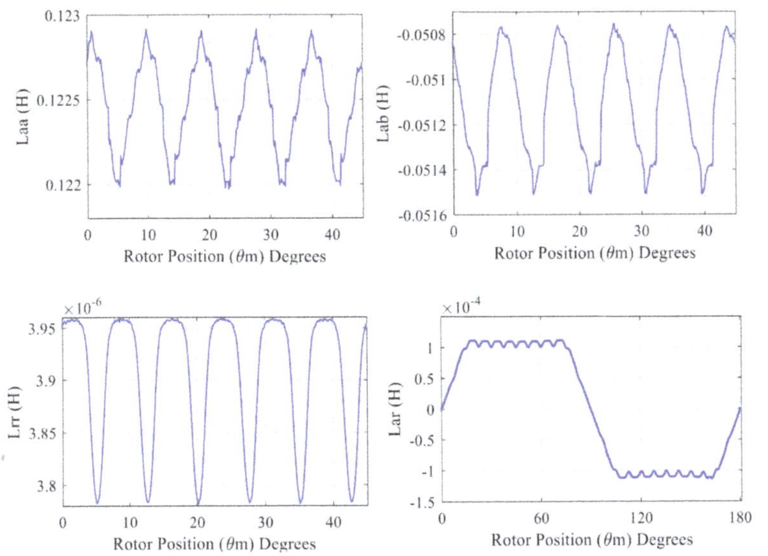

Figure 6. The calculated inductances as a function of the rotor position.

6. Test Setup

The test rig consists of two identical motors with specifications given in Table 1. Both machines are attached back to back on the same mechanical foundation as shown in Figure 7. The first motor is the test motor while the second acts as a loading machine. The grid supplies the test machine while the loading motor is controlled using ABB ACS-880 industrial inverter for better controllability of slip. The measurement time of the stator current of the test machine under healthy and broken bar cases is 100 sec with a sampling frequency of 10 kHz.

Table 1. Motor specifications.

Sr. No.	Parameter	Symbol	Value
1	Rated speed	N_r	1400 rpm@50 Hz
2	Rated power	P_r	18 kW@50 Hz
3	Connection	Y, Δ	Star (Y)
4	Power factor	$\cos\varphi$	0.860
5	Number of poles	P	4
6	Number of rotor bars	N_{rb}	40
7	Number of stator slots	Ns	48

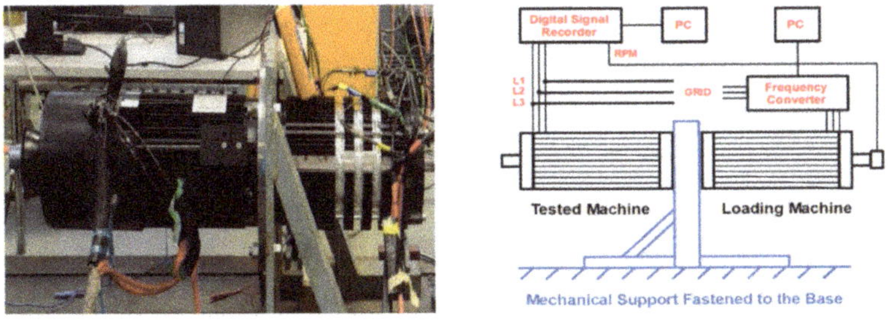

Figure 7. The test rig and its block diagram.

7. Results and Discussion

7.1. Stator Current Spectrum under Healthy and Broken Rotor Bar Cases

The varying inductances give rise to the harmonics in the stator voltage and currents. The most prominent of them are supply based and spatial harmonics. A comparison of the frequency spectrum of stator current obtained from the proposed model and the measurements taken from the laboratory test rig is shown in Figure 8. The only additional harmonics in the practical signal are the third harmonics coming from the supply side. Another major source of harmonics in the signal spectrum is the fault, which can act as a definition component for condition monitoring. Figure 9 shows the development of the left side harmonics (LSH) due to the broken bars at the rated load.

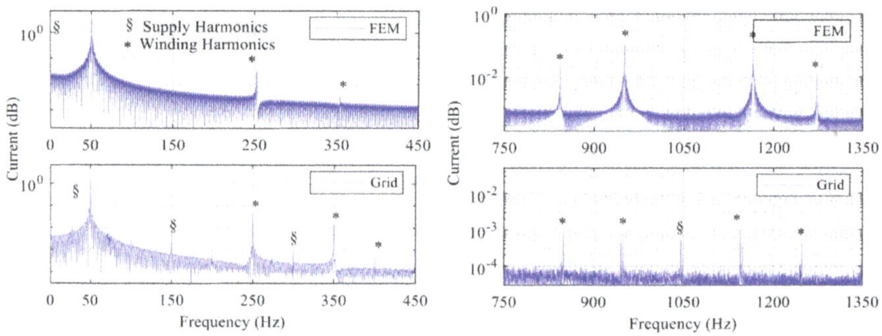

Figure 8. The frequency spectrum of stator current obtained from the proposed model and laboratory-based measurements.

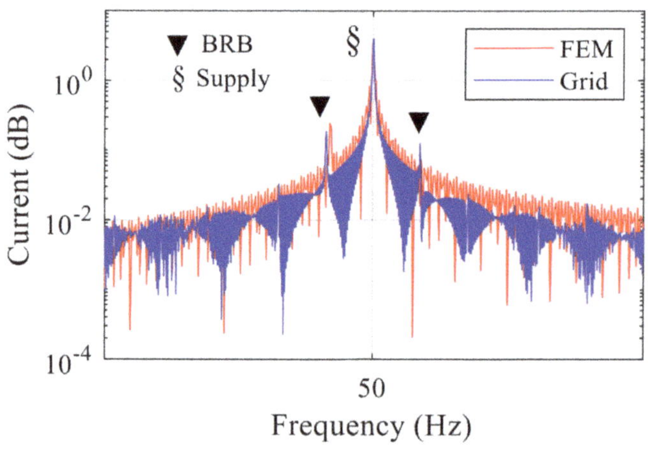

Figure 9. The broken rotor bars-based side-band frequencies.

7.2. Time Comparison

The overall computational speed of each worker (computer) can be increased by using all cores in a parallel fashion. This increase in speed is because of increased cache, which reduces memory latency; the parallel handling of the independent instructions; and the improved performance of the processor power wall. The per-step computational time for the calculation of inductances relevant to a single phase (*Laa*, *Lab*, *Lac*, and *Lar*) at different rotor positions by using the different number of cores is shown in Figure 10a. Since, with the increase in the number of cores, more processing power is being utilized, the per-step calculation time increases. Moreover, the variation in the calculation time with the increase in the number of cores also increases, which depends upon the processor being utilized by auxiliary programs like Windows, etc. Figure 10b shows the mean per-step simulation time, which increases from 100 to 200 s per step with the increase in the utilization of processing power from about 25% (one core) to 94% (four cores). The mean simulation time is calculated because each step takes a slightly different time for calculation because of the change in the number of mesh elements and the other programs running in parallel. The overall calculation time for all inductances for 1920 rotor positions is shown in Figure 10c. It is obvious that even the per step simulation time with an increase in the parallel processing increases, but the total calculation time decreases dramatically. Meanwhile, Figure 10d shows the time taken by the cluster of four computers working as a LAN. The simulation time decreases considerably and the non-linear decrease in time is due to the latency of the network. It is worth mentioning that the job scheduler prefers cores of different computers to work in parallel. This is the reason why the computational time in Figure 10d with four cores of different computers working in parallel is considerably less than the time taken by one computer with all four cores engaged in parallel as shown in Figure 10c.

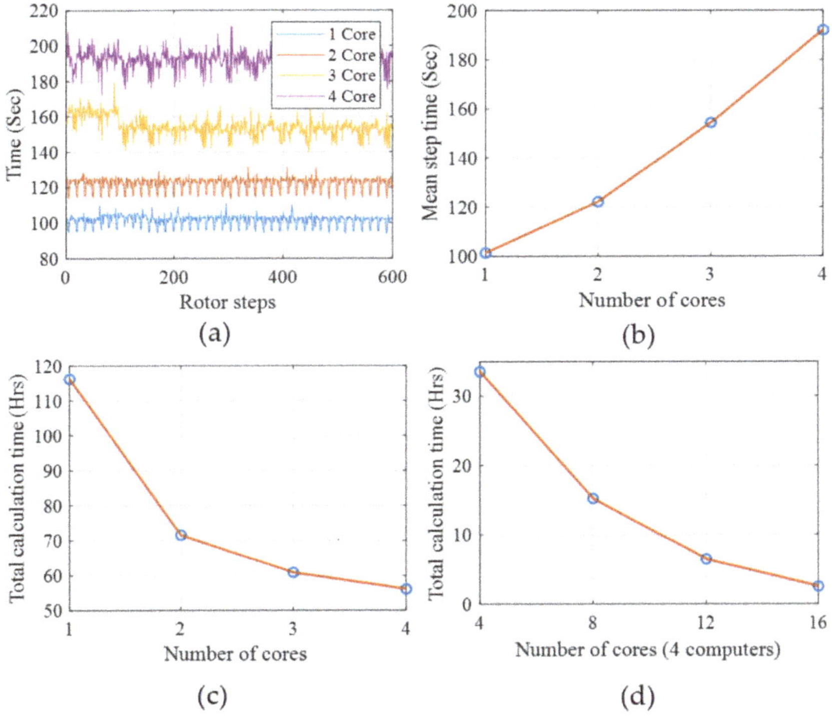

Figure 10. The simulation time, (**a**) per step with different number of single computer workers (cores) in parallel, (**b**) the mean per-step time with different number of cores of a single computer working in parallel, (**c**) overall time on one computer with different cores engaged in parallel, (**d**) overall time with different number of computers making a cluster with distributed cores in parallel.

8. Conclusions

This paper presents a hybrid FEM–analytical model solved in a parallel fashion on a cluster of computers for the reduction of computational time. The artificial intelligence (AI) based condition-monitoring techniques for predictive maintenance of electrical machines is gaining heightened popularity. This is not only because of their ability to detect faults at the incipient stage but also because of their aptitude for faults segregation. The accurate model of the electrical machines is the key element of these techniques, which is crucial for the collection of big data under various healthy and faulty conditions. This data is essential for the training of the diagnostic systems and for making the safety rules. Among several modeling techniques, the FEM-based models have proved their accuracy in the field of electrical machine design for the past few decades. The FEM models have very few approximations and can deal with almost any kind of geometrical complexities of the system as compared to their analytical counterparts. The analytical models have their own attractions, such as the reduced simulation time and the development of analytical equations, which are integral parts of drives and inverse problem theory, etc. The biggest challenge for FEM-based models is their complexity in the forms of computational time and required memory. While on the analytical side, the approximations are fatal for any reliable diagnostic algorithm. The world is witnessing the exponentially increasing trend in the power of processors and sophisticated IT networks, which leads toward cloud computation and industry 4.0 standards.

By exploiting the benefits of analytical, FEM models, and cloud computation, this paper proposes a hybrid analytical–FEM model for the simulation of the machine under healthy and faulty conditions with reduced calculation time. Most of the techniques dealing with the reduction in the simulation time of FEM models fail when the machine is in a faulty condition, which makes it purely unsymmetrical. Moreover, any approximation for the sake of reduced complexity can decrease the reliability of the model-dependent diagnostic algorithm. With the development of more sophisticated processors and industry 4.0 standards, the complete models of the system can be solved in very much less time as compared to the conventional techniques. In this paper, the model is first divided into offline and online portions. All the inductances and other necessary parameters are calculated in the offline section, and the results are saved in 3D matrices as a function of rotor position. Once the inductances are calculated, they can be used in an online dynamic model where the performance parameters such as speed, torque, flux, and currents can be investigated under different conditions. In the case of some faults such as broken bars or broken end rings, the inductance matrices need not be calculated again but the fault can be manipulated in the online portion. This can be achieved by changing the values of the corresponding elements in the resistance matrix. For reducing the calculation time, the FEM model is divided into several computers before dividing the specific portion into the cores of any particular processor. For making a computer cluster resembling a cloud, four computers having the same specifications were used as a local area network. The total rotor steps were then divided into several cores working in parallel and the results were collected on the main computer. This technique reduces the simulation time drastically without any need for model approximation. For the validation of results, the frequency spectrum of simulated stator current is compared with the one measured in the laboratory setup under healthy and broken rotor bar cases.

Author Contributions: Conceptualization, B.A., T.V., and A.B.; methodology, B.A., T.V., and A.B.; validation, A.K., A.R., and A.B.; data curation, B.A.; writing—original draft preparation, B.A.; writing—review and editing, A.R., and M.N.I.; visualization, T.V.; supervision, T.V., and A.B. All authors have read and agreed to the published version of the manuscript.

Funding: This research received no external funding.

Conflicts of Interest: The authors declare no conflict of interest.

References

1. Saidur, R. A review on electrical motors energy use and energy savings. *Renew. Sustain. Energy Rev.* **2010**, *14*, 877–898. [CrossRef]
2. Sahraoui, M.; Cardoso, A.J.M.; Ghoggal, A. The use of a modified prony method to track the broken rotor bar characteristic frequencies and amplitudes in three-phase induction motors. *IEEE Trans. Ind. Appl.* **2015**, *51*, 2136–2147. [CrossRef]

3. Naha, A.; Samanta, A.K.; Routray, A.; Deb, A.K. Low Complexity Motor Current Signature Analysis Using Sub-Nyquist Strategy with Reduced Data Length. *IEEE Trans. Instrum. Meas.* **2017**, *66*, 3249–3259. [CrossRef]
4. Asad, B.; Vaimann, T.; Belahcen, A.; Kallaste, A.; Rassõlkin, A.; Iqbal, M.N. Broken rotor bar fault detection of the grid and inverter-fed induction motor by effective attenuation of the fundamental component. *IET Electr. Power Appl.* **2019**, *13*, 2005–2014. [CrossRef]
5. Zhen, D.; Wang, T.; Gu, F.; Ball, A. Fault diagnosis of motor drives using stator current signal analysis based on dynamic time warping. *Mech. Syst. Signal Process.* **2013**, *34*, 191–202. [CrossRef]
6. Garcia-Ramirez, A.G.; Morales-Hernandez, L.A.; Osornio-Rios, R.A.; García-Pérez, A.; Romero-Troncoso, R.D.J. Thermographic Technique as a Complement for MCSA in Induction Motor Fault Detection. In Proceedings of the 2014 International Conference on Electrical Machines (ICEM), ICEM 2014, Berlin, Germany, 2–5 September 2014; pp. 1940–1945.
7. Taheri-Garavand, A.; Ahmadi, H.; Omid, M.; Mohtasebi, S.S.; Mollazade, K.; Smith, A.J.R.; Carlomagno, G.M. An intelligent approach for cooling radiator fault diagnosis based on infrared thermal image processing technique. *Appl. Therm. Eng.* **2015**, *87*, 434–443. [CrossRef]
8. Glowacz, A.; Glowacz, Z. Diagnostics of stator faults of the single-phase induction motor using thermal images, MoASoS and selected classifiers. *Measurement* **2016**, *93*, 86–93. [CrossRef]
9. Luong, P.; Wang, W. Smart Sensor-Based Synergistic Analysis for Rotor Bar Fault Detection of Induction Motors. *IEEE/ASME Trans. Mechatronics* **2020**, *25*, 1067–1075. [CrossRef]
10. Caesarendra, W.; Kosasih, B.; Tieu, A.K.; Zhu, H.; Moodie, C.A.; Zhu, Q. Acoustic emission-based condition monitoring methods: Review and application for low speed slew bearing. *Mech. Syst. Signal Process.* **2016**, *72–73*, 134–159. [CrossRef]
11. Vaimann, T.; Sobra, J.; Belahcen, A.; Rassõlkin, A.; Rolak, M.; Kallaste, A. Induction machine fault detection using smartphone recorded audible noise. *IET Sci. Meas. Technol.* **2018**, *12*, 554–560. [CrossRef]
12. Henao, H.; Demian, C.; Capolino, G.-A. A frequency-domain detection of stator winding faults in induction machines using an external flux sensor. *IEEE Trans. Ind. Appl.* **2003**, *39*, 1272–1279. [CrossRef]
13. Frosini, L.; Harlisca, C.; Szabo, L. Induction Machine Bearing Fault Detection by Means of Statistical Processing of the Stray Flux Measurement. *IEEE Trans. Ind. Electron.* **2015**, *62*, 1846–1854. [CrossRef]
14. Stone, G.C.; Sedding, H.G.; Chan, C. Experience With Online Partial-Discharge Measurement in High-Voltage Inverter-Fed Motors. *IEEE Trans. Ind. Appl.* **2018**, *54*, 866–872. [CrossRef]
15. Mirzaeva, G.; Saad, K.I.; Jahromi, M.G. Comprehensive Diagnostics of Induction Motor Faults Based on Measurement of Space and Time Dependencies of Air Gap Flux. *IEEE Trans. Ind. Appl.* **2017**, *53*, 2657–2666. [CrossRef]
16. Liu, R.; Yang, B.; Zio, E.; Chen, X. Artificial intelligence for fault diagnosis of rotating machinery: A review. *Mech. Syst. Signal Process.* **2018**, *108*, 33–47. [CrossRef]
17. Konar, P.; Chattopadhyay, P. Bearing fault detection of induction motor using wavelet and Support Vector Machines (SVMs). *Appl. Soft Comput.* **2011**, *11*, 4203–4211. [CrossRef]
18. Li, X.; Wang, K.; Jiang, L. The Application of AE Signal in Early Cracked Rotor Fault Diagnosis with PWVD and SVM. *J. Softw.* **2011**, *6*, 1969–1976. [CrossRef]
19. Soualhi, A.; Medjaher, K.; Zerhouni, N. Bearing Health Monitoring Based on Hilbert–Huang Transform, Support Vector Machine, and Regression. *IEEE Trans. Instrum. Meas.* **2015**, *64*, 52–62. [CrossRef]
20. Sadeghian, A.; Ye, Z.; Wu, B. Online Detection of Broken Rotor Bars in Induction Motors by Wavelet Packet Decomposition and Artificial Neural Networks. *IEEE Trans. Instrum. Meas.* **2009**, *58*, 2253–2263. [CrossRef]
21. Bin, G.F.; Gao, J.J.; Li, X.J.; Dhillon, B.S. Early fault diagnosis of rotating machinery based on wavelet packets—Empirical mode decomposition feature extraction and neural network. *Mech. Syst. Signal Process.* **2012**, *27*, 696–711. [CrossRef]
22. Prieto, M.D.; Cirrincione, G.; Espinosa, A.G.; Ortega, J.A.; Henao, H. Bearing Fault Detection by a Novel Condition-Monitoring Scheme Based on Statistical-Time Features and Neural Networks. *IEEE Trans. Ind. Electron.* **2013**, *60*, 3398–3407. [CrossRef]
23. Wang, J.; Liu, S.; Gao, R.X.; Yan, R. Current envelope analysis for defect identification and diagnosis in induction motors. *J. Manuf. Syst.* **2012**, *31*, 380–387. [CrossRef]
24. Palácios, R.H.C.; Da Silva, I.N.; Goedtel, A.; Godoy, W.F. A comprehensive evaluation of intelligent classifiers for fault identification in three-phase induction motors. *Electr. Power Syst. Res.* **2015**, *127*, 249–258. [CrossRef]

25. Wan, X.; Wang, D.; Tse, P.W.; Xu, G.; Zhang, Q. A critical study of different dimensionality reduction methods for gear crack degradation assessment under different operating conditions. *Measurement* **2016**, *78*, 138–150. [CrossRef]
26. Ali, M.Z.; Shabbir, N.S.K.; Zaman, S.M.K.; Liang, X. Single- and Multi-Fault Diagnosis Using Machine Learning for Variable Frequency Drive-Fed Induction Motors. *IEEE Trans. Ind. Appl.* **2020**, *56*, 2324–2337. [CrossRef]
27. Shao, S.; Yan, R.; Lu, Y.; Wang, P.; Gao, R.X. DCNN-Based Multi-Signal Induction Motor Fault Diagnosis. *IEEE Trans. Instrum. Meas.* **2020**, *69*, 2658–2669. [CrossRef]
28. Jung, U.; Koh, B.-H. Wavelet energy-based visualization and classification of high-dimensional signal for bearing fault detection. *Knowl. Inf. Syst.* **2015**, *44*, 197–215. [CrossRef]
29. Pandya, D.H.; Upadhyay, S.H.; Harsha, S.P. Fault diagnosis of rolling element bearing with intrinsic mode function of acoustic emission data using APF-KNN. *Expert Syst. Appl.* **2013**, *40*, 4137–4145. [CrossRef]
30. He, D.; Li, R.; Zhu, J. Plastic Bearing Fault Diagnosis Based on a Two-Step Data Mining Approach. *IEEE Trans. Ind. Electron.* **2013**, *60*, 3429–3440. [CrossRef]
31. Baccarini, L.M.R.; De Menezes, B.R.; Caminhas, W.M. Fault induction dynamic model, suitable for computer simulation: Simulation results and experimental validation. *Mech. Syst. Signal Process.* **2010**, *24*, 300–311. [CrossRef]
32. Cunha, C.C.M.; Oliveira, P.; Lyra, R.; Filho, B.C. Simulation and analysis of induction machines with rotor asymmetries. *IEEE Trans. Ind. Appl.* **2005**, *41*, 18–24.
33. Liang, J.; Qiu, Y.; Zhao, M.; Kang, S.; Lu, H. The modeling and numerical simulations of wind turbine generation system with free vortex method and simulink. *Energy Convers. Manag.* **2015**, *103*, 762–777.
34. Lubin, T.; Hamiti, T.; Razik, H.; Rezzoug, A. Comparison Between Finite-Element Analysis and Winding Function Theory for Inductances and Torque Calculation of a Synchronous Reluctance Machine. *IEEE Trans. Magn.* **2007**, *43*, 3406–3410.
35. Raziee, S.M.; Misir, O.; Ponick, B. Winding Function Approach for Winding Analysis. *IEEE Trans. Magn.* **2017**, *53*, 1–9.
36. Nandi, S. A Detailed Model of Induction Machines With Saturation Extendable for Fault Analysis. *IEEE Trans. Ind. Appl.* **2004**, *40*, 1302–1309.
37. Asad, B.; Vaimann, T.; Belahcen, A.; Kallaste, A.; Rassõlkin, A.; Iqbal, M.N. Modified winding function-based model of squirrel cage induction motor for fault diagnostics. *IET Electr. Power Appl.* **2020**, *14*, 1722–1734.
38. Sudhoff, S.D.; Kuhn, B.T.; Corzine, K.A.; Branecky, B.T. Magnetic Equivalent Circuit Modeling of Induction Motors. *IEEE Trans. Energy Convers.* **2007**, *22*, 259–270.
39. Apsley, J.M.; Williamson, S. Analysis of Multi-Phase Induction Machines with Winding Faults. In Proceedings of the IEEE International Conference on Electric Machines and Drives, San Antonio, TX, USA, 15 May 2005; pp. 249–255.
40. Wang, L.; Jatskevich, J.; Pekarek, S.D. Modeling of induction machines using a voltage-behind-reactance formulation. *IEEE Trans. Energy Convers.* **2008**, *23*, 382–392.
41. Sapena-Bano, A.; Martinez-Roman, J.; Puche-Panadero, R.; Pineda-Sanchez, M.; Perez-Cruz, J.; Riera-Guasp, M. Induction machine model with space harmonics for fault diagnosis based on the convolution theorem. *Int. J. Electr. Power Energy Syst.* **2018**, *100*, 463–481.
42. Ding, Q.; Yang, Z.; Sun, X.; Zhao, Q.; Zhu, H. Analysis of rotor slot width influence on a bearingless induction motor. *Comput. Electr. Eng.* **2020**, *81*, 106534.
43. Belahcen, A.; Rasilo, P.; Arkkio, A. Segregation of Iron Losses From Rotational Field Measurements and Application to Electrical Machine. *IEEE Trans. Magn.* **2014**, *50*, 893–896. [CrossRef]
44. Zhang, H. Online Thermal Monitoring Models for Induction Machines. *IEEE Trans. Energy Convers.* **2015**, *30*, 1–9. [CrossRef]
45. Zhang, Y.; Ruan, J.; Huang, T.; Yang, X.; Zhu, H.; Yang, G. Calculation of Temperature Rise in Air-cooled Induction Motors Through 3-D Coupled Electromagnetic Fluid-Dynamical and Thermal Finite-Element Analysis. *IEEE Trans. Magn.* **2012**, *48*, 1047–1050. [CrossRef]
46. Bourchas, K.; Stening, A.; Soulard, J.; Broddefalk, A.; Lindenmo, M.; Dahlen, M.; Gyllensten, F. Quantifying Effects of Cutting and Welding on Magnetic Properties of Electrical Steels. *IEEE Trans. Ind. Appl.* **2017**, *53*, 4269–4278. [CrossRef]

47. Sapena-Bano, A.; Chinesta, F.; Pineda-Sanchez, M.; Aguado, J.; Borzacchiello, D.; Puche-Panadero, R. Induction machine model with finite element accuracy for condition monitoring running in real time using hardware in the loop system. *Int. J. Electr. Power Energy Syst.* **2019**, *111*, 315–324. [CrossRef]
48. Ghahfarokhi, P.S.; Belahcen, A.; Kallaste, A.; Vaimann, T.; Gerokov, L.; Rassolkin, A. Thermal Analysis of a SynRM Using a Thermal Network and a Hybrid Model. In Proceedings of the 2018 23rd International Conference on Electrical Machines, ICEM 2018, Alexandroupoli, Greece, 3–6 September 2018; pp. 2682–2688.
49. Mukhrejee, V.; Far, M.F.; Martin, F.; Belahcen, A. Constrained Algorithm for the Selection of Uneven Snapshots in Model Order Reduction of a Bearingless Motor. *IEEE Trans. Magn.* **2017**, *53*, 1–4. [CrossRef]
50. Far, M.F.; Martin, F.; Belahcen, A.; Rasilo, P.; Awan, H.A.A. Real-Time Control of an IPMSM Using Model Order Reduction. *IEEE Trans. Ind. Electron.* **2020**, *1*. [CrossRef]
51. Far, M.F.; Martin, F.; Belahcen, A.; Montier, L.; Henneron, T. Orthogonal Interpolation Method for Order Reduction of a Synchronous Machine Model. *IEEE Trans. Magn.* **2018**, *54*, 1–6.
52. Borzacchiello, D.; Aguado, J.V.; Chinesta, F. Non-intrusive Sparse Subspace Learning for Parametrized Problems. *Arch. Comput. Methods Eng.* **2019**, *26*, 303–326. [CrossRef]
53. Villalonga, A.; Beruvides, G.; Castano, F.; Haber, R.E. Cloud-Based Industrial Cyber–Physical System for Data-Driven Reasoning: A Review and Use Case on an Industry 4.0 Pilot Line. *IEEE Trans. Ind. Informatics* **2020**, *16*, 5975–5984. [CrossRef]
54. Chen, Z.; Quan, W.; Wen, M.; Fang, J.; Yu, J.; Zhang, C.; Luo, L. Deep Learning Research and Development Platform: Characterizing and Scheduling with QoS Guarantees on GPU Clusters. *IEEE Trans. Parallel Distrib. Syst.* **2020**, *31*, 34–50. [CrossRef]
55. Gao, Y.; Ai, Y.; Tian, B.; Chen, L.; Wang, J.; Cao, D.; Wang, F.-Y. Parallel End-to-End Autonomous Mining: An IoT-Oriented Approach. *IEEE Internet Things J.* **2020**, *7*, 1011–1023. [CrossRef]
56. Bandyopadhyay, I.; Purkait, P.; Koley, C. A combined image processing and Nearest Neighbor Algorithm tool for classification of incipient faults in induction motor drives. *Comput. Electr. Eng.* **2016**, *54*, 296–312. [CrossRef]
57. Guesmi, H.; Ben Salem, S.; Bacha, K. Smart wireless sensor networks for online faults diagnosis in induction machine. *Comput. Electr. Eng.* **2015**, *41*, 226–239. [CrossRef]
58. Chen, J.; Li, K.; Bilal, K.; Zhou, X.; Li, K.; Yu, P.S. A Bi-layered Parallel Training Architecture for Large-Scale Convolutional Neural Networks. *IEEE Trans. Parallel Distrib. Syst.* **2019**, *30*, 965–976. [CrossRef]
59. You, Y.; Zhang, Z.; Hsieh, C.-J.; Demmel, J.; Keutzer, K. Fast Deep Neural Network Training on Distributed Systems and Cloud TPUs. *IEEE Trans. Parallel Distrib. Syst.* **2019**, *30*, 2449–2462. [CrossRef]
60. Huang, S.-R.; Huang, K.-H.; Chao, K.-H.; Chiang, W.-T. Fault analysis and diagnosis system for induction motors. *Comput. Electr. Eng.* **2016**, *54*, 195–209. [CrossRef]

applied sciences

MDPI

Article

Transient Modeling and Recovery of Non-Stationary Fault Signature for Condition Monitoring of Induction Motors

Bilal Asad [1,2,*], Toomas Vaimann [1], Anouar Belahcen [1,2], Ants Kallaste [1], Anton Rassõlkin [1], Payam Shams Ghafarokhi [3] and Karolina Kudelina [1]

1 Department of Electrical Power Engineering and Mechatronics, Tallinn University of Technology, 19086 Tallinn, Estonia; toomas.vaimann@taltech.ee (T.V.); Anouar.Belahcen@aalto.fi (A.B.); ants.kallaste@taltech.ee (A.K.); anton.rassolkin@taltech.ee (A.R.); karolina.kudelina@taltech.ee (K.K.)
2 Department of Electrical Engineering and Automation, Aalto University, FI-00076 Espoo, Finland
3 Department of Electrical Machine and Apparatus, Riga Technical University, LV-1658 Riga, Latvia; payam.shams-ghahfarokhi@rtu.lv
* Correspondence: bilal.asad@aalto.fi

Abstract: This paper presents the modeling and the broken rotor bar fault diagnostics by time–frequency analysis of the motor current under an extended startup transient time. The transient current-based nonstationary signal is retrieved and investigated for its time–frequency response to segregate the rotor faults and spatial harmonics. For studying the effect of reduced voltage on various parameters and the theoretical definition of the fault patterns, the winding function analysis (WFA)-based model is presented first. Moreover, an algorithm to improve the spectrum legibility is proposed. It is shown that by efficient utilization of the attenuation filter and consideration of the area containing the maximum power spectral density, the diagnostic algorithm gives promising results. The results are based on the machine's analytical model and the measurements taken from the laboratory setup.

Keywords: condition monitoring; fault diagnosis; Fourier transform; induction motors; modeling; wavelet transform

Citation: Asad, B.; Vaimann, T.; Belahcen, A.; Kallaste, A.; Rassõlkin, A.; Ghafarokhi, P.S.; Kudelina, K. Transient Modeling and Recovery of Non-Stationary Fault Signature for Condition Monitoring of Induction Motors. *Appl. Sci.* **2021**, *11*, 2806. https://doi.org/10.3390/app11062806

Academic Editor: Mohamed Benbouzid

Received: 17 February 2021
Accepted: 18 March 2021
Published: 21 March 2021

1. Introduction

Electrical machines have been showing their influential role in industrial and domestic applications since the second industrial revolution. This role is evident in electricity generation, such as wind power plants or electrical to mechanical energy converters, which are driving the industry. In various electrical machines, induction motors are ubiquitous because of their simple structure, good efficiency and easy maintenance. A variety of applications makes them consume more than fifty percent of the total energy generated worldwide.

The mechanically moving parts and harsh industrial environment make them vulnerable to faults. The electrical faults are mostly related to the stator, such as inter-turn short circuits, phase drop, voltage imbalance, earthing and inverter-related defects. However, the mechanical faults are mainly associated with the rotor, such as broken bars, bad bearings, eccentricity, broken end rings or inadequate foundations. All those faults are degenerative, making it crucial to detect them at the developing stage to avoid catastrophic situations. Various fault diagnostic techniques are available in the literature, such as vibration analysis, thermal analysis, acoustic analysis, electromagnetic field inference, leakage flux, infrared light detection and chemical analysis.

However, motor current signature analysis (MCSA)-based fault diagnostic techniques are gaining heightened popularity because of their noninvasive nature and lower complexity. Moreover, a vast field of data processing techniques improves their flexibility and makes them more reliable.

It is a well-studied fact that almost all rotor faults modulate the stator currents with specific frequency components. The detection of those frequency components can lead to the cause of the fault. The Fourier transform can be considered a foundation stone for all advanced signal processing techniques. The majority of MCSA-based methods depend on a fast Fourier transform (FFT) of the signal; for example, in [1], the authors used the FFT on the active and reactive currents of a motor to investigate the broken rotor bars and load oscillations. The authors of [2] used the FFT in conjunction with Park's vector to make an artificial ants clustering technique for the fault diagnostics of an induction motor. The Park's vector makes an entire domain of condition monitoring of electrical equipment [3–7], giving promising results if used in conjunction with FFT. In [8], the autoregressive method relied on a discrete-time Fourier transform (DTFT) and a notch filter. The researchers in [9] used the FFT to prove that slot harmonics could be used as potential indicators to detect broken rotor bars. In [10], the authors used an adoptive notch filter and FFT for broken rotor bar fault diagnostics of an induction motor. In [11], the FFT was used on simulations and practical results to investigate broken rotor bars and mechanical vibrations. In [12], Nandi used the FFT to study the stator current frequency spectrum for different fault conditions, and in [13], the FFT was used along with a band-stop filter for the detection of broken rotor bar frequencies.

However, there are certain limitations of the FFT, putting a question mark on its reliability for fault diagnostics. These limitations include spectral leakage, which is the power of the main components leaking into the subsequent frequency bins. If the acquired signal's length and sampling frequency are not good enough, the faults representing frequency components are highly likely to be buried under the corresponding primary frequency bins. This problem becomes worse when the motor is working under no-load or lesser-load conditions, as the faulty frequencies are the function of slip. The other issue of the FFT is that the signal should be in a steady state regime, be stationary and not have any discontinuities. These problems are becoming worse as the inverters are coming forward as an integral part of the drive system. The inverter-fed voltage is full of harmonics, which makes the frequency spectrum hazy to understand. Moreover, the drive control algorithms can also have an impact on the amplitude of the harmonics. For example, in the case of direct torque control (DTC) motors, the drive directly influences the current signal carrying all the information about the motor's health [14].

Researchers have tried several different techniques to cope with these problems. The use of Hilbert transforms to extract the signal's envelope, which possesses considerable information regarding the electrical machine's health, can be found in [15]. The authors in [16] used a fractional Fourier transform to recover the faulty frequencies from a non-stationary signal. In [17], the authors used a sliding discrete Fourier transform for the detection of broken rotor bars, while [18–21] used the wavelet technique to improve the accuracy. These techniques' capability to handle nonstationary signals opens a new paradigm in the field of fault diagnostics [22–24]. However, the poor time–frequency resolution is still a challenge.

This paper presents the following attractive features:

- For studying the impact of broken bars on different performance parameters, a WFA-based analytical model is developed. This model takes less simulation time to study the effect of low voltage and asymmetry on the machine's performance. It also helps to plot the faulty theoretical patterns used as a benchmark for differentiation among fault and spatial harmonics;
- Unlike most of the papers where a high inertia load or variable transformer is used, an industrial inverter-based low voltage test to extend the motor's transient interval is proposed. It is also described how an industrial inverter can reduce the voltage while maintaining a constant rated frequency. Increasing the transient length of the motor startup current improves the time–frequency resolution of the spectrum;

- An algorithm to improve the spectrum's legibility is proposed, which helps segregate various frequency patterns in the transient regime;
- The wavelet transform is preferred over the short-time Fourier transform (STFT) to avoid inherited FFT drawbacks. Moreover, a band-stop infinite impulse response (IIR) filter is used to attenuate the fundamental component, which improves the spectrum's legibility;
- It is proposed that the selection of time–frequency regions with a 95% confidence interval (CI) in the form of a contour plot gives a more unambiguous indication of faulty patterns. By adjusting the level of the CI, the spatial and switching frequency-based patterns can be avoided. To the best of the author's knowledge, this technique is not presented in the literature so far.

2. Theoretical Background

2.1. The Modeling of Induction Motor

Electrical machines' modeling is the first milestone in machine design, control, analysis and diagnostics. The famous modeling techniques are divided into two major categories: analytic and numerical. The commonly used analytic methods are two-axis theory [25], winding function method (WFM) [26–28], reluctance network [29] and convolution theorem-based models [30]. The finite element method (FEM)-based models are typical examples of numerical modeling of electrical machines. Although the numerical models have the least number of approximations, the complexity and computational time is the biggest drawback [31]. It becomes worst in the field of fault diagnostics, where the motor's symmetry cannot be exploited to reduce the simulation time. For the fault diagnostic in the transient regime, the V-shaped pattern made by the sideband frequencies is significant. This pattern generates because of the varying slip until the steady state region. This nonstationary signal moves in a bandwidth from 0 Hz to 50 Hz. Thus, the higher-order slotting harmonics become less significant for the analysis of the fault-based, nonstationary signal. The winding function-based model with much less complexity than the corresponding FEM-based models is essential to differentiate among several harmonics. Hence, the air gap is considered constant by neglecting the slot openings with the magnetic material's infinite permeability. The stator winding is approximated as a pure sinusoid, while the rotor turn function represents its actual cage structure. Various self and mutual inductances can be calculated using the winding function formula as in Equation (1). The analytical formulas for the calculation of different inductances given in Equations (2)–(6) are the solution of (1) for the winding functions, as described in [32]:

$$L_{ij}(\theta) = \mu_o rl \int_0^{p\pi} g^{-1}(\varphi,\theta) N_i(\varphi,\theta) N_j(\varphi,\theta) d\theta \tag{1}$$

$$L_m = \frac{\mu_o rl}{g}\left(\frac{p}{2}\right)\left(\frac{N_s}{p}\right)^2 \pi \tag{2}$$

$$L_{AB} = L_{BC} = L_{CA} = -\frac{L_m}{2} \tag{3}$$

$$L_{rkk} = \left(\frac{\mu_o rl}{g}\right)\alpha_r\left[1-\frac{\alpha_r}{p\pi}\right]\left(\frac{p}{2}\right) \tag{4}$$

$$L_{rki} = \left(\frac{\mu_o rl}{g}\right)\left[-\frac{\alpha_r{}^2}{p\pi}\right]\left(\frac{p}{2}\right) \tag{5}$$

$$L_{sri}(\theta_r) = \left(\frac{\mu_o rlN_s}{g}\right)\sin\left(\frac{\alpha_r}{2}\right)\cos\left(\theta_r+(i-1)\alpha_r+\frac{\alpha_r}{2}\right) \tag{6}$$

where μ_o is the permeability of the free space; r is the machine's radius, which is taken as the center of the air gap; l is the machine's effective length; g is the air gap width; α_r is the angular displacement between two consecutive rotor bars; p is the number of the poles; Ns is the number of effective turns per phase in the stator; and θ and φ are the rotor and stator angular positions from a reference point. In Equation (6), the inductance is multiplied with a cosine function of θr to emulate the rotating rotor, where integer i changes the angle of the respective rotor bar from the stator reference point. Lm is the magnetization inductance, L_{AB}, L_{BC} and L_{CA} are stator mutual inductances, L_{rkk} and L_{rki} are rotor self and mutual inductances and L_{sri} is the mutual inductance between the stator and rotor.

The performance parameters, such as torque, speed, currents and voltages, can be calculated using Equations (7)–(9):

$$\begin{bmatrix} Vs \\ 0 \end{bmatrix} = \begin{bmatrix} \boldsymbol{R}_s & 0 \\ 0 & \boldsymbol{R}_r \end{bmatrix}\begin{bmatrix} \boldsymbol{I}_s \\ \boldsymbol{I}_r \end{bmatrix} + \frac{d}{dt}\begin{bmatrix} \boldsymbol{L}_{ss} & \boldsymbol{L}_{sr} \\ \boldsymbol{L}_{rs} & \boldsymbol{L}_{rr} \end{bmatrix} \tag{7}$$

$$T_e = I_s^T(\frac{d}{d\theta}L_{rs})I_r \tag{8}$$

$$J\frac{d}{dt}\omega_m = T_e - T_L - B_f\omega_m \tag{9}$$

where $\boldsymbol{Vs}$ is the stator input voltage vector and $\boldsymbol{Rs}$ and $\boldsymbol{Rr}$ are the stator and rotor resistance matrices of orders (3 × 3) and (40 × 40), respectively. Similarly, $\boldsymbol{Lss}$, $\boldsymbol{Lsr}$, $\boldsymbol{Lrs}$ and $\boldsymbol{Lss}$ are the matrices containing stator–stator, stator–rotor, rotor–stator, and rotor–rotor self and mutual inductances at a specific rotor position. Moreover, T_e is the generated torque, $\boldsymbol{I}_s$ is the stator current, $\boldsymbol{I}_r$ is the rotor current, T_L is the loading torque, B_f is friction coefficient, ω_m is rotor angular velocity and J is the moment of inertia. The detailed model with slot openings and the actual winding functions can be found in [33].

2.2. The Fault Signature Equations and Modulation

Almost all rotor-related faults produce frequency components in the induction motor's current and the voltage spectrum as a slip function. The most common rotor faults can be diagnosed by detecting the frequencies represented by the equations shown in Table 1.

Table 1. Fault definition frequencies.

Fault	Modulating Frequencies
Broken Rotor Bars	$f_{BR} = f_s \pm 2ksf_s,\ k = 1, 2, 3, \ldots$
PSH and Eccentricity	$f_{ecce} = \left[(kn_b \pm n_d)\left(\frac{1-s}{p}\right) \pm v\right]fs$ More precisely: $f_{ecce} = \left[1 \pm k\left(\frac{1-s}{p}\right)\right]fs$ $f_{ecce} = f_s \pm kf_r,\ k = 1, 2, 3, \ldots$

where k is the harmonic order, ν is the supply fed harmonics, f_s is the supply frequency, f_{br} are the broken bar frequencies, n_d is the dynamic eccentricity (0 for static and 1, 2, 3 . . . . for dynamic), f_r is rotor frequency, s is the slip and p is the number of poles.

The phase current of an ideal and symmetrical machine can be defined as

$$i_a(t) = I_m\, sin(\omega t + \alpha) \tag{10}$$

where Im is the peak current, ω is the supply frequency in the angular domain and α is the phase angle. The unsymmetrical rotor with broken rotor bars, broken end rings or bad bearings starts modulating the current with a frequency dependent upon the rotor's speed and the modulation index, which depends upon the severity of the fault:

$$i_{af}(t) = [1 + m(t)]i_a(t) \tag{11}$$

where $m(t)$ is the modulating signal having a modulation index M, which depends on the number of broken bars (Nb) and the total number of rotor bars (Nt). If the rotor is rotating and the winding distributions are considered as sinusoidal, the modulating signal is also a sinusoid, such that

$$m(t) = M\cos(\omega_o t + \varphi) \tag{12}$$

where $\omega_o = 2\pi f$ is the fault characteristic frequency and depends on its nature and the machine's slip. In the case of the broken rotor bars, the characteristic fault frequency is at $2sf_s$:

$$i_{af}(t) = [1 + M\cos(4\pi s f_s t + \varphi)]i_a(t) \tag{13}$$

$$f_o = 2sf_s \quad and \quad \omega_o = 2\pi(2sf_s) \tag{14}$$

$$m(t) = M\cos(4\pi s f_s t + \varphi) \tag{15}$$

$$i_{af}(t) = [1 + M\cos(4\pi s f_s t + \varphi)]I_m \sin(\omega t) \tag{16}$$

$$i_{af}(t) = I_m \sin(2\pi f_s t) + \left(\frac{MI_m}{2}\right)[\sin(2\pi f_s(1+2s)t + \varphi) + \sin(2\pi f_s(1-2s)t + \varphi)] \tag{17}$$

The modulation index M can be approximated as a ratio between the number of broken and total rotor bars [34]:

$$M \approx \frac{N_b}{N_t} \tag{18}$$

The sine terms with frequencies $fs(1 + 2s)$ and $fs(1 - 2s)$ in Equation (17) correspond to the right-hand side (RHS) and left-hand side (LHS) harmonics, which makes a pattern during the transient regime as the slip changes from the maximum value to some specific value, depending upon the load.

In the steady state current, each fault is represented with a particular frequency value, as described previously. In the transient region, the frequency changes continuously. The rapidly evolving slip in the transient regime gives rise to nonstationary signals making specific patterns. The most common causes are the rotor asymmetries, such as the inherent eccentricity, the principal slotting harmonics and the broken bars or end rings. Figure 1 presents the theoretical development of those patterns for a harmonic order of two.

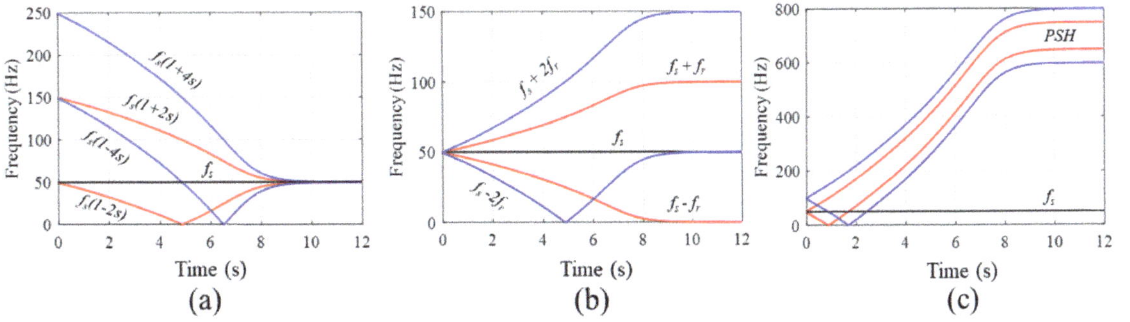

Figure 1. The development of frequency patterns in a transient regime due to (**a**) broken bars, (**b**) dynamic eccentricity and (**c**) principal slotting harmonics.

2.3. The Recovery of the Nonstationary Signal and Related Signal Processing

The frequency spectrum's legibility in either a steady state or transient regime depends on the signal's length and sampling frequency. The fault-representing frequencies are the function of motor slip, as described in Table 1. During the transient interval, the slip is not constant, but decreases with an increase in speed as the motor goes toward a steady state regime. This changing slip results in the creation of nonstationary signals in the transient domain. With a rated supply, this transient interval is so small that those patterns do not become visible. It becomes worse when the motor operates at a low load. The transient interval can be extended by decreasing the supply voltage with a high inertia load, while the input voltage can be changed using a variable three-phase transformer or an industrial inverter. The increasing use of the frequency converter as a substantial part of the drive system increases the test's feasibility. By reducing the supply voltage while maintaining a constant frequency, the rotor's inertia can control the transient time of the motor. Moreover, with reduced voltage, the leakage inductance also takes more time to reach the steady state point, which also supports an extended transient interval. For any reliable signal processing-based diagnostic algorithm, the role of digital filters is undeniable.

Digital filters are the mathematical algorithms capable of reducing or enhancing specific parameters of a signal. They are used for either the separation of combined signals or the restoration of a distorted signal. Their diversified nature has many types, and they are being used extensively in almost every signal processing-based application.

They are the potential tool to remove the strong supply-based frequency components for improving the time–frequency spectrum's legibility. This is necessary because the fundamental component is powerful in amplitude compared with the fault-based harmonics, which are very weak, particularly at the fault's embryonic stage. Those low-amplitude frequency components become barely visible in the presence of the fundamental component.

To attenuate the fundamental component, the selection of an appropriate band-stop filter becomes very crucial. The infinite impulse response (IIR) band-stop filters, such as the Chebyshev II, can give promising results because of their steep transition band and low passband ripples among various digital filters. The excellent transition band and low passband ripples can reduce the filter's impact on the recovered signal.

In light of the mentioned theoretical aspects, the frequency spectrum resolution for transient and steady signals is improved in the following way, as presented in Figure 2:

- Measure the signal with the sampling frequency, meeting the Nyquist criterion. The sampling frequency can be improved later using data interpolation. This is necessary for the accurate detection of zero-crossing points. Moreover, any small constant offset due to data acquisition devices should be removed to avoid a high-amplitude 0 Hz component in the spectrum. This is achieved by subtracting the mean value of the signal from itself;
- Most signal processing techniques, such as DTFT, are sensitive to signal discontinuities and the fractional number of acquired cycles. This problem is solved by counting the integral number of cycles using zero-crossing detection. Saving the signal from the first until the last zero crossings will remove the starting and ending fractional portions of the signal;
- Each sinusoidal signal has three zero crossings, which can be exploited to get the integral number of cycles. If the number of zero crossings is odd, then the signal consists of an integral number of cycles; otherwise, the samples from the second-last zero-crossing until the end should be discarded;
- The band-stop IIR filter is then tuned to suppress the fundamental component;
- The recovered signal shows good spectral legibility both for frequency (steady state) and time–frequency (transient) analysis.

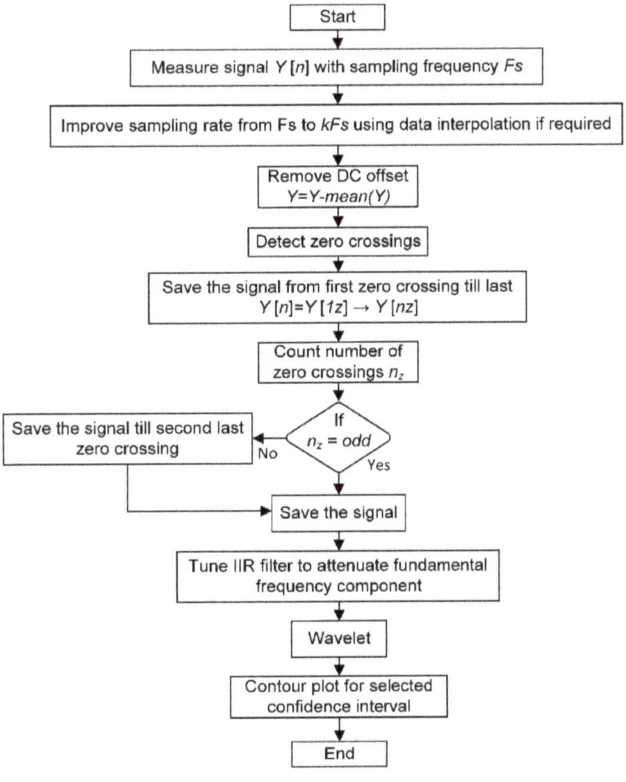

Figure 2. Flowchart for better frequency resolution and development of contour plot.

3. Simulation Results

The winding function model described in Section 2 was used for the transient and steady-state analysis of the motor's performance parameters under healthy and broken bar conditions. The broken bars were simulated by increasing the value of corresponding entries in the resistance matrix. Figure 3b shows the speed–torque dynamic response under healthy and faulty conditions. The zoomed window shows the shift of the operating point at the load application with a slight increase in the slip. The speed and torque ripples due to broken bars are visible in Figure 3a,b.

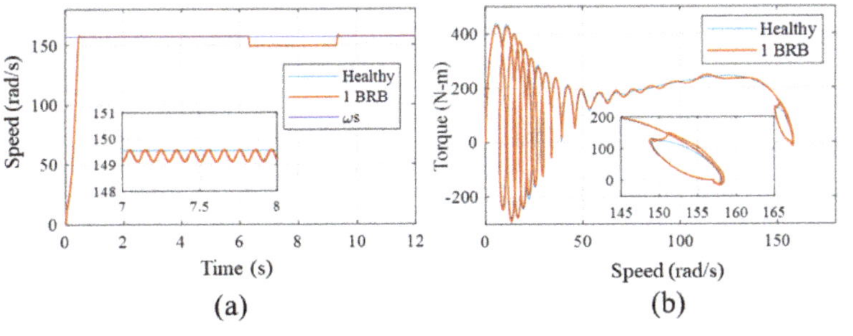

Figure 3. (**a**) The speed dynamic response under healthy and faulty cases. (**b**) The speed–torque dynamic response under the healthy and broken bar cases with shifting of the operating point upon the application of a load in the zoomed window.

These ripples were because of the fault-based induced voltage on the stator side. The frequency of those harmonics depended upon the slip, as discussed previously. During the transient interval, the slip changed dramatically from one to the nominal value, generating a nonstationary signal in the current. This transient interval was so narrow that any specific pattern was hardly visible with rated voltage conditions because of the low-frequency resolution. This phenomenon is described in Figure 4. The motor's simulated transient current at 10%, 15%, 25%, and the rated voltage is presented. The envelope is highlighted by plotting the absolute value of the analytical function given below in (19) and (20). In all subfigures, the *x*-axis represents the time in seconds while the y-axis represents the amplitude in amperes. Moreover, as the constant air-gap is considered in the simulation model, the non-stationary signal contains only broken bar based pattern.

$$i_A(t) = i(t) + j\,\hat{i}(t) \tag{19}$$

where *i(t)* is the current, *j* is the imaginary unit representing the complex number and $\hat{i}(t)$ is its Hilbert-transformed signal. This can be achieved by the convolution of the signal with $1/\pi t$, as shown by the following equation:

$$H(t) = \frac{1}{\pi}\int_{-\infty}^{\infty}\left(\frac{f(x)}{t-x}\right)dx \tag{20}$$

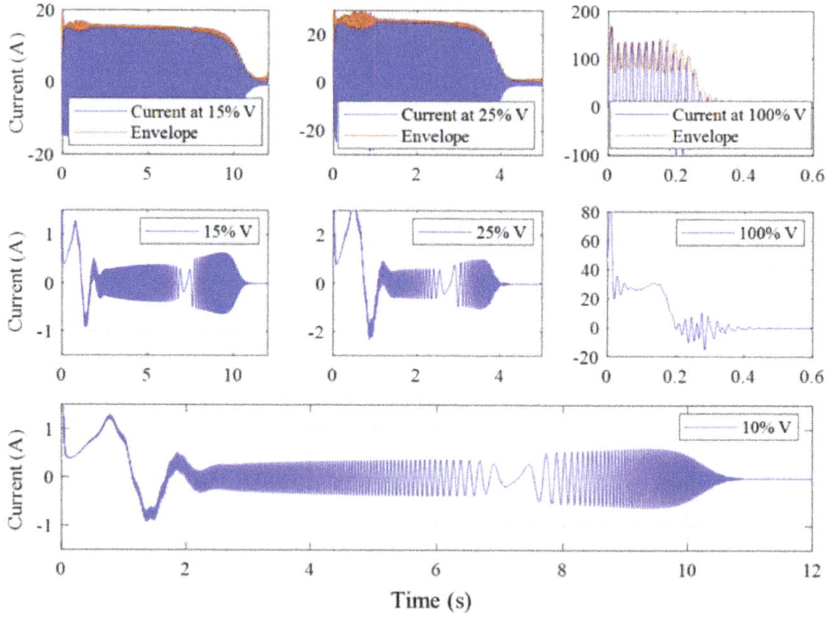

Figure 4. The motor's simulated startup currents at different voltage levels (**top row**), the recovered nonstationary signal (**middle row**) and the nonstationary signal at 10% of the rated voltage representing the changing frequency pattern (**bottom**).

All these steps can be achieved in Matlab by using the simple command ab (hilbert(i[t])). The envelope contains all harmonics other than the fundamental component. For a steady state signal, the Hilbert filter is the right choice for attenuation of the fundamental component. However, in the transient regime, the phase shift of all frequency components poses a problem.

Hence, the nonstationary current signatures were recovered by attenuating the fundamental component using the IIR band-stop filter, as shown in Figure 4 (middle and last row). With a decrease in the applied voltage, the fault-based nonstationary signal increased in length and became more legible. The time–frequency analysis of those non-stationary signals at rated voltage and 10% of the rated voltage is shown in Figure 5. With reduced voltage and the corresponding increase in the signal length, the V-shaped pattern became visible, being hardly present in the current at the rated voltage. Moreover, the supply frequency component (50 Hz) was attenuated using the proposed filter. The fault pattern became more legible by using filter and contour plots together. The contour plot shows the areas with a 95% confidence interval. With the rated voltage, the total power was confined in the region until 0.5 s without any pattern (Figure 5b), while with extended time, the pattern became visible as in Figure 5d. In the magnitude scalogram, the color bar represents the amplitude of the specific frequency component in amperes.

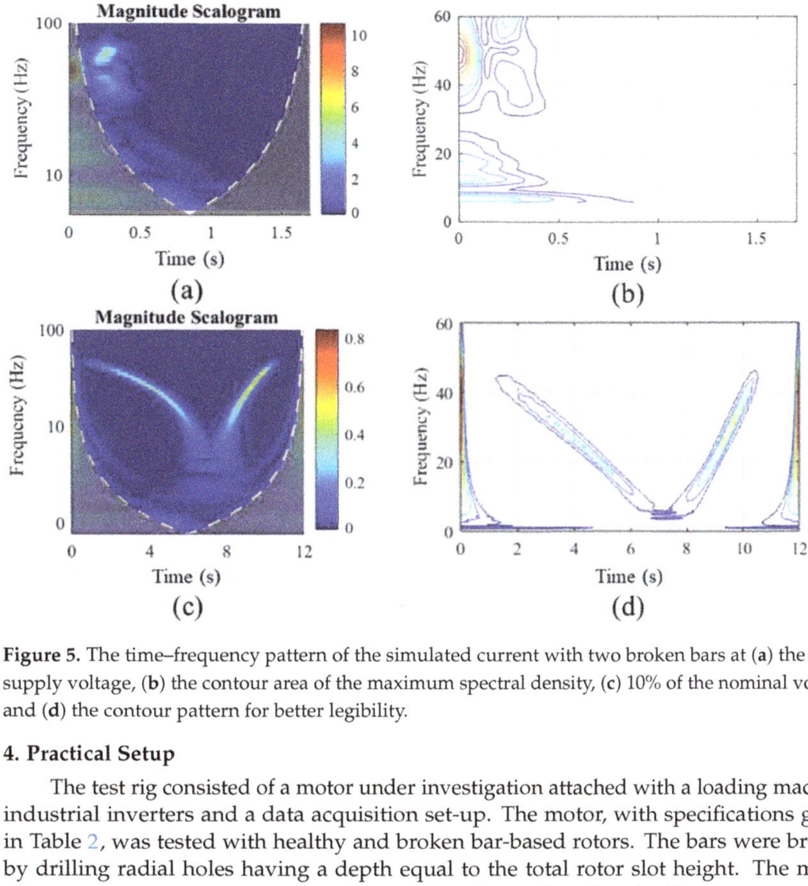

Figure 5. The time–frequency pattern of the simulated current with two broken bars at (**a**) the rated supply voltage, (**b**) the contour area of the maximum spectral density, (**c**) 10% of the nominal voltage and (**d**) the contour pattern for better legibility.

4. Practical Setup

The test rig consisted of a motor under investigation attached with a loading machine, industrial inverters and a data acquisition set-up. The motor, with specifications given in Table 2, was tested with healthy and broken bar-based rotors. The bars were broken by drilling radial holes having a depth equal to the total rotor slot height. The motor phase currents were measured under transient intervals using the Dewetron transient recorder. The measured signal's sampling frequency was 10 kHz, which was good enough for better frequency resolution. The test set-up is shown in Figure 6. In the industrial inverter case, the variable voltage at a constant frequency could be achieved by changing the rated parameters. Since the voltage/Hz ratio in the scalar control remained constant, by changing the rated frequency with a constant rated voltage, the output voltage could be varied, as shown in Table 3. Moreover, the acceleration and deceleration time should be equal to zero to avoid the drive controller's influence on the transient interval.

Table 2. The machine specifications.

Parameter	Symbol	Value
Number of poles	P	4
Number of phases	φ	3
Connection	-	Delta
Stator slots	Ns	36, non-skewed
Rotor bars	Nb	28, skewed
Rated voltage	V	400 V@50 Hz
Rated current	I	8.8 A
Rated power	Pr	7.5 kW @ 50 Hz

(a) (b)

Figure 6. (**a**) The rotor with two broken bars. (**b**) The test bench with the loading motor on the right side and the test motor on the left side.

Table 3. The setting of the industrial inverter to achieve the desired voltage.

V_{rated} (V)	f_{rated} (Hz)	V/Hz	f_{set} (Hz)	V_{out} (V)
300	300	1	50	50
300	150	2	50	100
300	100	3	50	150
300	75	4	50	200
300	60	5	50	250
300	50	6	50	300

5. Results and Discussion

Figure 7 shows the motor's transient current at 15%, 25% and 100% of the rated supply voltage with two broken bars. The recovered nonstationary signal was very legible at 10% of the supply voltage. Unlike the simulation-based results, these signals contained several higher-order harmonics. The most prominent causes of those harmonics were the non-sinusoidal stator and rotor winding distribution, the stator, the rotor slot openings and the magnetic material's nonlinear behavior. However, those frequency components were of less significance, as the most prominent fault-based pattern remained in the bandwidth of 0–50 Hz.

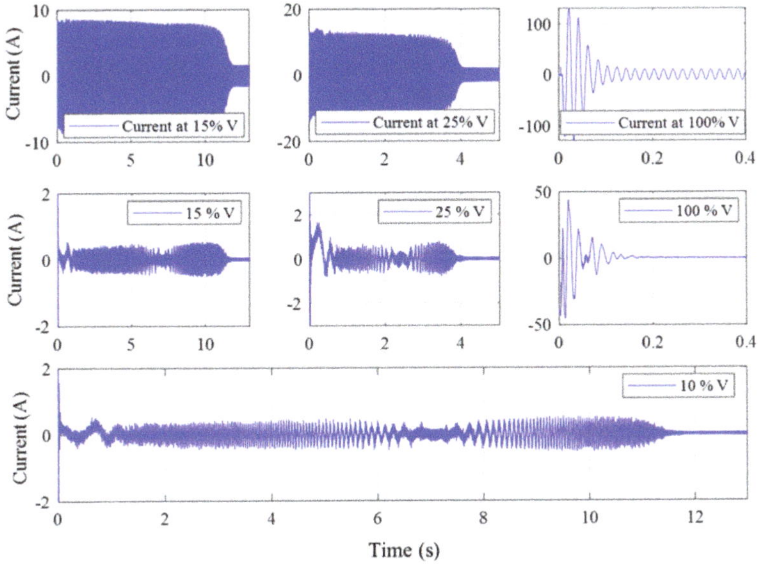

Figure 7. The motor's startup currents measured at different voltage levels (**top row**), the corresponding recovered nonstationary signals (**middle row**) and the nonstationary signal at 10% of the rated voltage representing the changing frequency pattern (**bottom**).

The frequency spectrum of the stator current during the steady state interval for healthy and broken rotor bars at no load and rated load conditions is shown in Figure 8. The spectrum in Figure 8b shows the evolution of the left-side band (LSB) and right-side band (RSB) harmonics. These harmonics are the function of slip, and the results are based on the measurements taken under rated load conditions.

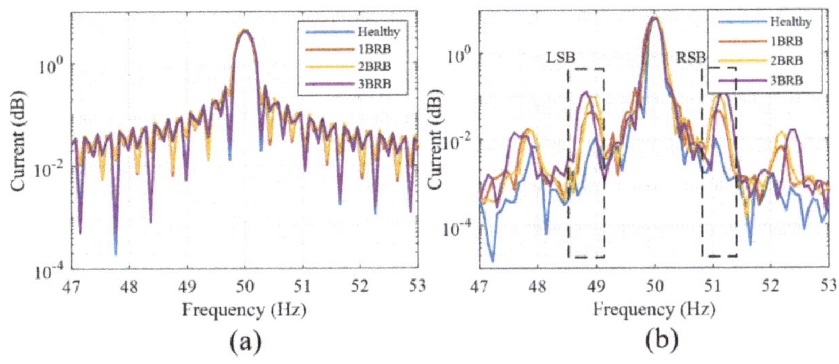

Figure 8. The frequency spectrum from 47 Hz to 53 Hz under healthy and broken rotor bar cases under steady state and (**a**) no-load and (**b**) rated-load conditions.

The sideband frequencies increased in amplitude with the increase in the number of broken bars. Besides that, the sideband fault frequencies tended to shift slightly away from the fundamental component with the rise in broken bars. With the increase in the number of broken bars, the average generated torque decreased with a slight rise in slip. The segregation of these components is very difficult from the rest of the harmonics if the entire spectrum is considered.

The severity of this problem increases with the decrease in load. As the load decreased, these harmonics started hiding under the fundamental component, with total disappearance at the no load condition as in Figure 8a. Moreover, the skewness in the rotor and stator slots affected their amplitudes, making them less detectable at the developing stage. The skews tended to attenuate the rotor harmonics for smooth speed torque features. It is also important to mention that these harmonics remain venerable when the inverter feeds the motor. This effect can be in the form of a colossal number of harmonics fed by the inverter, making faulty frequencies even more difficult for segregation. Moreover, the impact of the drive controller cannot be ignored. This impact is worst in DTC-controlled motors, where the controller of the drive tries to eliminate the current harmonics to reduce torque ripples [14].

The leading causes of the speed and torque ripples were the current harmonics, due to non-sinusoidal winding distribution on the stator and rotor side, the supply-based odd multiples of the fundamental component even if the machine was grid-fed, the inherent dynamic eccentricity and the magnetic material's nonlinear behavior. Those harmonics are inevitable, even if the machine is in a healthy condition. Figure 9 shows those harmonics in the stator current spectrum of a grid-fed healthy motor working under 50% of the rated load. All the harmonics except the supply-fed one were the function of slip and tended to move away from the parent supply component with the increase in load. As discussed earlier, the legibility of those harmonics in the current spectrum not only depended upon slip but the fluctuating load, and the drive's controller also made the detection of desired components challenging.

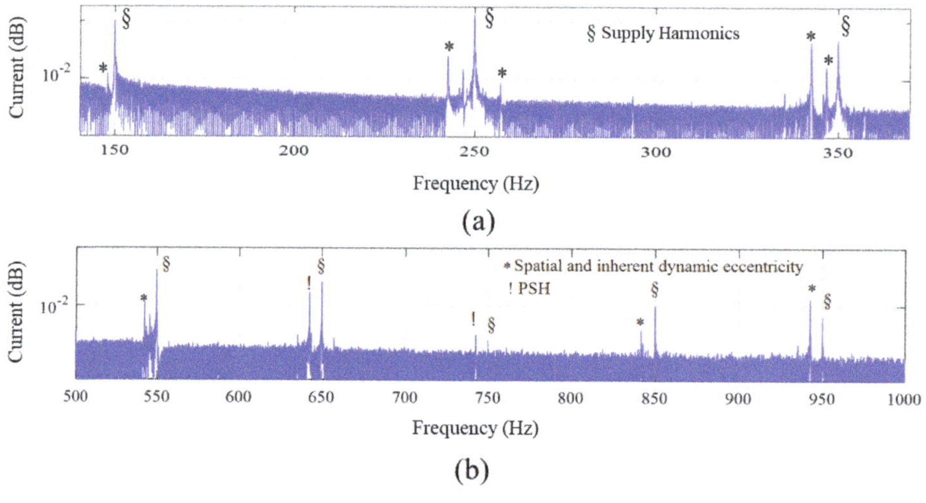

Figure 9. The generation of spatial harmonics in the stator current of the healthy machine (**a**) under a steady state regime with 50% of the rated load in the bandwidth of 100–400 Hz and (**b**) under a steady state regime with 50% of the rated load in the bandwidth of 500–1000 Hz.

To avoid all those problems, the inspection of the motor current under the transient regime showed promising results. For an increase in the transient time for better time–frequency resolution, the supply voltage could be reduced using a three-phase transformer or the industrial inverter, as discussed previously. With the increase in the applied voltage, the motor took less time to reach a steady state interval. Under the transient regime, the continuously decreasing slip moved the RHS and LHS frequency components toward the fundamental component. It made a V-shaped pattern whose width depended upon the transient time. An exceptional intention was needed here, as there were several slip-dependent frequency components, which could make similar patterns in the transient region. Those frequencies are the spatial and inherent dynamic eccentricity-based harmonics, which develop near all supply-fed higher-order harmonics.

Figure 10 shows the harmonics' development in the transient and steady state regime under the healthy condition with the attenuated fundamental component. It is evident that in the transient region, the rotor slotting and eccentricity-based harmonics made the V-shaped patterns, as in Figure 10a. Those patterns started from 50 Hz and vanished completely as they approached the parent supply component, as in Figure 10b. It is essential to differentiate them from the actual faulty patterns to avoid false alarms.

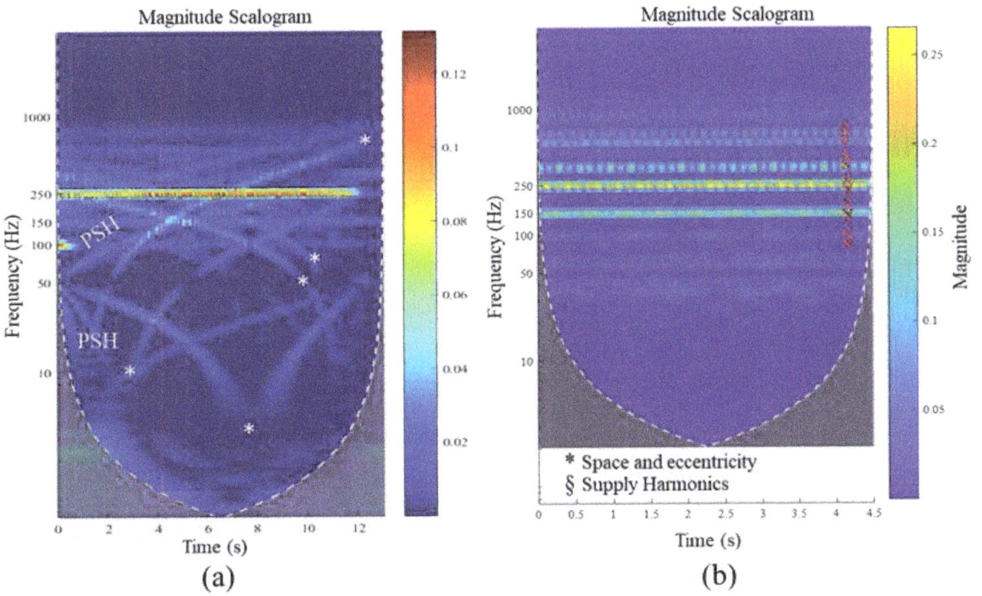

Figure 10. The generation of spatial harmonics in the stator current of the healthy machine. (**a**) The development of time–frequency patterns due to the space harmonics. (**b**) The time-frequency spectrum under a steady state regime under 50% of the rated load.

Figure 11 shows the time–frequency response, with the respective regions containing 85% of the total spectral energy in the contour plot for the healthy and broken bar cases. In the presence of the fundamental component, the faulty pattern is barely visible even in the extended current, as in Figure 11c. Moreover, the influence of the most potent supply component on the visibility of the fault pattern is evident in the contour plot. Although the faulty pattern is visible to some extent in the magnitude scalogram, it is absent in the contour plot, as the maximum spectral energy remained in the region near the fundamental component.

This pattern was not very legible because the sideband frequencies were lower in amplitude than the fundamental component. This pattern can be made very clear by attenuating the fundamental component, as shown in Figure 12. The IIR band-stop filter, having a filter order of two and a stopband attenuation of 40 showed promising results, while attenuating a narrow bandwidth of frequencies from 49.9 Hz to 50.1 Hz. Figure 12a shows the transient current envelope, while the evolution of spatial frequencies in the transient regime is presented in Figure 12b. The development of a faulty pattern with good legibility is shown in Figure 12c,d for one and two broken bars, respectively.

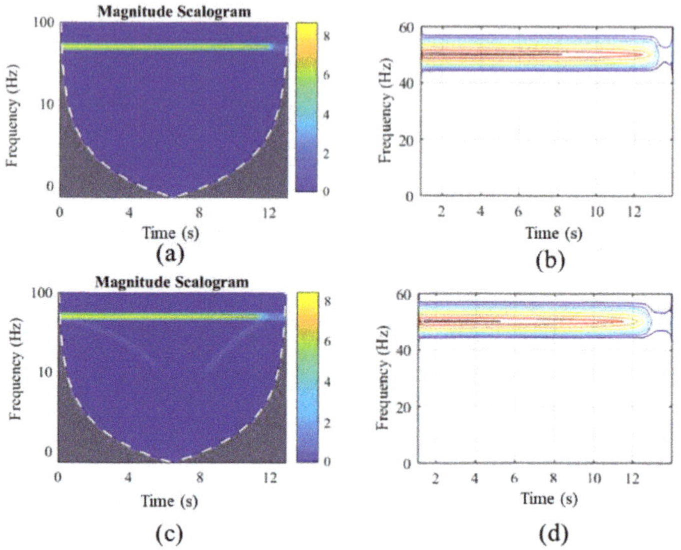

Figure 11. The results based on the measurements taken at 10% of the rated supply voltage. (**a**) The time–frequency response of the healthy motor's phase current in the transient regime. (**b**) The respective contour plot. (**c**) The time–frequency response of the faulty motor's phase current in the transient regime. (**d**) The respective contour plot with a 95% confidence interval.

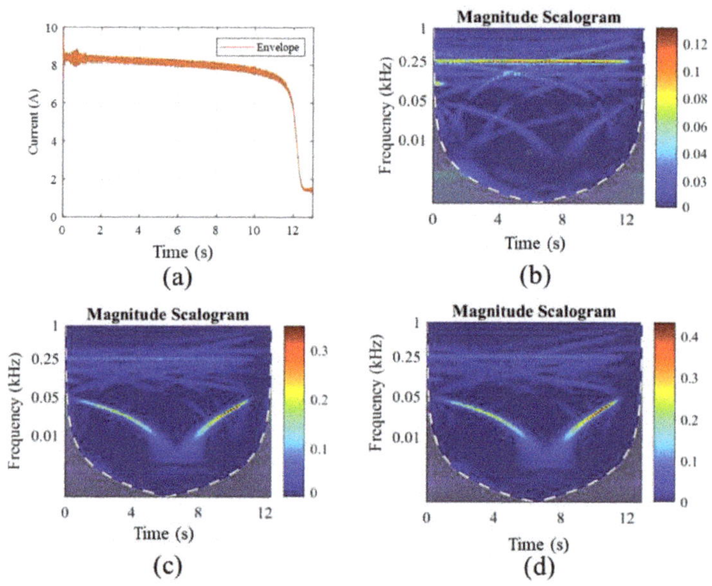

Figure 12. The results based on the measurements taken at 10% of the rated supply voltage. (**a**) The envelope of the motor's phase current in the transient regime. (**b**) Time–frequency spectrum in the healthy case with an attenuated fundamental component. (**c**) Time–frequency spectrum in the case of one broken bar with the attenuated fundamental component. (**d**) Time–frequency spectrum in the case of two broken bar with an attenuated fundamental component.

The contour plot with a bump mother wavelet for the healthy and one broken rotor bar (BRB) cases is shown in Figure 13. Since the spatial harmonics were weaker than the fault frequencies, the contour plot shows the evolution of the sideband frequencies, which avoids confusion with the slotting-related patterns.

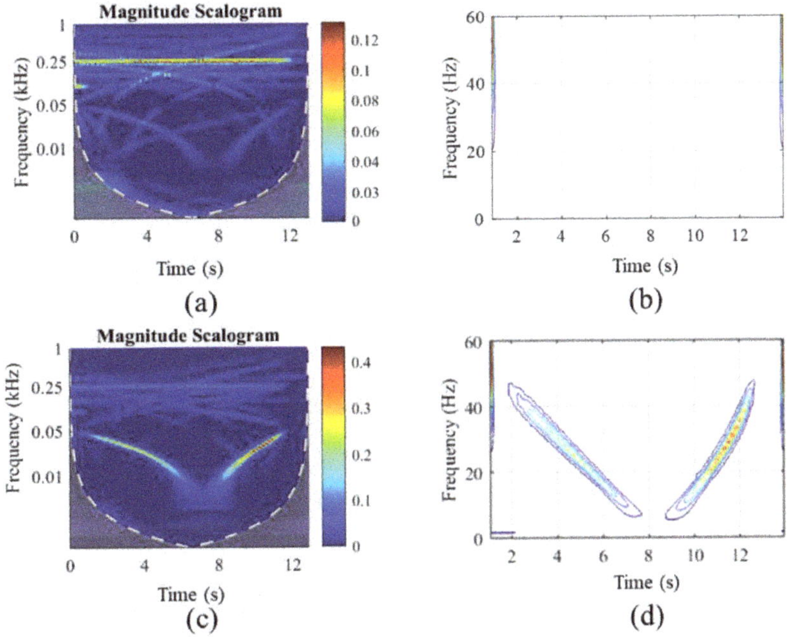

Figure 13. (**a**) The time–frequency plot of the healthy motor with 10% of the rated voltage. (**b**) The time–frequency contour plot. (**c**) The time–frequency plot for 1 BRB motor with 10% of the rated voltage. (**d**) The corresponding contour plot.

6. Conclusions

The broken rotor bars fault diagnostics by time–frequency analysis of the motor current under the transient period was investigated in this paper. The detection of fault frequencies in the steady state interval was prevalent in the literature but possessed several difficulties. Since the sideband frequencies were the function of slip, they became challenging to discover under low-load conditions. The spectral leakage of the prevailing supply components was hazardous for the visibility of faulty components, as they were fragile in amplitude. The spectral leakage was the function of the signal length, the sampling frequency and the type of window used to compute the FFT or wavelet.

Moreover, the inclusion of a vast bandwidth of inverter-fed frequencies made the detection of faulty frequencies hazier. The drive controller's impact on reducing the current ripples was another fact, making fault diagnosis under a steady state regime difficult.

These drawbacks can be resolved by investigating the health in the transient regime. The transient interval's biggest flaw is the little time, which reduces the resolution of the spectrum. This transient time can be increased by lowering the applied voltage without any external load. By doing so, the broken bars-based V-shaped pattern can be seen with great accuracy. Moreover, the FFT-based drawbacks are considerably reduced by using the wavelet approach. The transient spectrum's legibility is further enhanced by attenuating the fundamental component with an IIR filter's help. The IIR band-stop filter has a sharp transition interval and low passband ripples, having less influence on the remaining frequency components.

Several spatial and supply-related frequency patterns can cause a false alarm if they are not differentiated accurately from the actual faulty components. Those components are much weaker in amplitude compared with the fault-related frequencies. The use of the filter to attenuate those components is nearly impossible, as they are not stationary. Hence, the regions with the maximum spectral energy in the form of a contour plot are used, which can be easily adjusted as a threshold to show the pattern. The results are presented using the simulations and the measurements taken from a laboratory-based test rig.

The proposed approach can be used for the pattern-based training of advanced artificially intelligent (AI) diagnostic algorithms.

Author Contributions: Conceptualization, B.A., T.V. and A.B.; methodology, B.A., T.V. and A.B.; validation, A.K., A.R. and A.B.; data curation, B.A. and K.K.; writing—original draft preparation, B.A.; writing—review and editing, B.A., K.K. and A.R.; visualization, T.V. and P.S.G.; supervision, T.V., A.K. and A.B. All authors have read and agreed to the published version of the manuscript.

Funding: The research leading to these results received funding from the EEA/Norway Grants 2014–2021, Industrial Internet methods for electrical energy conversion systems monitoring and diagnostics.

Institutional Review Board Statement: Not applicable.

Informed Consent Statement: Not applicable.

Conflicts of Interest: The authors declare no conflict of interest.

References

1. Bossio, G.R.; De Angelo, C.H.; Bossio, J.M.; Pezzani, C.M.; Garcia, G.O. Separating Broken Rotor Bars and Load Oscillations on IM Fault Diagnosis Through the Instantaneous Active and Reactive Currents. *IEEE Trans. Ind. Electron.* **2009**, *56*, 4571–4580. [CrossRef]
2. Soualhi, A.; Clerc, G.; Razik, H. Detection and Diagnosis of Faults in Induction Motor Using an Improved Artificial Ant Clustering Technique. *IEEE Trans. Ind. Electron.* **2013**, *60*, 4053–4062. [CrossRef]
3. Marques Cardoso, A.J.; Saraiva, E.S. On-line diagnostics of three-phase induction motors by Park's vector. In Proceedings of the ICEM, Pisa, Italy, 12–14 September 1988; pp. 231–234.
4. Oliveira, L.M.R.; Cardoso, A.J.M. Extended Park's vector approach-based differential protection of three-phase power transformers. *IET Electr. Power Appl.* **2012**, *6*, 463. [CrossRef]
5. Freire, N.M.A.; Estima, J.O.; Marques Cardoso, A.J. Open-Circuit Fault Diagnosis in PMSG Drives for Wind Turbine Applications. *IEEE Trans. Ind. Electron.* **2013**, *60*, 3957–3967. [CrossRef]
6. Cruz, S.M.A.; Cardoso, A.J.M. Stator winding fault diagnosis in three-phase synchronous and asynchronous motors, by the extended Park's vector approach. *IEEE Trans. Ind. Appl.* **2001**, *37*, 1227–1233. [CrossRef]
7. Cruz, M.A.; Marques Cardoso, S.A.J. Rotor Cage Fault Diagnosis in Three-Phase Induction Motors by Extended Park's Vector Approach. *Electr. Mach. Power Syst.* **2000**, *28*, 289–299.
8. Ayhan, B.; Trussell, H.J.; Chow, M.Y.; Song, M.H. On the Use of a Lower Sampling Rate for Broken Rotor Bar Detection With DTFT and AR-Based Spectrum Methods. *IEEE Trans. Ind. Electron.* **2008**, *55*, 1421–1434. [CrossRef]
9. Khezzar, A.; Kaikaa, M.Y.; El Kamel Oumaamar, M.; Boucherma, M.; Razik, H. On the Use of Slot Harmonics as a Potential Indicator of Rotor Bar Breakage in the Induction Machine. *IEEE Trans. Ind. Electron.* **2009**, *56*, 4592–4605. [CrossRef]
10. Malekpour, M.; Phung, B.T.; Ambikairajah, E. Stator current envelope extraction for analysis of broken rotor bar in induction motors. In Proceedings of the 2017 IEEE 11th International Symposium on Diagnostics for Electrical Machines, Power Electronics and Drives (SDEMPED), Tinos, Greece, 29 August–1 September 2017; pp. 240–246.
11. Belahcen, A.; Martinez, J.; Vaimann, T. Comprehensive computations of the response of faulty cage induction machines. In Proceedings of the 2014 International Conference on Electrical Machines (ICEM), Berlin, Germany, 2–5 September 2014; IEEE: Piscataway, NJ, USA, 2014; pp. 1510–1515.
12. Nandi, S.; Toliyat, H.A.; Li, X. Condition Monitoring and Fault Diagnosis of Electrical Motors—A Review. *IEEE Trans. Energy Convers.* **2005**, *20*, 719–729. [CrossRef]
13. Asad, B.; Vaimann, T.; Kallaste, A.; Belahcen, A. Harmonic Spectrum Analysis of Induction Motor With Broken Rotor Bar Fault. In Proceedings of the 2018 IEEE 59th International Scientific Conference on Power and Electrical Engineering of Riga Technical University (RTUCON), Riga, Latvia, 12–13 November 2018; pp. 1–7.
14. Asad, B.; Vaimann, T.; Belahcen, A.; Kallaste, A.; Rassõlkin, A.; Iqbal, M.N. Broken rotor bar fault detection of the grid and inverter-fed induction motor by effective attenuation of the fundamental component. *IET Electr. Power Appl.* **2019**, *13*, 2005–2014. [CrossRef]

15. Puche-Panadero, R.; Pineda-Sanchez, M.; Riera-Guasp, M.; Roger-Folch, J.; Hurtado-Perez, E.; Perez-Cruz, J. Improved Resolution of the MCSA Method Via Hilbert Transform, Enabling the Diagnosis of Rotor Asymmetries at Very Low Slip. *IEEE Trans. Energy Convers.* **2009**, *24*, 52–59. [CrossRef]
16. Pineda-Sanchez, M.; Riera-Guasp, M.; Antonino-Daviu, J.A.; Roger-Folch, J.; Perez-Cruz, J.; Puche-Panadero, R. Diagnosis of Induction Motor Faults in the Fractional Fourier Domain. *IEEE Trans. Instrum. Meas.* **2010**, *59*, 2065–2075. [CrossRef]
17. Moussa, M.A.; Boucherma, M.; Khezzar, A. A Detection Method for Induction Motor Bar Fault Using Sidelobes Leakage Phenomenon of the Sliding Discrete Fourier Transform. *IEEE Trans. Power Electron.* **2017**, *32*, 5560–5572. [CrossRef]
18. Kia, S.H.; Henao, H.; Capolino, G.-A. Diagnosis of Broken-Bar Fault in Induction Machines Using Discrete Wavelet Transform Without Slip Estimation. *IEEE Trans. Ind. Appl.* **2009**, *45*, 1395–1404. [CrossRef]
19. Singh, S.; Kumar, N. Detection of Bearing Faults in Mechanical Systems Using Stator Current Monitoring. *IEEE Trans. Ind. Inform.* **2017**, *13*, 1341–1349. [CrossRef]
20. Kang, M.; Kim, J.-M. Reliable Fault Diagnosis of Multiple Induction Motor Defects Using a 2-D Representation of Shannon Wavelets. *IEEE Trans. Magn.* **2014**, *50*, 1–13. [CrossRef]
21. Sapena-Bano, A.; Pineda-Sanchez, M.; Puche-Panadero, R.; Martinez-Roman, J.; Matic, D. Fault Diagnosis of Rotating Electrical Machines in Transient Regime Using a Single Stator Current's FFT. *IEEE Trans. Instrum. Meas.* **2015**, *64*, 3137–3146. [CrossRef]
22. Antonino-Daviu, J. Electrical monitoring under transient conditions: A new paradigm in electric motors predictive maintenance. *Appl. Sci.* **2020**, *10*, 6137. [CrossRef]
23. Gyftakis, K.N.; Spyropoulos, D.V.; Mitronikas, E. Advanced Detection of Rotor Electrical Faults in Induction Motors at Start-up. *IEEE Trans. Energy Convers.* **2020**. [CrossRef]
24. Gyftakis, K.N.; Panagiotou, P.A.; Lee, S. Bin Generation of Mechanical Frequency Related Harmonics in the Stray Flux Spectra of Induction Motors Suffering from Rotor Electrical Faults. *IEEE Trans. Ind. Appl.* **2020**, *56*, 4796–4803. [CrossRef]
25. Cunha, C.C.M.; Lyra, R.O.C.; Filho, B. Simulation and Analysis of Induction Machines With Rotor Asymmetries. *IEEE Trans. Ind. Appl.* **2005**, *41*, 18–24. [CrossRef]
26. Faiz, J.; Ojaghi, M. Unified winding function approach for dynamic simulation of different kinds of eccentricity faults in cage induction machines. *IET Electr. Power Appl.* **2009**, *3*, 461. [CrossRef]
27. Nandi, S. Modeling of Induction Machines Including Stator and Rotor Slot Effects. *IEEE Trans. Ind. Appl.* **2004**, *40*, 1058–1065. [CrossRef]
28. Marfoli, A.; Bolognesi, P.; Papini, L.; Gerada, C. Mid-Complexity Circuital Model of Induction Motor with Rotor Cage: A Numerical Resolution. In Proceedings of the 2018 XIII International Conference on Electrical Machines (ICEM), Alexandroupoli, Greece, 3–6 September 2018; pp. 277–283.
29. Sudhoff, S.D.; Kuhn, B.T.; Corzine, K.A.; Branecky, B.T. Magnetic Equivalent Circuit Modeling of Induction Motors. *IEEE Trans. Energy Convers.* **2007**, *22*, 259–270. [CrossRef]
30. Sapena-Bano, A.; Martinez-Roman, J.; Puche-Panadero, R.; Pineda-Sanchez, M.; Perez-Cruz, J.; Riera-Guasp, M. Induction machine model with space harmonics for fault diagnosis based on the convolution theorem. *Int. J. Electr. Power Energy Syst.* **2018**, *100*, 463–481. [CrossRef]
31. Sapena-Bano, A.; Chinesta, F.; Pineda-Sanchez, M.; Aguado, J.V.; Borzacchiello, D.; Puche-Panadero, R. Induction machine model with finite element accuracy for condition monitoring running in real time using hardware in the loop system. *Int. J. Electr. Power Energy Syst.* **2019**, *111*, 315–324. [CrossRef]
32. Asad, B.; Vaimann, T.; Kallaste, A.; Rassolkin, A.; Belahcen, A. Winding Function Based Analytical Model of Squirrel Cage Induction Motor for Fault Diagnostics. In Proceedings of the 2019 26th International Workshop on Electric Drives: Improvement in Efficiency of Electric Drives (IWED), Moscow, Russia, 30 January–2 February 2019; pp. 1–6.
33. Asad, B.; Vaiman, T.; Belahcen, A.; Kallaste, A.; Rassolkin, A.; Iqbal, M.N. Modified Winding Function-based Model of Squirrel Cage Induction Motor for Fault Diagnostics. *IET Electr. Power Appl.* **2020**, *14*, 1722–1734. [CrossRef]
34. Bellini, A.; Filippetti, F.; Franceschini, G.; Tassoni, C.; Kliman, G.B. Quantitative evaluation of induction motor broken bars by means of electrical signature analysis. *IEEE Trans. Ind. Appl.* **2001**, *37*, 1248–1255. [CrossRef]

applied sciences

MDPI

Article

Transfer Learning-Based Fault Diagnosis under Data Deficiency

Seong Hee Cho [1], Seokgoo Kim [1,2] and Joo-Ho Choi [3,*]

1 Department of Aerospace and Mechanical Engineering, Korea Aerospace University, Goyang 10540, Korea; shcho0923@kau.kr (S.H.C.); seokgoo.kim@ufl.edu (S.K.)
2 Department of Mechanical and Aerospace Engineering, University of Florida, Gainesville, FL 32611, USA
3 School of Aerospace and Mechanical Engineering, Korea Aerospace University, Goyang 10540, Korea
* Correspondence: jhchoi@kau.ac.kr

Received: 4 September 2020; Accepted: 30 October 2020; Published: 3 November 2020

Abstract: In the fault diagnosis study, data deficiency, meaning that the fault data for the training are scarce, is often encountered, and it may deteriorate the performance of the fault diagnosis greatly. To solve this issue, the transfer learning (TL) approach is employed to exploit the neural network (NN) trained in another (source) domain where enough fault data are available in order to improve the NN performance of the real (target) domain. While there have been similar attempts of TL in the literature to solve the imbalance issue, they were about the sample imbalance between the source and target domain, whereas the present study considers the imbalance between the normal and fault data. To illustrate this, normal and fault datasets are acquired from the linear motion guide, in which the data at high and low speeds represent the real operation (target) and maintenance inspection (source), respectively. The effect of data deficiency is studied by reducing the number of fault data in the target domain, and comparing the performance of TL, which exploits the knowledge of the source domain and the ordinary machine learning (ML) approach without it. By examining the accuracy of the fault diagnosis as a function of imbalance ratio, it is found that the lower bound and interquartile range (IQR) of the accuracy are improved greatly by employing the TL approach. Therefore, it can be concluded that TL is truly more effective than the ordinary ML when there is a large imbalance between the fault and normal data, such as smaller than 0.1.

Keywords: transfer learning; fault diagnosis; data deficiency; imbalanced data; linear motion guide

---

1. Introduction

In recent years, fault diagnosis of mechanical components has been studied very actively and many important advances have been made primarily based on the machine learning techniques such as k-nearest neighbor (KNN) [1], support vector machine (SVM) [2], convolutional neural network (CNN) [3], and others. However, these research techniques have dealt with situations where training and test data lie in the same domain with enough amounts of fault data. While this unusual situation enables the diagnosis performance more superior, it does not reflect the circumstances of the real field where the fault diagnosis should be carried out under a lack of fault data due to cost and safety issues.

In the study of Prognostics and Health Management (PHM), the lack of fault or failure data, which is referred to as the "data deficiency", has been considered as the main obstacle that deteriorates the performance of PHM. To tackle this issue, there have been several efforts with different perspectives and strategies. For example, Kim et al. [4] proposed data augmentation prognostics (DAPROG) by augmenting the run-to-fail (RTF) data obtained in the past for various operating conditions to the current one using dynamic time warping (DTW). Sobie et al. [5] generated virtual training data exploiting the high-resolution simulations based on the roller bearing dynamics and improved the machine learning classification performance. A data fusion approach was also proposed as a means to compensate the absence of data in certain classes, which are addressed by Diez-Olivan et al. [6], with reviews and challenges for the industrial prognosis; Azamfar et al. [7], for the gearbox fault diagnosis using the convolutional neural network (CNN) and motor current signature analysis, Huang et al. [8] for the mechanical fault diagnosis based on IoT; Luwei et al. [9], for the rotating machine fault classification, to name a few.

Over the years, transfer learning (TL) has emerged as one of the solutions. TL is a method of extracting knowledge obtained in one domain (source domain) to solve problems in another domain (target domain), combining with machine learning techniques. The fundamental idea is that if the source and target domain are similar, using the trained weights of the source domain is more efficient than the training of the target domain from scratch. As a result, the TL algorithm compensates for data deficiency in the target domain, such as the insufficient amount and unlabeled samples. Several studies have been made in this direction. For example, Zhang et al. [10] proposed transfer learning based on the neural networks (NN) for multiclass fault diagnosis of rolling bearings. Parameters of the NN in the source domain are obtained through the training phase and transferred to the target task which has more classes to classify. Classification accuracy and training time were improved compared to those by the traditional NN. Wen et al. [11] proposed a new deep transfer learning method for fault diagnosis of different working conditions. They extracted domain-invariant features by using sparse auto-encoder and the maximum mean discrepancy between domains. Compared to the results of traditional methods, such as the deep belief network (DBN), SVM, and artificial neural network (ANN), the proposed method has made good improvement. They have also conducted a comparative study for different ratios of samples between the source and target dataset. Qian et al. proposed a transfer learning-based fault diagnosis network combined with high-order Kullback–Leibler (HKL) divergence [12]. The proposed network has extracted features through the sparse filtering and HKL divergence for discrimination and generalization ability. They have validated their method by both the bearing and gearbox datasets. Lu et al. [13] proposed domain adaptation combined with a deep convolutional generative adversarial network (DA-DCGAN) for the DC series arc fault diagnosis in a photovoltaic (PV) system. They used normal and fault data of the source domain (PV emulator in the laboratory) and only normal data of the target domain (PV system in the field) for an adversarial training process. By generating fault data of the target domain artificially, high detection accuracy was achieved. Guo et al. [14] proposed a deep convolutional transfer learning network (DCTLN) for intelligent fault diagnosis of machines with unlabeled data. They optimized their network models in the direction of minimizing the maximum mean discrepancy (MMD) between the source and target domain datasets for learning domain-invariant features from raw signals.

The above-mentioned papers have shown several improvements in the fault diagnosis by using TL, but the imbalance issue in the sample size between the normal and fault data was not studied, which may affect the performance of TL greatly. Even though a few papers [11,12] have addressed a similar issue, it was about the sample imbalance between the source and target domain, in which the size of fault data is the same or greater than the normal. Recalling that the fault does not occur often in the real operation conditions because of the periodic maintenance or the costly effort to make by intention, the imbalance issue between the normal and fault data is of great importance for practical use. For this reason, the effect of transfer learning in the fault diagnosis of imbalanced data is investigated in this study.

2. Transfer Learning

Transfer learning is to extract knowledge from one domain task and utilize (or transfer) it to solve another domain task. The domain in which knowledge is extracted is referred to as the source domain, and the domain in which the knowledge is transferred is referred to as the target domain. Depending on the type of knowledge and the way it is transferred, transfer learning is divided by several approaches [15]. In this paper, parameter transfer is employed, in which a neural network is trained by the source domain dataset and the pre-trained model is transferred to the target domain to fine-tune its parameters by the target domain dataset. In the parameter transfer-based approach, the source datasets are usually relatively large in sample size while the target datasets are small. The approach assumes that the parameters of the source and target domains share a common part shown as follows [10]:

$$\theta_s = \theta + \theta_1 \tag{1}$$

$$\theta_t = \theta + \theta_2 \tag{2}$$

where θ denotes parameters, such as weight and bias, of a neural network, and the subscript s and t denote the source and target domain, respectively. The idea is that by initializing the parameters with those of the source domain, the accuracy and training efficiency can be improved substantially [10].

In this paper, the influence of the imbalance ratio in the target domain, defined as the proportion of sample size between normal and fault, is investigated in the transfer learning framework, and compared with the results by the traiditional neural network. As stated before, the motivation of this study is due to the lack of fault data in the real field. The flowchart is presented in Figure 1. As shown in Figure 1a, in the traditional machine learning-based fault diagnosis, datasets of different states (e.g., normal and fault) are divided into the training and test. Training data are used to train a model for the diagnosis and test data to measure the performance of the model. Note in the figure that the degraded dataset is smaller in sample size than that of the normal dataset, which is usually the case in the field, and may be responsible for the poor classification performance. In the transfer learning (TL)-based diagnosis, there are two groups of datasets from the different domains: the source and target as shown in Figure 1b. The former is used to extract knowledge and the latter to utilize the knowledge for its diagnosis. In this case, the degraded dataset in the source domain is relatively larger in sample size than those in the target domain, which is going to be exploited for the diagnosis of the target domain. In this paper, parameter transfer is used so that a neural network (NN) is trained in the source domain and the pre-trained model is transferred to fine-tune the parameter in the target domain. The concept is that the model parameters on the left in the source domain are knowledge-transferred to serve as the initial values of the model on the right in the target domain, which can improve the accuracy and efficiency of training greatly.

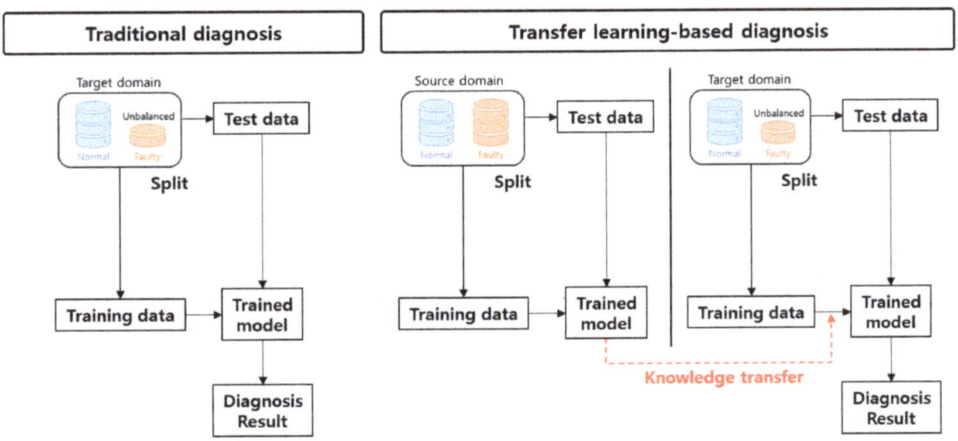

(a) (b)

Figure 1. Flowchart of fault diagnosis: (**a**) traditional machine learning-based approach, and (**b**) transfer learning-based approach [15].

3. Application

3.1. Linear Motion Guide Dataset

A linear motion (LM) guide is a frequently used mechanical part in the manufacturing field because it moves heavy workpieces while maintaining precision at high speeds. The LM guide in the experiment is a ball guide type with a rectangular LM block whose height and width are 24 mm and 34 mm. Only one LM block is installed in a single rail. The seal type of the LM block is an end seal for dust prevention. The workpiece is fixed to the surface of the LM block, which travels along a linear rail with dozens of balls circulating inside of the block as shown in Figure 2. Since the balls move along the track in the LM block, flaking may occur due to the rolling contact fatigue loads. Four LM guides are prepared, and test rigs are made, which consist of the motor and belt drive (not shown here), rail, workpiece, and LM guides as shown in Figure 3a. The LM block reciprocates along a 450 mm-long rail with a sequence of acceleration for 0.23 s, a constant speed of 1.667 m/s for 0.04 s, and deceleration for 0.23 s, which is shown in Figure 4a. A three-axis accelerometer with a sampling rate of 10.2 kHz is installed at the LM block as shown in Figure 3b. The vibration signal is measured intermittently at the 0, 20 k, 50 k, and 100 k'th cycles for the early period, and at every 100 k cycle for the subsequent period. At each recording cycle, the signal is recorded twice with different speeds: first at the original 1.667 m/s (high) and next at 0.1 m/s (low). This is to simulate the on-line operation during the production and the off-line inspection operation during the maintenance, respectively. The speed profile for the latter case is shown in Figure 4b. In the recording cycle, the measurement is repeated nine or ten times at random under each speed. The experiment is terminated when one of the LM guides encounters the flaking appearance during the inspection, which has taken about two months. As a result, the total number of datasets is given by 680 and 723 for the high and low speeds, respectively, where a dataset refers to the measured data for a single trip. Figure 5a shows the plot of the vibration signal under the low speed over the whole period acquired for one of the LM guides. Figure 5b is its three-dimensional spectrogram, which is the stack of the frequency spectrum made by fast Fourier transform at each cycle. The figure shows that the amplitudes at around 3 kHz increase from a certain time period, which is about the last 25% of the whole cycle. The same phenomena are found in the other LM guides. The authors believe that this represents the fault progression of LM guides, eventually leading to the flaking failure.

Since the data can be easily acquired during the maintenance, the latter is defined as the source domain, whereas the former being the target domain. Then the imbalance condition is considered where the fault data in the target domain are much less available than those in the source domain due to the infrequent occurrence in the real operation. Among the datasets in the source domain (low speed), those in the first six.

Note that 25% and the last 25% of the whole period are chosen as the normal and fault at our convenience to make the number of datasets in each class the same. The results are 152 datasets in common for the normal and fault conditions. Then, the NN is trained using the all the datasets (152 normal samples and 152 fault samples) in the source domain. In the target domain (high speed), the datasets of the normal and fault conditions are chosen in the same way, and the results are 160 datasets in common. Then, the datasets are divided into the training and test, with the number being 100 and 60, respectively. The training datasets are used to train the NN which is pre-trained and transferred from the source domain. The test datasets are used to examine the accuracy of the trained NN. The numbers in each case are summarized in Table 1.

To simulate the imbalance condition where the fault data in the target domain are much less available than those in the source domain, only a part of the fault datasets are taken as shown in Table 1, which is given by multiplying the imbalance ratio (IR). For example, when the IR is set at 0.1, only 10 percent of the fault datasets are used in the target domain for the training.

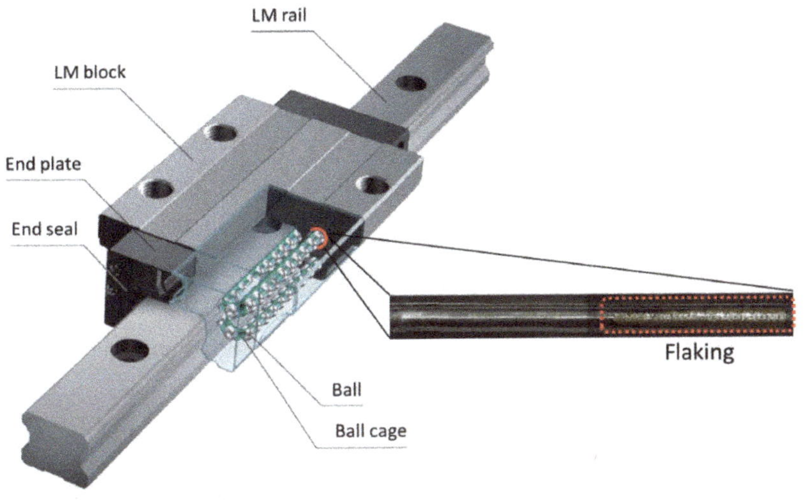

Figure 2. Linear motion (LM) guide structure and flaking inside the LM block.

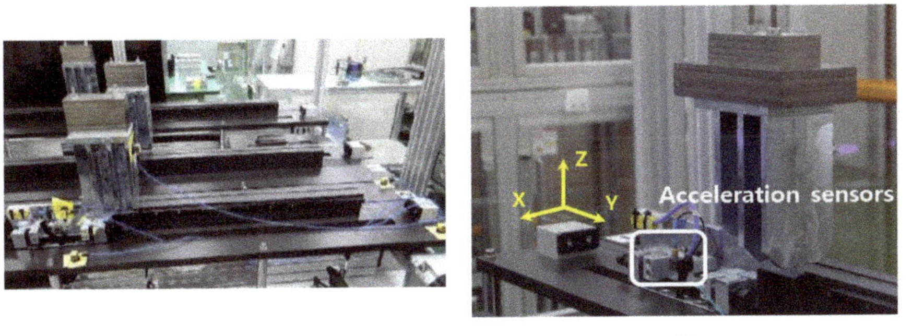

(**a**) (**b**)

Figure 3. Experimental setup: (**a**) linear motion guide test rigs; (**b**) installed accelerometer.

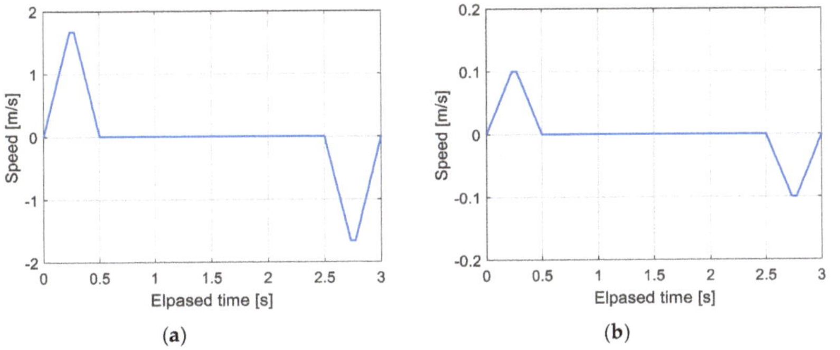

(**a**) (**b**)

Figure 4. Speed profile of LM guide: (**a**) high speed during production (target domain); (**b**) low speed during an inspection (source domain).

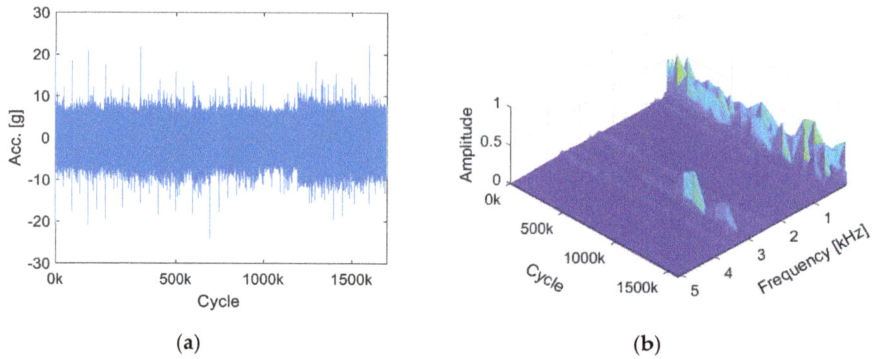

Figure 5. Vibration signal of low speed for one of the LM guides: (**a**) raw signal in time domain; (**b**) frequency spectrum by Fourier transform.

Table 1. Details of LM dataset.

	Purpose	Normal Dataset	Fault Dataset
Source Domain	Training	152	152
Target Domain	Training	100	100*IR [1]
	Test	60	60

[1] Imbalance ratio (IR) denotes the imbalance ratio.

3.2. Training Neural Networks of Source Domain

As shown in Figure 1, the transfer learning is preceded by the training of the NN in the source domain using the 152 datasets of the normal and fault as given in Table 1. As is addressed in many works of literature (e.g., see [16,17]), the training process consists of the training and validation, in which the training is to determine the parameters (weights and bias) under a given architecture, whereas the validation is to find out the optimum architecture or hyper-parameters that give the best performance.

Thirteen time-domain statistical features as listed in Table 2 are extracted from the raw signals in the three directions, which gives the thirty-nine features. These are used as the input nodes of the NN. In terms of the architecture, a single hidden layer is considered with the transfer function being the hyperbolic tangent sigmoid. The optimum number of nodes in the hidden layer is then sought by the validation process, more specifically the five-fold cross-validation in practice. To this end, the datasets are divided into the five folds, in which the four are used for the training (determining the weights and bias of the network), while the remaining one is used for the validation. The diagnosis accuracy is then calculated by the ratio of the number of correct classifications to the total number of validation datasets. Since there are five folds, this is repeated five times, from which the average accuracy is obtained. Furthermore, since the NN solutions can vary widely due to the random initial conditions, this is repeated 300 times, and the grand average is obtained accordingly. The process is repeated by increasing the number of nodes in the hidden layer from 2 to 100 by increments of 2 (e.g., 2, 4, 6 . . . , 100). The result of validation in the source domain is given in Figure 6a, where the optimum number of hidden nodes is chosen at 22 by visual inspection. Once this is done, the NN is trained again to obtain the parameters (weights and bias) under the optimum architecture (namely, 22 hidden nodes) using the training datasets with initial values of the parameters given at random. Repeating this by 300 times, 300 sets of parameters are obtained for the NN in the source domain, which we call the pre-trained NN parameters to be used in the target domain. The reason for the repetition is due to the arbitrary nature of the NN solution.

Table 2. Equations for statistical features.

Feature	Formula	Feature	Formula
Mean	$\frac{1}{N}\sum_{i}^{N} x_i$	Crest factor	$\frac{\max(\lvert x_i \rvert)}{\sqrt{\frac{1}{N}\sum_i^N x_i^2}}$
Root amplitude	$\left(\frac{1}{N}\sum_{i}^{N}\sqrt{\lvert x_i \rvert}\right)^2$	Clearance factor	$\frac{\max(\lvert x_i \rvert)}{\left(\frac{1}{N}\sum_i^N \sqrt{\lvert x_i \rvert}\right)^2}$
RMS	$\sqrt{\frac{1}{N}\sum_{i}^{N} {x_i}^2}$	Shape factor	$\frac{\sqrt{\frac{1}{N}\sum_i^N {x_i}^2}}{\frac{1}{N}\sum_i^N \lvert x_i \rvert}$
Standard deviation	$\sqrt{\frac{1}{N-1}\sum_{i}^{N}(x_i-\overline{x})^2}$	Impulse factor	$\frac{\max(\lvert x_i \rvert)}{\frac{1}{N}\sum_i^N \lvert x_i \rvert}$
Peak	$\max(\lvert x_i \rvert)$	Peak-to-peak	$\max(x_i)-\min(x_i)$
Skewness	$\frac{\sum_i^N (x_i-\overline{x})^3}{(N-1)\sigma^3}$	Root sum of squares	$\sqrt{\sum_{i}^{N} x_i^2}$
Kurtosis	$\frac{\sum_i^N (x_i-\overline{x})^4}{(N-1)\sigma^4}$		

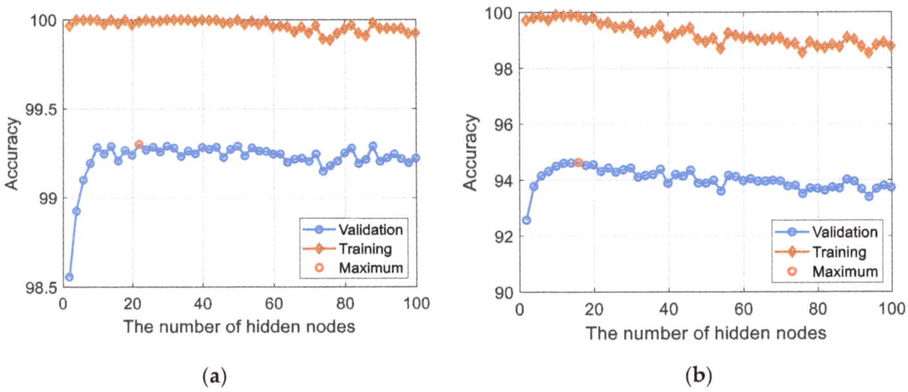

Figure 6. Diagnosis accuracy as a function of the number of hidden nodes: (**a**) the source domain; (**b**) the target domain with IR of 0.5.

3.3. *Training Neural Networks of Target Domain by Transfer Learning*

Once the NN is trained in the source domain, it can be exploited by TL in the target domain. There are two ways for this: One is named TL1, which uses the same architecture in the target domain as that of the source domain. The other is TL2, which, in addition to the same architecture, initial values for the training are given by the pre-trained NN parameters of the source domain. Whichever it is, the training is carried out using the architecture of the source domain by 300 times for the training datasets in the target domain as given in Table 1. In the training, the initial values of the parameters are assumed at random in TL1 whereas they are given by the pre-trained NN parameters in TL2. The trained NNs are applied to the test datasets of Table 1, from which the diagnosis accuracies are obtained as defined previously in the source domain. Since the 300 number of accuracy is obtained as a result, the accuracy is represented by the distribution, which occurs due to the randomness of NN solutions. Furthermore, the number of fault datasets vary by IR. So this is repeated for each IR from 0.1 to 1.0 by the increment of 0.1. Note that the smaller the IR, the more the data deficiency becomes greater.

To make a comparison of the TL-based approach, the ordinary machine learning (ML) approach is also performed. In this case, the training and validation are carried out solely by the training datasets in the target domain to determine the optimum architecture. The procedure is the same as that addressed in the source domain. The process is repeated for each IR from 0.1 to 1.0 by the increment of 0.1. Figure 6b gives the result of validation when IR is 0.5, where the optimum number of hidden nodes is found at 16. The results for other IRs are found in Table 3. Once this is done, the NN has trained again by 300 times to obtain the 300 sets of parameters under the architecture using the training datasets with the initial values assumed at random. Finally, the trained NNs are applied to the test datasets of Table 1, from which the diagnosis accuracies are obtained, representing the distribution due to the randomness of NN solutions.

Table 3. The number of nodes in hidden layer determined in the target domain.

IR	0.1	0.2	0.3	0.4	0.5	0.6	0.7	0.8	0.9	1.0
The number of nodes in hidden layer	10	18	14	10	16	10	12	16	12	12

The results of accuracy distribution in TL1, TL2, and ML are given as a function of IR as shown in Figure 7a–c, respectively. In the figures, the bottom, top, and red lines of the box indicate the 25th, 75th percentile, and median of the distribution, respectively. The top and bottom of the extended dotted lines are the 5th and 95th percentiles. In Figure 8a,b, the lower bound (fifth percentile) and the IQR between 25th and 75th percentiles are shown, which indicate the lower confidence limit of accuracy and the degree of statistical dispersion, respectively [18]. When the IR is 0.1, the lower bound of TL2 is 69.2, which is greater than 57.5 (TL1) and 53.3 (ML), and the IQR is 5.83, which is only about 60% of 9.58 (TL1) and 10.41 (ML). The higher the bound and the lower the IQR, the better the performance. This can be identified more clearly in Figure 9, where the histograms of diagnosis accuracy and its lower bound are displayed for TL1, TL2, and ML, respectively. While the histograms of TL1 and ML are similar and show much larger variance, TL2 shows smaller variance with greater lower confidence limit. However, as the IR increases greater than 0.1, the lower bound and IQR of TL2 are not superior to the others. The authors believe that these results are due to the negative transfer [10], which happens when the fault characteristics between the source and target domain are quite different. In that case, adding the fault data in the target domain does not help to improve the TL performance. Overall observations are summarized as follows: (1) TL1 and ML behave similarly, whereas TL2 is different from the others. That is, the initialization of NN parameters is of more importance than the adoption of source domain architecture in making the difference of TL from ordinary ML in the target domain. (2) TL2 is superior to ML and TL1 only when the number of fault data is very small such as when IR is 0.1 or less as is evident in Figure 8, in which the lower bound of accuracy for TL2 is higher and the IQR for TL2 is shorter than the other two. This is why TL is necessary for the case when the data are too few to train by their own. (3) It is of surprise that TL2 is soon found less effective than the other two as the IR gets greater than 0.1. Nevertheless, it should not be overlooked that the proposed approach is truly useful in the case of significant data deficiency, which is the highlight of the study.

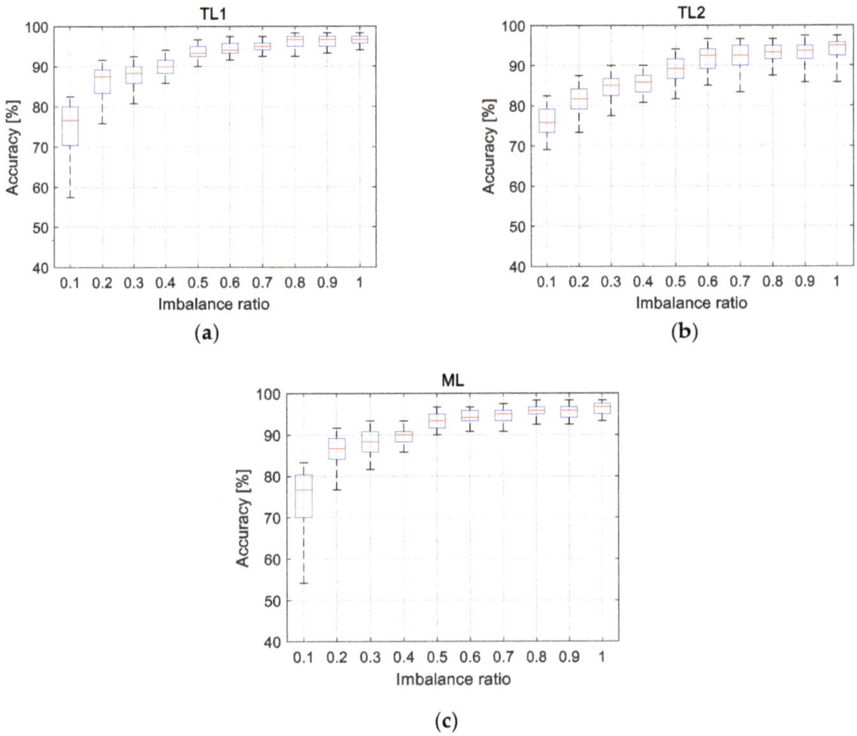

Figure 7. Diagnosis accuracy of the target domain: (**a**) transfer learning (TL)1; (**b**) TL2; (**c**) machine learning (ML).

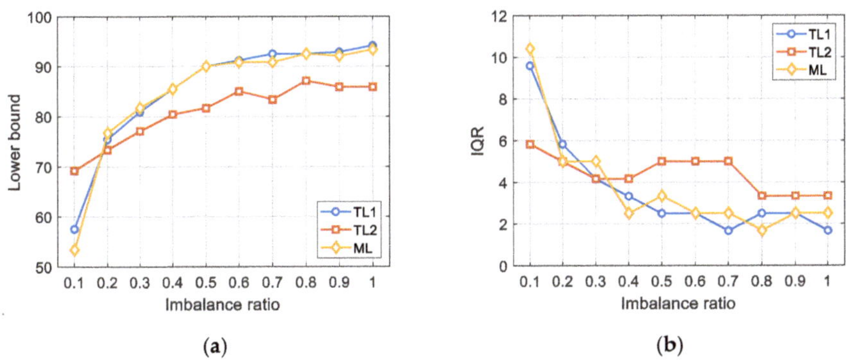

Figure 8. Two measures of diagnosis accuracy: (**a**) lower bound; (**b**) interquartile range (IQR).

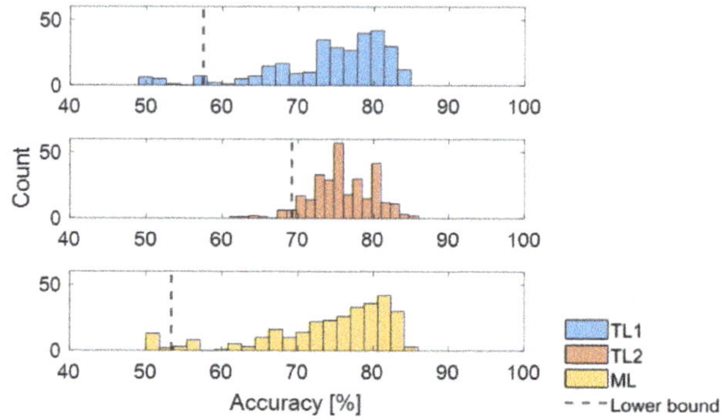

Figure 9. Histogram of diagnosis accuracy when IR is 0.1.

4. Conclusions

Collecting the fault data during the real operation is difficult because of safety and cost issues, which results in the imbalance issue in the number of normal versus fault data, and even data deficiency. The performance of fault diagnosis is then greatly deteriorated due to this. On the other hand, a good number of normal and fault data can be acquired in a similar condition such as the lab test or maintenance inspection. In this study, a transfer learning-based fault diagnosis is proposed to solve this problem, which is to pre-train the NN using the large datasets as the source domain, and transfer the knowledge to train the NN using the imbalanced datasets in the real target domain. To illustrate this, normal and fault datasets are acquired from the run-to-fail test of the LM guides, in which the data at high and low speeds are regarded as those for the real operation (target) and maintenance inspection (source), respectively. To study the effect of imbalance, the number of fault data in the target domain is reduced by multiplying the imbalance ratio (IR), and the accuracy of diagnosis is explored as a function of IR. From the study, it is concluded that TL is truly more effective than ordinary ML when there is a large imbalance between the fault and normal such as smaller than 0.1. When the imbalance ratio is 0.1, the variance is improved substantially such that the IQR is about 60% of the others. However, to our surprise, it soon becomes less so, reversing the performance as the IR is increased, which is a too small benefit to employ TL as the solution. This seems to be attributed to the negative transfer arising when the fault characteristics differ too much between the source and target domain. How to figure out and overcome this will be another future research topic, which involves numerous implementations of various experimental datasets.

Author Contributions: Conceptualization, S.H.C. and S.K.; methodology, S.H.C. and S.K.; writing—original draft preparation, S.H.C.; writing—review and editing, J.-H.C.; supervision, J.-H.C. All authors have read and agreed to the published version of the manuscript.

Funding: This work was supported by the National Research Foundation of Korea (NRF) grant funded by the Korea government(MSIT) (No. 2020R1A4A407990411).

Conflicts of Interest: The authors declare no conflict of interest.

References

1. Bagheri, B.; Ahmadi, H.; Labbafi, R. Application of Data Mining and feature extraction on Intelligent Fault Diagnosis by Artificial Neural Network and K-Nearest Neighbor. In Proceedings of the XIX International Conference on Electrical Machines—ICEM 2010, Rome, Italy, 6–8 September 2010. [CrossRef]
2. Hu, Q.; He, Z.; Zhang, Z.; Zi, Y. Fault diagnosis of rotating machinery based on improved wavelet package transform and SVMs ensemble. *Mech. Syst. Signal. Process.* **2007**, *21*, 688–705. [CrossRef]
3. Eren, L.; Ince, T.; Kiranyaz, S. A Generic Intelligent Bearing Fault Diagnosis System Using Compact Adaptive 1D CNN Classifier. *J. Signal. Process. Syst.* **2019**, *91*, 179–189. [CrossRef]

4. Kim, S.; Kim, N.H.; Choi, J.-H. Prediction of remaining useful life by data augmentation technique based on dynamic time warping. *Mech. Syst. Signal. Process.* **2020**, *136*, 106486. [CrossRef]
5. Sobie, C.; Freitas, C.; Nicolai, M. Simulation-driven machine learning: Bearing fault classification. *Mech. Syst. Signal. Process.* **2018**, *99*, 403–419. [CrossRef]
6. Diez-Olivan, A.; Del Ser, J.; Galar, D.; Sierra, B. Data fusion and machine learning for industrial prognosis: Trends and perspectives towards Industry 4.0. *Inf. Fusion* **2019**, *50*, 92–111. [CrossRef]
7. Azamfar, M.; Singh, J.; Bravo-Imaz, I.; Lee, J. Multisensor data fusion for gearbox fault diagnosis using 2-D convolutional neural network and motor current signature analysis. *Mech. Syst. Signal. Process.* **2020**, *144*, 106861. [CrossRef]
8. Huang, M.; Liu, Z.; Tao, Y. Mechanical fault diagnosis and prediction in IoT based on multi-source sensing data fusion. *Simul. Model. Pr. Theory* **2020**, *102*, 101981. [CrossRef]
9. Luwei, K.C.; Yunusa-Kaltungo, A.; Sha'Aban, Y.A. Integrated Fault Detection Framework for Classifying Rotating Machine Faults Using Frequency Domain Data Fusion and Artificial Neural Networks. *Machines* **2018**, *6*, 59. [CrossRef]
10. Zhang, R.; Tao, H.; Wu, L.; Guan, Y. Transfer Learning With Neural Networks for Bearing Fault Diagnosis in Changing Working Conditions. *IEEE Access* **2017**, *5*, 14347–14357. [CrossRef]
11. Wen, L.; Gao, L.; Li, X. A New Deep Transfer Learning Based on Sparse Auto-Encoder for Fault Diagnosis. *IEEE Trans. Syst. Man Cybern. Syst.* **2017**, *49*, 136–144. [CrossRef]
12. Qian, W.; Li, S.; Wang, J. A New Transfer Learning Method and its Application on Rotating Machine Fault Diagnosis under Variant Working Conditions. *IEEE Access* **2018**, *6*, 69907–69917. [CrossRef]
13. Lu, S.; Sirojan, T.; Phung, B.T.; Zhang, D.; Ambikairajah, E. DA-DCGAN: An Effective Methodology for DC Series Arc Fault Diagnosis in Photovoltaic Systems. *IEEE Access* **2019**, *7*, 45831–45840. [CrossRef]
14. Guo, L.; Lei, Y.; Xing, S.; Yan, T.; Li, N. Deep Convolutional Transfer Learning Network: A New Method for Intelligent Fault Diagnosis of Machines with Unlabeled Data. *IEEE Trans. Ind. Electron.* **2018**, *66*, 7316–7325. [CrossRef]
15. Zheng, H.; Wang, R.; Yang, Y.; Yin, J.; Li, Y.; Li, Y.; Xu, M. Cross-Domain Fault Diagnosis Using Knowledge Transfer Strategy: A Review. *IEEE Access* **2019**, *7*, 129260–129290. [CrossRef]
16. Bin, G.; Gao, J.; Li, X.; Dhillon, B. Early fault diagnosis of rotating machinery based on wavelet packets—Empirical mode decomposition feature extraction and neural network. *Mech. Syst. Signal. Process.* **2012**, *27*, 696–711. [CrossRef]
17. Saravanan, N.; Ramachandran, K. Incipient gear box fault diagnosis using discrete wavelet transform (DWT) for feature extraction and classification using artificial neural network (ANN). *Expert Syst. Appl.* **2010**, *37*, 4168–4181. [CrossRef]
18. Buch, K.D. Decision Based Non-Linear Filtering Using Interquartile Range Estimator for Gaussian Signals. In Proceedings of the 2014 Annual IEEE India Conference (INDICON), Pune, India, 11–13 December 2014. [CrossRef]

Publisher's Note: MDPI stays neutral with regard to jurisdictional claims in published maps and institutional affiliations.

applied sciences

MDPI

Article

Feature Extraction for Bearing Fault Detection Using Wavelet Packet Energy and Fast Kurtogram Analysis

Xiaojun Zhang *, Jirui Zhu, Yaqi Wu, Dong Zhen and Minglu Zhang

Tianjin Key Laboratory of Power Transmission and Safety Technology for New Energy Vehicles, School of Mechanical Engineering, Hebei University of Technology, Tianjin 300401, China; jir_zhu@163.com (J.Z.); wuyaqi159753@163.com (Y.W.); d.zhen@hebut.edu.cn (D.Z.); zhangml@hebut.edu.cn (M.Z.)

* Correspondence: xjzhang@hebut.edu.cn

Received: 28 September 2020; Accepted: 28 October 2020; Published: 31 October 2020

Abstract: An integrated method for fault detection of bearing using wavelet packet energy (WPE) and fast kurtogram (FK) is proposed. The method consists of three stages. Firstly, several commonly used wavelet functions were compared to select the appropriate wavelet function for the application of WPE. Then the analyzed signal is decomposed using WPE and the energy of each decomposed signal is calculated and selected for signal reconstruction. Secondly, the reconstructed signal is analyzed by FK to select the best central frequency and bandwidth for the band-pass filter. Finally, the filtered signal is processed using the squared envelope frequency spectrum and compared with the theoretical fault characteristic frequency for fault feature extraction. The procedure and performance of the proposed approach are illustrated and estimated by the simulation analysis, proving that the proposed method can effectively extract the weak transients. Moreover, the analysis results of gearbox bearing and rolling bearing cases show that the proposed method can provide more accurate fault features compared with the individual FK method.

Keywords: wavelet packet energy (WPE); fast kurtogram (FK); wavelet packet parameters; rolling element bearing

1. Introduction

The condition monitoring of rolling bearings is very important in ensuring the safety of the mechanical system as it is one of the basic components and widely applied in various rotating machines. The measured dynamic signal of bearing has the characteristics of non-linear and non-stationary with various background noise. Moreover, the original signal is too low in energy to extract fault characteristics. As a result, some progress has been made in the research of effective health monitoring for mechanical system and key components in recent years [1–3]. The vibration signal produced by the rotating machine contains significant information related to the state of the machine; therefore, the vibration-based analysis has been widely used as an effective method to identify the machine faults [4]. However, the measured vibration signal is consisted of stationary, non-stationary, and background noise. It is necessary to select appropriate signal processing technology to obtain effective fault features for rolling bearings fault detection [5–7].

In the fault diagnosis of rolling bearings, the analysis method based on the time domain and frequency domain are commonly used for stationary signal processing, and the time-frequency domain analysis method has been widely used to analyze the non-stationary signals for fault features extraction in both the time and frequency domains. The popular time-frequency analysis methods include short time Fourier transform (STFT), empirical mode decomposition (EMD), Wigner-Ville distribution (WVD), wavelet pack transform (WPT), etc. Although these methods have been widely used in fault detection, they also have their own shortcomings. For instance, the STFT cannot satisfy the localization of time and frequency at the same time. The WVD is affected by cross-interference items. The EMD can effectively decompose amplitude modulation signals into a series of intrinsic mode functions (IMFs), but it is easily affected by modal aliasing, which makes IMFs lose their specific physical meaning. Compared with the above-mentioned method, WPT can perform orthogonal decomposition on the signal of the whole frequency band to obtain more detailed and comprehensive information. Therefore, it has been widely used in fault diagnosis of gearboxes, rolling bearings, and rotors [8–10]. On the basis of WPT, wavelet packet energy (WPE) is an improved time-frequency analysis method which can get the energy change of the signal in each frequency band and then extract the characteristic frequency band signal to reflect the impact vibration phenomenon of bearing. In recent years, WPE has become a prevalent method in the field of machinery condition monitoring and fault diagnosis [11–14].

Spectra kurtosis (SK) is one of the effective approaches for transient signal enhancement. It was originally developed by Dwyer [15] to highlight the transient components. The spectra kurtosis method uses the properties of kurtosis, is very sensitive to the transient components in a signal, and can provide an index for indicating the frequency band for transient components extraction. The definition of the "kurtogram" was first introduced by Antoni [16]. The fast kurtogram (FK) overcomes the limitation of pre-determining the parameters for the band-pass filter and can optimize the parameters adaptively according to the characteristics of the analyzed signal. Therefore, FK has been favored in the field of rotating machinery fault detection, and has been widely adopted and developed in recent years. Smith et al. applied the FK to analyze the vibration signal obtained from a planetary gearbox test bench for detecting planet bearings faults [17]. He et al. proposed a new multi-faults detection method using the FK and Minimum entropy deconvolution (MED) to extract weak impulsive signal [18]. Zhang et al. presented an effective method for band-filter parameter optimization in rolling element bearing fault diagnosis based on the genetic algorithms and FK [19]. The above research shows that the FK can be applied to bearing fault diagnosis and extract the fault features effectively.

To effectively extract fault features from background noise, this paper proposes a hybrid method based on WPE and FK for bearing fault diagnosis. Firstly, in order to select the optimized wavelet packet basis function and obtain a reconstructed signal with sufficient fault information, various wavelet functions are compared and the energy distribution were analyzed. Secondly, FK analysis is carried out on the reconstructed signal to select the optimal center frequency and bandwidth of the band-pass filter. Finally, the filtered signal is processed with the square envelope spectrum, and the fault features are extracted for bearing fault detection. Compared with the conventional FK analysis, the diagnosis result shows that the proposed method not only can extract the fault characteristic frequency but also optimize the wavelet packet parameters. Therefore, the effectiveness of the proposed method in the fault diagnosis of rolling bearings is verified.

The remainder of this paper is outlined as follows. A brief review of the theoretical background of wavelet packet transform and fast kurtogram are presented in Sections 2 and 3. In Section 4, the performance evaluation of the proposed method based on simulation study is introduced. The procedure of the proposed method is demonstrated in Section 5. Section 6 validates the practical applications of the detector via application case studies. Finally, conclusions are drawn in Section 7.

2. Wavelet Packet Transform

2.1. WPT

Compared with wavelet decomposition, wavelet packet (WPT) makes up for the shortcomings of high-frequency part decomposition of the measurement signal. It separates the frequency band of the signal in multiple levels and refines the fault features of the measurement signal in each frequency band [20,21]. Suppose the original signal is $x_j^n(t) \in u_j^n$, then $x_j^n(t)$ can be decomposed by wavelet packet as:

$$x_j^n(t) = \sum_t d_l^{j,n} u_n\left(2^j t - l\right) \quad (1)$$

Among them, j represents the scale and l represents the translation factor. $u_n\left(2^j t - l\right)$ represents the basis function in u_j^n, and $d_l^{j,n}$ represents the WPT coefficient. $\{d_l^{j,n}\}$ can be decomposed into $\{d_l^{j,\,2n}\}$ and $\{d_l^{j,\,2n+1}\}$, namely:

$$\begin{cases} d_l^{j,2n} = \sum\limits_k a_{k-2l} d_k^{j+1,n} \\ d_l^{j,2n+1} = \sum\limits_k a_{k-2l} d_k^{j+1,n} \end{cases} \tag{2}$$

where k represents the translation parameter; n represents the oscillation parameter.

2.2. Frequency Band Energy Feature Extraction

Suppose the highest frequency of the time signal $x(t)$ is $f_{\max}$, and perform J-layer WPT decomposition on $x(t)$ to get 2^J groups of WPT coefficients, respectively $w_{Ji}\left(i = 0, 1, \cdots, 2^J - 1\right)$. The frequency band corresponding to these 2^J coefficients is $\left[\frac{i}{2^J} f_{\max}, \frac{i+1}{2^J} f_{\max}\right]$.

Take the three-layer WPT as an example. First, find the signal energy of each frequency band. Suppose the corresponding energy of $w_{3i}(i = 0,\ 1, \cdots, 7)$ is $E_{3i}(i = 0,\ 1, \cdots, 7)$, then:

$$E_{3i} = \|w_{3i}\|^2 = \int_{-\infty}^{+\infty} |w_{3i}|^2 dt \tag{3}$$

Secondly, the feature vector is constructed with E_{3i} as the element:

$$T = [E_{30}, E_{31}, \cdots, E_{37}] \tag{4}$$

When the signal energy is large, $E_{3i}(i = 0,\ 1,\ \cdots,\ 7)$ represents a large value, so it will bring certain difficulties in signal processing. Therefore, the feature vector T will be normalized:

$$E = \left|\sum_{i=0}^{7} |E_{3i}|^2\right|^{\frac{1}{2}} \tag{5}$$

Finally, get the normalized vector $R = [E_{30}/E, E_{31}/E,\ \cdots, E_{37}/E]$.

3. Kurtogram Calculation

Antoni clarified the theoretical basis of spectral kurtosis (SK) for diagnosing mechanical faults, and successfully applied the SK method in actual mechanical fault diagnosis [22]. Subsequently, some researchers combined SK with other methods for machinery fault detection and achieved certain results [23,24]. Considering the Wold-Cramer decomposition of non-stationary measurement signals, and define $Y(t)$ as the system excited by the signal $x(t)$, then $Y(t)$ can be presented as:

$$Y(t) = \int_{-\infty}^{+\infty} e^{j2\pi f t} H(t, f) dX(f) \tag{6}$$

where $dX(f)$ is a spectral process, $H(t, f)$ is the complex envelope of process $Y(t)$ under frequency f. The SK using the fourth-order spectrum cumulate is defined as:

$$C_{4Y}(f) = S_{4Y}(f) - S_{2Y}^2(f) \quad (f \neq 0) \tag{7}$$

where $S_{2nY}(f)$ indicates the instantaneous moment of order $2n$, which can be expressed in Equation (8). It has the ability of presenting the complex envelope energy.

$$S_{2nY}(t, f) = E\{\left|H(t, f) dX(f)\right|^{2n}\}/df \tag{8}$$

Hence the SK in the energy-normalized fourth-order spectral can be cumulated:

$$SK_Y(f) = \frac{C_{4Y}(f)}{S_{2Y}^2(f)} = \frac{S_{4Y}(f)}{S_{4Y}^2(f)} - 2,\ f \neq 0 \tag{9}$$

A simple model of the measured signal can be represented $x(t) = z(t) + n(t)$, $x(t)$ is the measured vibration signal, $n(t)$ is the random noise signal, and $z(t)$ is the pure measurement signal. $SK_X(f)$ and $SK_Z(f)$ are used to represent the measured vibration signal and the SK of the pure measurement signal respectively, then SK is written as

$$SK_x(f) = \frac{SK_z(f)}{[1+\rho(f)]^2} \tag{10}$$

where $\rho(f) = S_{2N}(f)/S_{2Z}(f)$ represents the power spectral density (PSD) ratio of the noise to the actual vibration signal. It can be seen that $SK_x(f)$ and $SK_z(f)$ are approximately equal at frequencies with a large signal-to-noise ratio, and $SK_x(f)$ is approximately equal to zero at frequencies with strong random noise.

4. Simulation Analysis

4.1. Data Acquisition

The vibration signal of a bearing with incipient faults can be represented as an amplitude modulation model. Considering the random sliding process, the modeled vibration signal can be expressed in Equation (11):

$$\begin{cases} x_1(t) = \sum\limits_{m=1}^{M} Ah_1\left(t - \frac{i}{f_i} - \tau_i\right) \\ h_1(t) = \exp(-C_1 t)\cos(2\times 3000\pi t) \end{cases} \tag{11}$$

where A = 1 is the amplitude of the modeled signal, C_i = 900 is attenuation coefficient, f_i = 100 Hz is the fault characteristic frequency, τ_i is the random fluctuation that occurs during the i-th impact. The sampling frequency of the data is 20 kHz, and the sampling points is 8192.

4.2. Optimization of the WPE Parameters

Figure 1 shows the waveform and the corresponding spectrum of the simulated outer race fault signals. It is not easy to recognize the fault features of the simulated signal from Figure 1.

In order to get better wavelet packet parameters, firstly, select several commonly used wavelet packet parameters (i.e., db1, db3, db5, db7, db10, coif1, coif2, coif3, coif4, coif5, sym1, sym3, sym5, sym7, sym10) from Table 1. Then, decomposed, energy calculated and reconstructed by using the above 15 wavelet functions respectively. Secondly, the reconstructed signal is filtered using FK and the filtered signal is square envelope spectrum analysis. Then, compare the amplitude of square envelope spectrum to determine the optimal wavelet base in Figure 2. It can be seen that db1, coif2, and sym1 decomposition effect is better than others. To further select the wavelet function, the above three wavelet functions are compared as shown in Figure 3. It can be seen that the decomposition of db1 and sym1 are similar, better than coif2. Finally, due to the db1 and sym1 decomposition effect being similar, then select the wavelet function (i.e., db2, db3, db4, db5) and (i.e., sym2, sym3, sym4, sym5) for comparison. Figure 4a,b depict that decomposition of db2, db3 and sym2, sym3 show a similar effect; however, the magnitude of db4, db5 is greater than the magnitude of sym4, sym5 in Figure 4c,d. Therefore, wavelet function dbN decomposition is best.

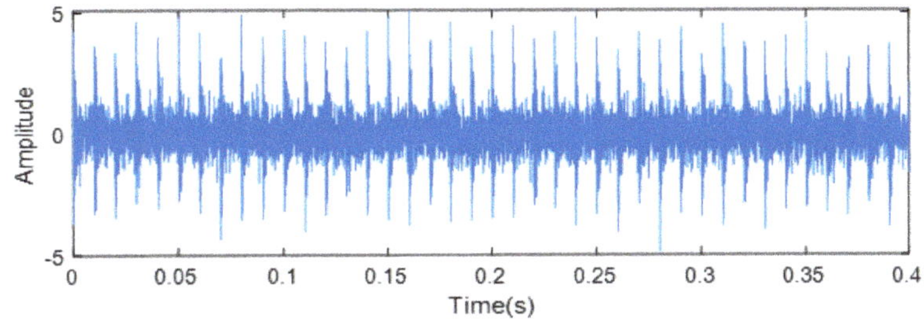

(**a**)

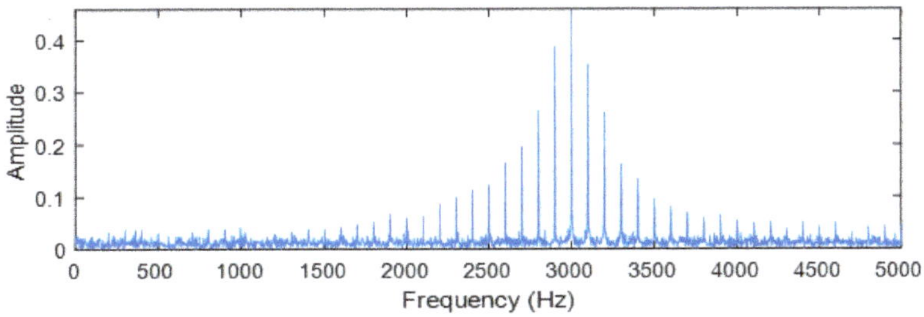

(**b**)

Figure 1. Simulated signal of a faulty bearing: (**a**) waveform; (**b**) spectrum.

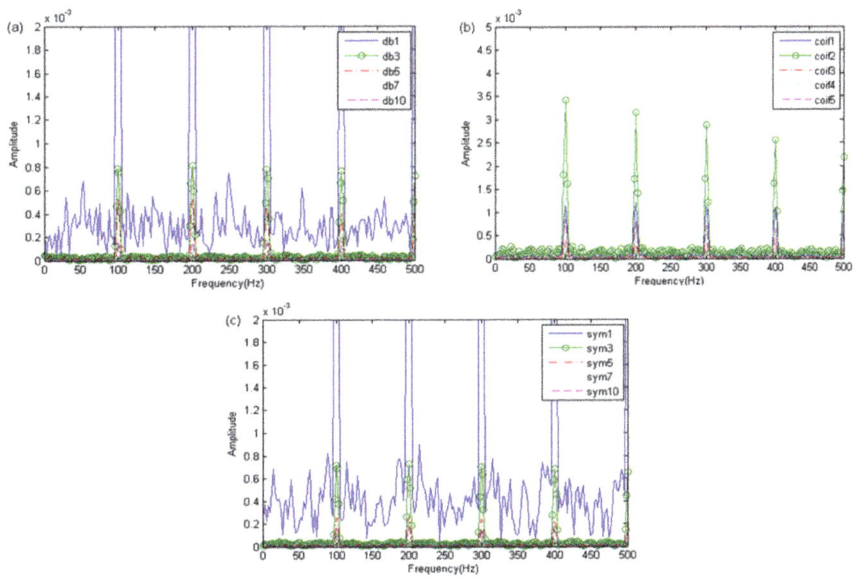

Figure 2. The results of the envelope analysis: (**a**) dbN; (**b**) coifN; (**c**) symN.

Table 1. Wavelet packet parameters.

Wavelet Function	Order Range
dbN	1–45
coifN	1–5
symN	1–45

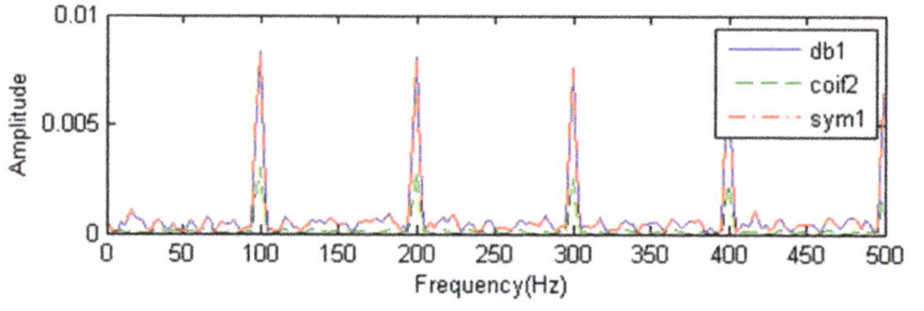

Figure 3. The results of the envelope analysis.

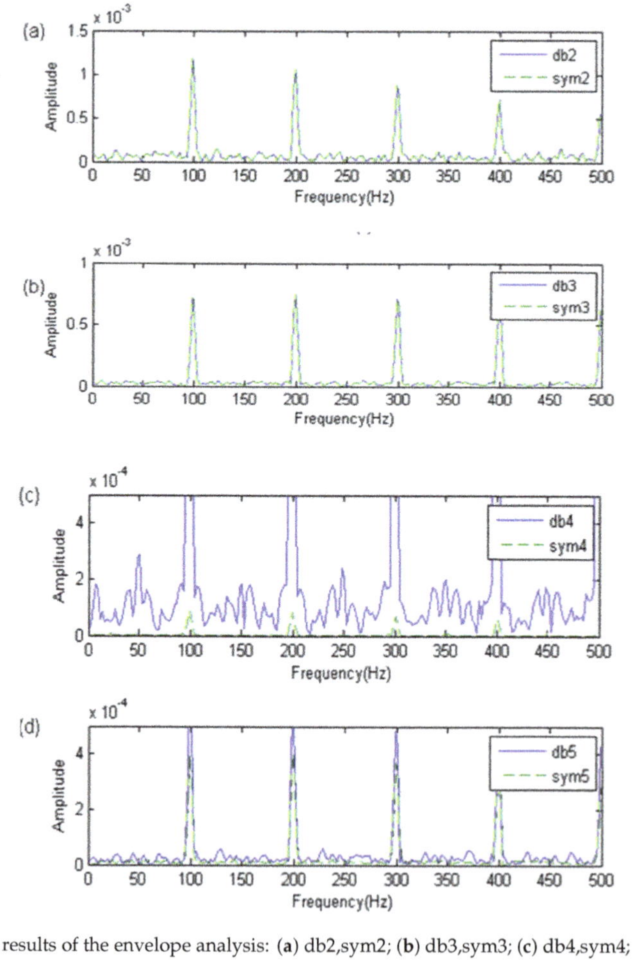

Figure 4. The results of the envelope analysis: (**a**) db2,sym2; (**b**) db3,sym3; (**c**) db4,sym4; (**d**) db5,sym5.

The above concludes that the decomposition of wavelet function dbN is the best. However, the order of the wavelet packet is not determined. In this paper, we applied the FK to optimize the wavelet packet order. The FK of the reconstructed signal is depicted in Figure 5. It can be found that the center frequency of FK utilize db3 is located in the subspace the frequency band. Therefore, db3 is selected as the wavelet function.

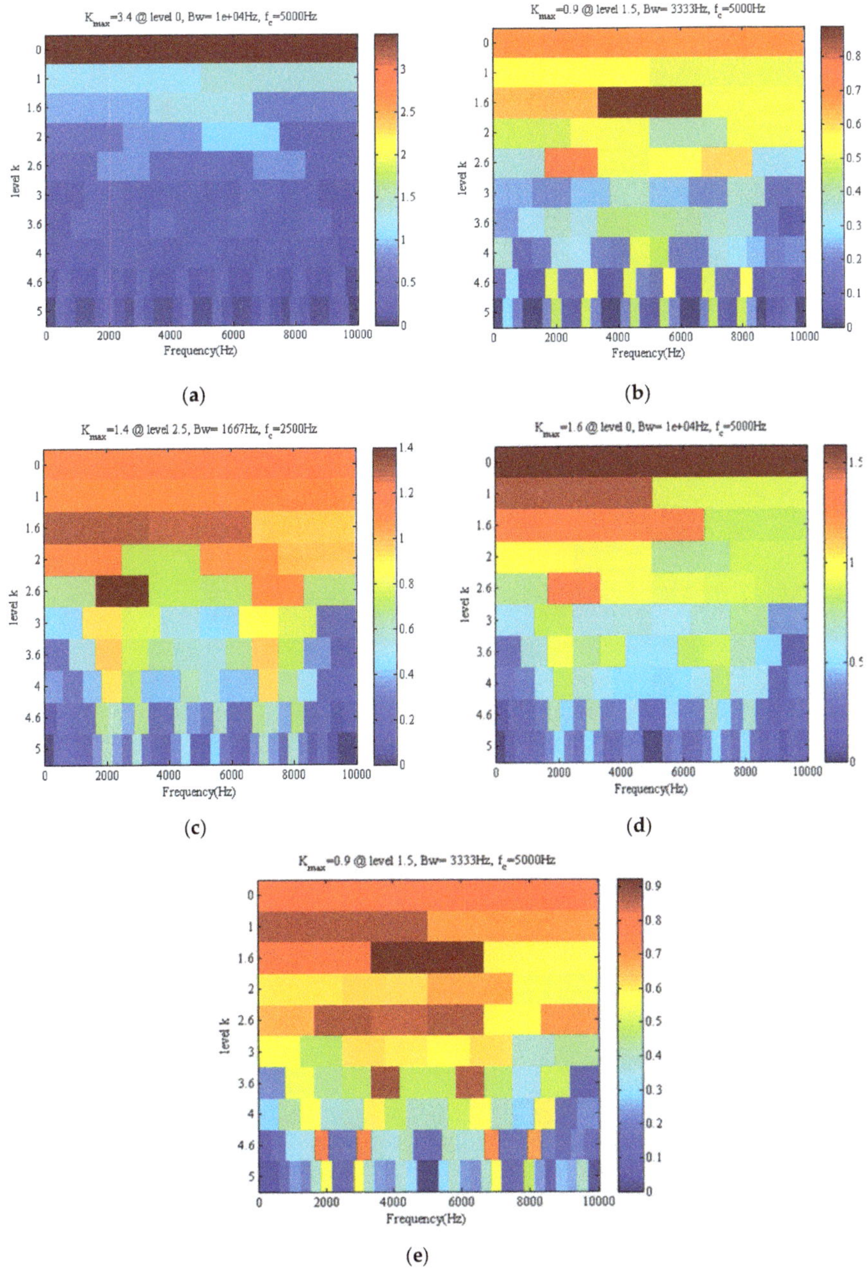

Figure 5. The results of the fast kurtogram (FK) (**a**) db1; (**b**) db2; (**c**) db3; (**d**) db4; (**e**) db5.

4.3. Results and Analysis

The WPE method is utilized to extract the fault characteristics of the faulty signal. The wavelet function db3 is used, decomposition layers are selected to be 3, then 8 decomposition bands are generated and the frequency range of 3-layer wavelet packet displayed in Table 2. The proportional relationships of the energy bands are analyzed and calculated; the comparison result is illustrated in Figure 6 by histograms. It can be seen that the changes in the WPE distribution is presented in the node [3,2] and the node [3,3]. Figure 7 illustrated the results of the square envelope spectrum the proposed method, and f_o represents the fault frequency of the bearing outer ring.

Table 2. The frequency range of three layer wavelet packet decomposition.

Serial Number	Node Situation	Frequency/Hz
1st	node[3,0]	0–1250 Hz
2nd	node[3,1]	1250–2500 Hz
3rd	node[3,2]	2500–3750 Hz
4th	node[3,3]	3750–5000 Hz
5th	node[3,4]	5000–6250 Hz
6th	node[3,5]	6250–7500 Hz
7th	node[3,6]	7500–8750 Hz
8th	node[3,7]	8750–10,000 Hz

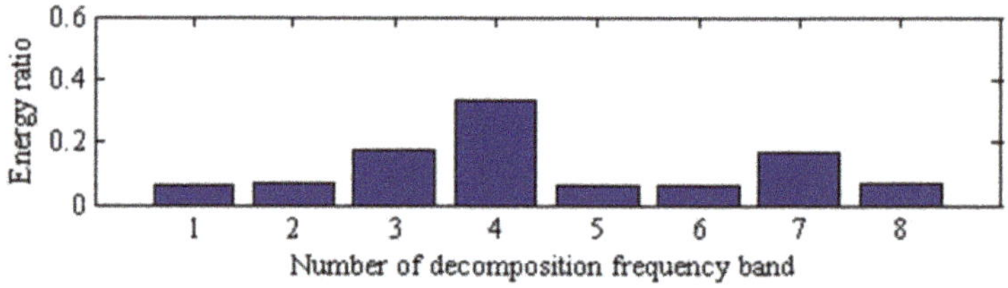

Figure 6. Energy distribution proportional histogram of simulated signal.

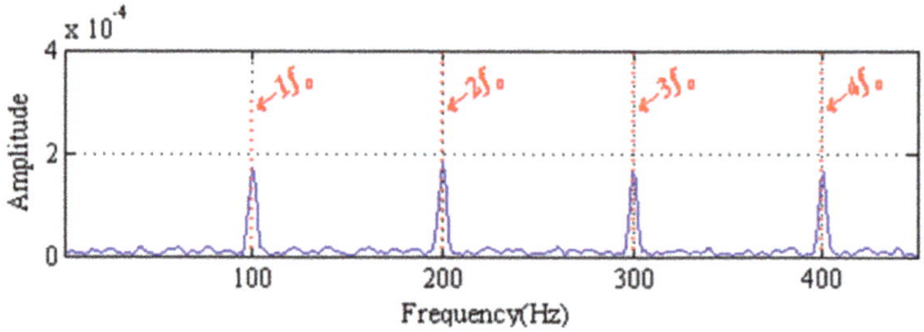

Figure 7. The results of the proposed method.

5. The Flowchart of the Fault Diagnosis Method

This paper proposes a new method using wavelet packet energy (WPE) and fast kurtogram (FK). The specific diagnosis steps are as follows, and the basic process is shown in Figure 8.

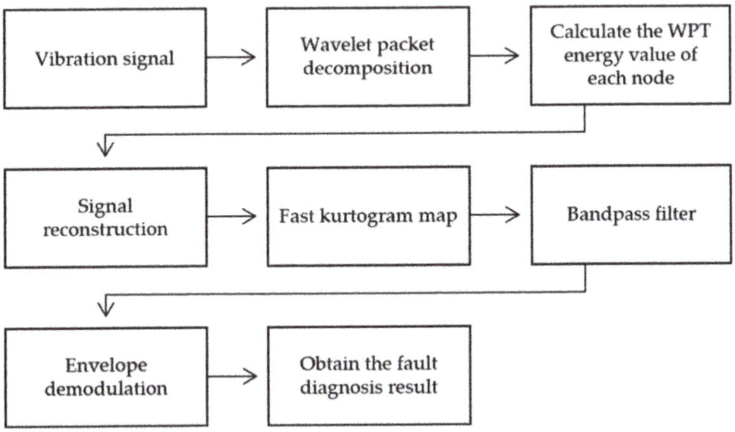

Figure 8. Rolling bearing fault diagnosis flowchart.

(1) By selecting the appropriate wavelet function and the number of decomposition layers, the measurement signal is decomposed by WPT and passed to obtain the WPT coefficient of each node;
(2) Calculate the WPT energy value of each node;
(3) Select the decomposed signal with the concentrated energy for reconstruction;
(4) Calculate the SK map of the acquired reconstructed signal and select the center frequency and bandwidth corresponding to the maximum kurtosis value;
(5) Filter the reconstructed signal using the optimized center frequency and bandwidth as the band-pass filter parameters;
(6) Perform square envelope demodulation on the filtered signal to obtain the envelope spectrum of the signal;
(7) Analyze the envelope spectrum to obtain the fault diagnosis result.

6. Experiments Validation

6.1. Data Acquisition

The first verification test case is fault detection for rolling bearing. The bearing test rig consists of a motor, coupling, intermediate shaft, supporting bearings, and electrical brake, as shown in Figure 9. A piezoelectric accelerometer was installed vertically on the drive end of the motor, and another piezoelectric accelerometer was positioned on the bearing housing with a sensitivity of 1.04 mV/ms^2. The standard load of the bearing test rig is 680 Nm, and the final output speed is 388 rpm. Tables 3 and 4 show the parameters of the experimental bearing, among them, f_o, f_i, f_b, and f_c are the fault characteristic frequencies of the outer ring, inner ring, rolling element, and cage of the rolling element bearing, respectively.

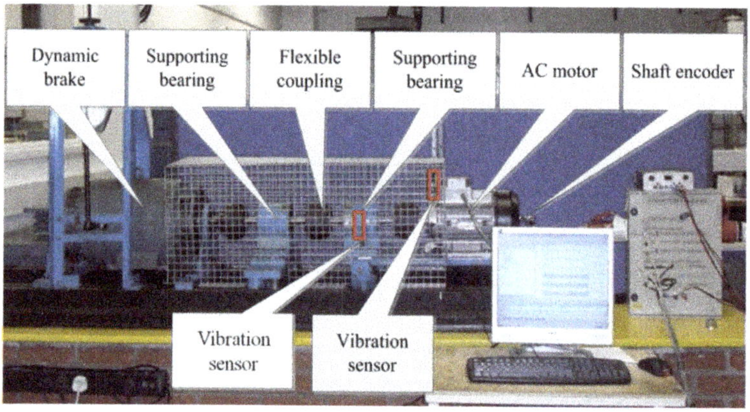

Figure 9. The motor bearing test rig.

Table 3. The parameters of the motor bearings.

Bearing Designation	Pitch Diameter D_m (mm)	Ball Diameter d (mm)	Ball Numbers z
6206ZZ	46.4	9.53	9
f_o	f_i	f_b	f_c
89.33	130.99	62.42	9.93

Table 4. Specifications of the supporting bearing.

Bearing Designation	Pitch Diameter D_m (mm)	Ball Diameter d (mm)	Ball Numbers z
6008	54	7.9	12
f_o	f_i	f_b	f_c
49.25	65.17	33.60	4.10

6.2. Motor Bearing Fault Detection

Figure 10 shows the waveform and its frequency spectrum. It is not easy to identify the fault of rolling bearing from Figure 10. Thus, it is necessary to conduct further analysis.

The proposed method is used to handle the measured signal of the gearbox bearing. In Figure 11, the changes in WPE distribution in the node [2,3] and the node [3,6] can be seen. Selecting the above two subspaces are applied as the eigenvalue of the outer race fault signals. To further prove the superiority of the proposed method, the individual FK method is applied for comparison. Figure 12 shows the analysis results of the proposed method and the individual FK method. Three obvious spectrum lines pointed at the characteristic frequencies of the bearing with outer race (i.e., f_o, $2f_o$, and $3f_o$) can be recognized in Figure 12a. In contrast, although three harmonics of fault frequency (i.e., f_o, $2f_o$, and $3f_o$) can be identified in Figure 12b, the noise is significantly more than the proposed method. The experimental results demonstrated that the proposed method can obtain more accurate recognized result than the individual FK method for bearing outer race fault detection.

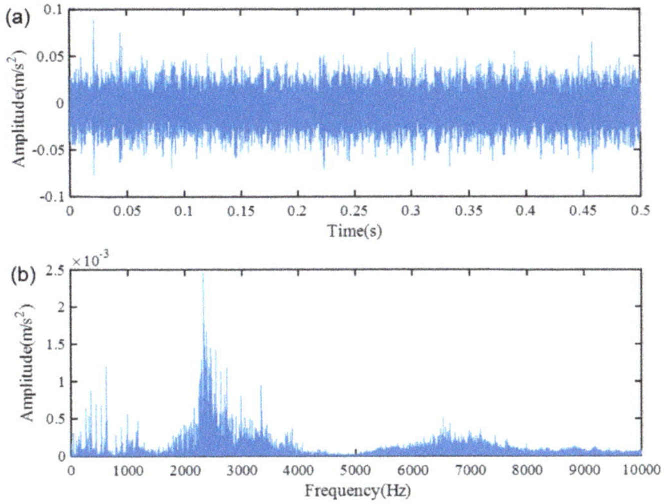

Figure 10. The motor vibration signal: (**a**) waveform; (**b**) spectrum.

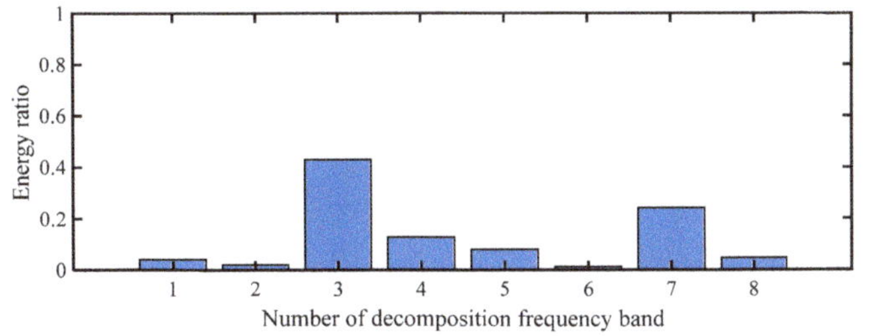

Figure 11. Energy distribution proportional histogram of the vibration signal.

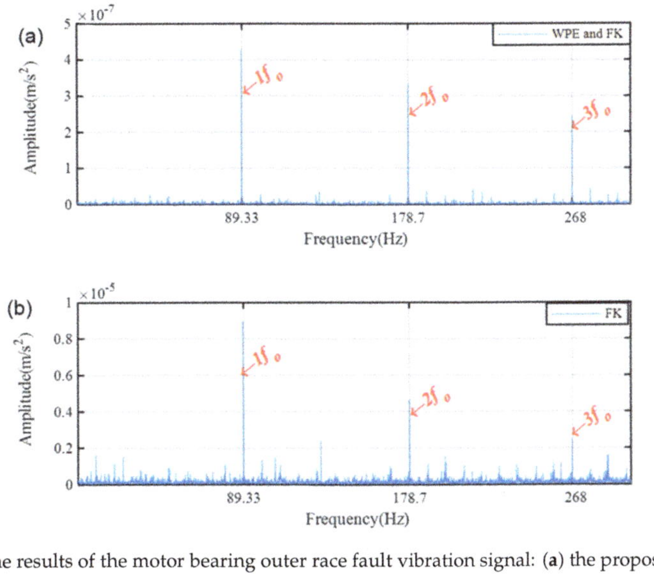

Figure 12. The results of the motor bearing outer race fault vibration signal: (**a**) the proposed method; (**b**) individual FK method.

6.3. Supporting Bearing Fault Detection

Figure 13 depicts the waveform and spectrum. Evidently, the fault feature information is merged with random noise. Thus, the diagnosis is barely possible.

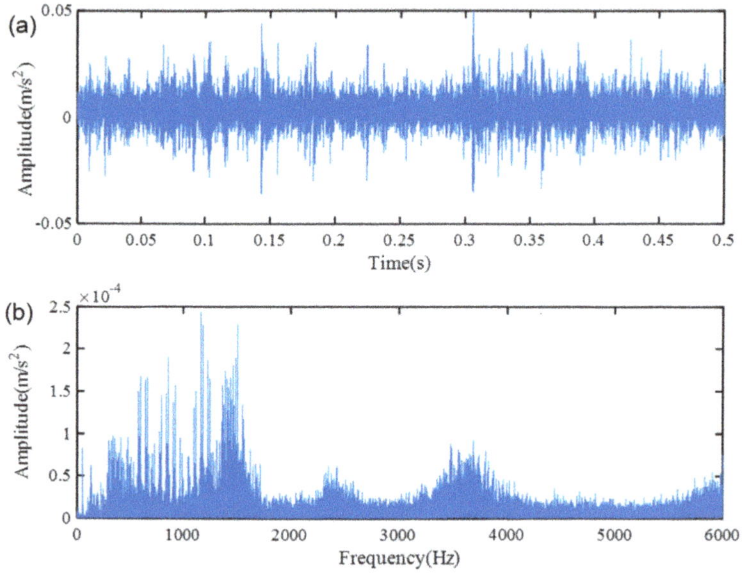

Figure 13. The supporting bearing vibration signal: (**a**) waveform; (**b**) spectrum.

The proposed method is utilized to diagnosis the fault information. In Figure 14, the changes in WPE distribution in the node [3,4] and the node [3,5] are shown. Figure 15 illustrates the analysis results of the individual FK and the proposed method, respectively. It can be found from Figure 15a that the result presented obvious peaks reflected the supporting bearing fault and its harmonics. In Figure 15b, although the inner race fault frequency and its harmonics can be recognized, the signal-to noise ratio is poor. The comparison results show that the proposed method is more accurate than the individual FK method in supporting bearing fault diagnosis.

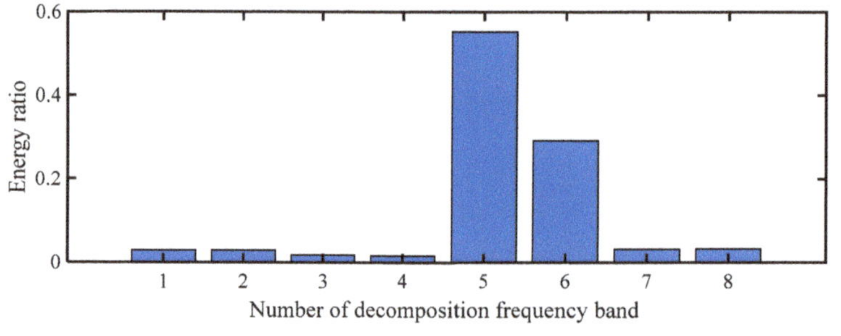

Figure 14. Energy distribution proportional histogram of the vibration signal.

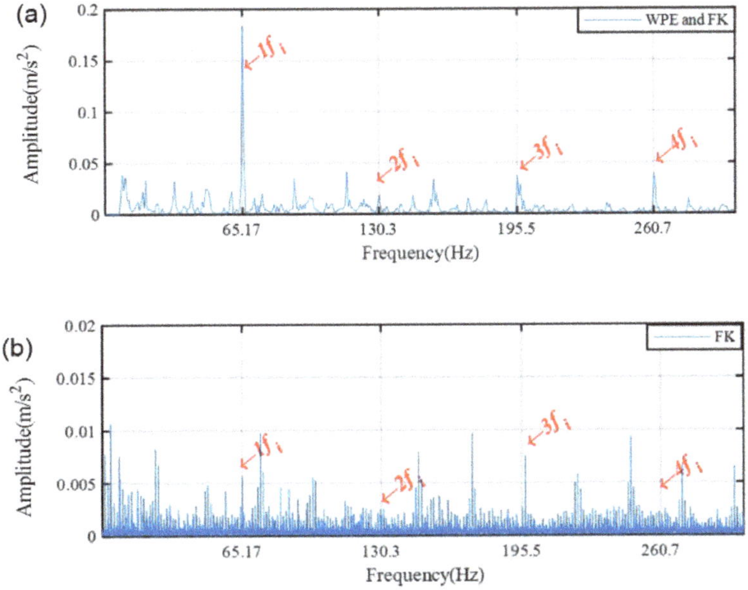

Figure 15. The results of the supporting bearing inner race fault vibration signal: (**a**) the proposed method; (**b**) individual FK.

7. Conclusions

Aiming at the problem of energy concentration of faulty bearing vibration signals, this paper proposed a new method combining wavelet packet energy (WPE) and fast kurtogram (FK) to extract the transient signals. In the simulation analysis, the wavelet packet parameters were optimized by comparison analysis. And it is verified that the wavelet function of db3 can obtain better decomposition results than others. Through experiments, even both the proposed method and the individual FK method can obtain the bearing faults characteristic frequency, but it can be seen that the results of the proposed method significantly exceeds the results obtained by the individual FK method. In addition, the signal-to-noise ratio of the envelope spectrum obtained by the proposed method is much higher than that of the individual FK method. Therefore, the experiment results demonstrate that the proposed method outperforms the individual FK method. Furthermore, it can effectively solve the energy concentration problem in feature extraction for bearing fault.

Author Contributions: X.Z. and J.Z. conceived and wrote the paper; Y.W. and J.Z. analyzed the data; D.Z. and M.Z. proposed the analysis method and revised the manuscript. All authors have read and agreed to the published version of the manuscript.

Funding: This research was funded by the National Key Research and Development Project of China, grant number 2018YFB1309401.

Conflicts of Interest: The authors declare no conflict of interest.

References

1. Li, Z.; Ming, A.; Zhang, W.; Liu, T.; Chu, F.; Li, Y. Fault feature extraction and enhancement of rolling element bearings based on maximum correlated kurtosis deconvolution and improved empirical wavelet transform. *Appl. Sci.* **2019**, *9*, 1876. [CrossRef]
2. Zhang, X.; Wang, J.; Liu, Z.; Wang, J. Weak feature enhancement in machinery fault diagnosis using empirical wavelet transform and an improved adaptive bistable stochastic resonance. *ISA Trans.* **2019**, *84*, 283–295. [CrossRef] [PubMed]

3. Cui, L.; Wang, X.; Wang, H.; Jiang, H. Remaining useful life prediction of rolling element bearings based on simulated performance degradation dictionary. *Mech. Mach. Theory* **2020**, *153*, 103967. [CrossRef]
4. Lei, Y.; Liu, Z.; Lin, J.; Lu, F. Phenomenological models of vibration signals for condition monitoring and fault diagnosis of epicyclic gearboxes. *J. Sound Vib.* **2016**, *369*, 266–281. [CrossRef]
5. Zhou, R.; Bao, W.; Li, N.; Huang, X.; Yu, D. Mechanical equipment fault diagnosis based on redundant second generation wavelet packet transform. *Digit. Signal Process.* **2010**, *20*, 276–288. [CrossRef]
6. Guo, J.; Zhen, D.; Li, H.; Shi, Z.; Gu, F.; Ball, A. Fault detection for planetary gearbox based on an enhanced average filter and modulation signal bispectrum analysis. *ISA Trans.* **2020**, *101*, 408–420. [CrossRef]
7. Cui, L.; Wang, X.; Wang, H.; Ma, J. Research on remaining useful life prediction of rolling element bearings based on time-varying Kalman filter. *IEEE Trans. Instrum. Meas.* **2020**, *69*, 2858–2867. [CrossRef]
8. Xing, Y.; Wang, Y.; Shi, L.; Guo, H.; Chen, H. Sound quality recognition using optimal wavelet-packet transform and artificial neural network methods. *Mech. Syst. Signal Process.* **2015**, *66*. [CrossRef]
9. Mavaddaty, S.; Ahadi, S.M.; Seyedin, S. Speech enhancement using sparse dictionary learning in wavelet packet transform domain. *Comput. Speech Lang.* **2017**, *44*, 22–47. [CrossRef]
10. Xu, X.; Luo, M.; Tan, Z.; Pei, R. Echo signal extraction method of laser radar based on improved singular value decomposition and wavelet threshold denoising. *Infrared. Phys. Tech.* **2018**, *92*, 327–335. [CrossRef]
11. Bianchi, D.; Mayrhofer, E.; Gröschl, M.; Betz, G.; Vernes, A. Wavelet packet transform for detection of single events in acoustic emission signals. *Mech. Syst. Signal. Process.* **2015**, *64*, 441–451. [CrossRef]
12. He, Q. Vibration signal classification by wavelet packet energy flow manifold learning. *J. Sound Vib.* **2013**, *332*, 1881–1894. [CrossRef]
13. Han, L.; Li, C.; Guo, S.; Su, X. Feature extraction method of bearing AE signal based on improved FAST-ICA and wavelet packet energy. *Mech. Syst. Signal Process.* **2015**, *62*, 91–99. [CrossRef]
14. Hoseinzadeh, M.S.; Khadem, S.E.; Sadooghi, M.S. Quantitative diagnosis for bearing faults by improving ensemble empirical mode decomposition. *ISA Trans.* **2018**, *83*, 261–275. [CrossRef]
15. Dwyer, R. Detection of non-gaussian signals by frequency domain kurtosis estimation. In Proceedings of the IEEE International Conference on Acoustics Speech and Signal Processing, Boston, MA, USA, 14–16 April 1983; Volume 8, pp. 607–610.
16. Antoni, J. Fast computation of the kurtogram for the detection of transient faults. *Mech. Syst. Signal. Process.* **2007**, *21*, 108–124. [CrossRef]
17. Smith, W.A.; Fan, Z.; Peng, Z.; Li, H.; Randall, R.B. Optimised spectral kurtosis for bearing diagnostics under electromagnetic interference. *Mech. Syst. Signal Process.* **2016**, *75*, 371–394. [CrossRef]
18. He, D.; Wang, X.; Li, S.; Lin, J.; Zhao, M. Identification of multiple faults in rotating machinery based on minimum entropy deconvolution combined with spectral kurtosis. *Mech. Syst. Signal Process.* **2016**, *81*, 235–249. [CrossRef]
19. Zhang, Y.; Randall, R.B. Rolling element bearing fault diagnosis based on the combination of genetic algorithms and fast kurtogram. *Mech. Syst. Signal Process.* **2009**, *23*, 1509–1517. [CrossRef]
20. Guo, J.; Zhen, D.; Li, H.; Shi, Z.; Gu, F.; Ball, A. Fault feature extraction for rolling element bearing diagnosis based on a multi-stage noise reduction method. *Measurement* **2019**, *139*, 226–235. [CrossRef]
21. Mishra, C.; Samantaray, A.K.; Chakraborty, G. Rolling element bearing fault diagnosis under slow speed operation using wavelet de-noising. *Measurement* **2017**, *103*, 77–86. [CrossRef]
22. Antoni, J. The spectral kurtosis: A useful tool for characterising non-stationary signals. *Mech. Syst. Signal Process.* **2006**, *20*, 282–307. [CrossRef]
23. Miao, Y.; Zhao, M.; Lin, J. Improvement of kurtosis-guided-grams via Gini index for bearing fault feature identification. *Meas. Sci. Technol.* **2017**, *28*, 1–14. [CrossRef]
24. Ren, Y.; Li, W.; Zhang, B.; Zhu, Z.; Jiang, F. Fault diagnosis of rolling bearings based on improved kurtogram in varying speed conditions. *Appl. Sci.* **2019**, *9*, 1157. [CrossRef]

applied sciences

MDPI

Article

Fatigue Life Analysis of Ball Bearings and a Shaft System Considering the Combined Bearing Preload and Angular Misalignment

Yu Zhang [1,*], Mengqi Zhang [2], Yawen Wang [3] and Liyang Xie [2,4]

1 School of Mechanical Engineering, Shenyang Jianzhu University, Shenyang 110168, China
2 School of Mechanical Engineering and Automation, Northeastern University, Shenyang 110819, China; mqiz0313@163.com (M.Z.); lyxie@mail.neu.edu.cn (L.X.)
3 Department of Mechanical and Aerospace Engineering, University of Texas at Arlington, Arlington, TX 76019, USA; yawen.wang@uta.edu
4 Ministry of Education Key Laboratory of Vibration and Control of Aero-Propulsion System, Northeastern University, Shenyang 110819, China
* Correspondence: yu.zhang@sjzu.edu.cn; Tel.: +86-139-4046-4015

Received: 18 March 2020; Accepted: 13 April 2020; Published: 16 April 2020

Abstract: Bearing preload significantly affects the running performance of a shaft-bearing system including the fatigue life, wear, and stiffness. Due to the mounting error, the bearing rings are often angularly misaligned. The effects of the combined bearing preload and angular misalignment on the fatigue life of ball bearings and a shaft-bearing system are analyzed in this paper. The contact force distribution of angular contact ball bearings in the shaft-bearing system is investigated based on the system model. The system model includes the bearing model, and the shaft model is verified by comparing with the manufacturer's manual and the results from other theoretical models, with the difference between the results from the present bearing model and manufacturer manual within 3%. The global optimization method is used to replace the Newton–Raphson algorithm to solve the ball elements' displacements and friction coefficients, which improves the computation efficiency of the system model. The fatigue life of each bearing is evaluated with the consideration of the two preload methods and two angular misalignment cases. The fatigue life results show that the system life at the optimal angular misalignment is more than 1.5 times that without angular misalignment at the low preload value, and this ratio decreases as the preload value increases. The optimal angular misalignment of both the shaft-bearing system and the misaligned bearing is not always consistent, which depends on the preload value and bearing life. Both the constant-displacement preload and constant-force preload do not cause a significant difference in the highest system life. The different misaligned bearings can lead to different highest system lives as the preload value is low.

Keywords: shaft-bearing system; angular contact ball bearing; bearing preload; angular misalignment; fatigue life

1. Introduction

The shaft-bearing system is a key part in mechanical transmissions, in which the rolling element bearing is commonly applied due to its low friction, low wear, and low energy consumption. With the high speed and high precision requirements for rotating machinery, the shaft is usually designed to be supported by multiple rolling element bearings. Many scholars have investigated the dynamic characteristics of the shaft-bearing system [1–3], and a comprehensive review on dynamic model development of a shaft-bearing system was also presented [4]. In addition to dynamic analysis, the fatigue life of both the rolling element bearing and shaft system has also attracted much attention.

Among various causes that affect the fatigue life of the rolling element bearing, both preload and angular misalignment are two common and important factors.

Many researchers have conducted numerous studies on the fatigue life of the rolling element bearings considering the preload and angular misalignment. Harris [5], as one of first researchers, presented the dependence of the fatigue life of a cylindrical roller bearing having crowned rolling elements on the angular misalignment. Hagiu [6] investigated the relation between the preload and service life for an angular contact ball bearing by theoretical and experimental methods. Hwang and Lee [7] reviewed the three categories of preload technologies and clarified that the determination of proper preload should take the fatigue life, stiffness, and temperature of rolling element bearings into account. Considering the significance of the pressure distribution of roller elements on the fatigue life estimation, Tong et al. [8] extended the 3D elastic contact method to simulate the contact pressure and analyzed the fatigue life of a tapered roller bearing with the consideration of the angular misalignment effect. Yang et al. [9] analyzed the effects of the combined external loads and angular misalignment on the double-row tapered roller bearing, and the results demonstrated that the external load, rotation speed, and angular misalignment had a significant influence on the fatigue life of a double-row tapered roller bearing. Warda et al. [10,11] investigated the effect of the correction parameters of roller generators and angular misalignment on the fatigue life of the radial cylindrical roller bearing, in which the bearing radial clearance and complex loads were both taken into account.

The above research works were mainly limited to a single bearing, and many works investigated the relations among the preload, fatigue life, temperature, and stiffness in a shaft-bearing system. Jiang et al. [12] investigated a variable preload technology for machine tool spindles working at different ranges of rotation speed, and the experimental results showed that the variable preload technology can reduce the temperature rise of the system in the high speed condition compared with the application of constant preload and improved the bearing stiffness at the low speed range. Xu et al. [13] developed an analytical method for determining the optimum preload based on the critical state between the skidding and rolling of ball bearings for different speed ranges, and their results were verified with the help of a spindle bearing experimental setup. Than and Huang [14] investigated the thermal effect of the spindle bearing system during high speed rotations when the preload was applied, and the time-varying thermal effects on the preload and stiffness of bearings was obtained. Zhang et al. [15] investigated the effect of external load and preload on the number of rolling elements in the contact zone based on the a quasi-dynamic model and determined an optimum preload for a simplified bearing-rotor system by taking the bearing fatigue life as the optimization target.

For the shaft-bearing system with high speed and high precision requirements, bearing preload is necessary to increase bearing stiffness and suppress vibration. In addition, the angular misalignment of the rolling bearing due to mounting error is common and unavoidable and would cause considerable variations in ball-raceways' contact force distribution, which affects the bearing fatigue life. Currently, very few studies have investigated the effects of the combined preload and angular misalignment on the fatigue life of rolling element bearings and the shaft-bearing system at high speed. The fatigue life variation of rolling element bearings and the shaft-bearing system with the combined preload and angular misalignment has not been well understood. In this sense, understanding the fatigue life variation considering the combined preload and angular misalignment is important for proper selection and assembly of rolling bearings in the shaft-bearing system, which can be achieved by analyzing the effects of the combined preload and angular misalignment on the fatigue life of rolling element bearings and the shaft-bearing system.

In this paper, a generic shaft-bearing system model combing the shaft model and the bearing model is introduced. A numerical method is presented to improve the computation efficiency of the system model. The ball-raceway contact forces of angular contact ball bearings (ACBBs) in shaft-bearing systems are evaluated under the complex operation conditions considering both the preload and angular misalignment. Based on the fatigue life theory, the fatigue life of each ACBB is calculated.

The effects of the combined preload and angular misalignment on the fatigue life of ACBBs and the shaft-bearing system are discussed. Finally, some useful conclusions are given.

2. Shaft-Bearing System Model

2.1. Quasi-Static Model of Angular Contact Ball Bearing

The scheme diagram of a shaft supported by multiple ACBBs is shown in Figure 1. O-xyz denotes the inertial coordinate system where the z-axis is coincident with the shaft axis, and O^k-$x^k y^k z^k$ denotes the local coordinate system for the kth bearing in which the forward direction of the z^k-axis is defined as from the small side of the bearing to its big side. Here, the quasi-static bearing model is presented in the bearing local coordinate system. A ball bearing is taken as an example to analyze the interactions between ball elements and raceways.

The kinematics of the ball element is shown in Figure 2. Here, the inner ring is assembled with a shaft that rotates at angular speed ω_i about the bearing axis while the outer ring is fixed. For a bearing with pitch diameter d_m and ball diameter D, one can get the angular speed ω_c at which the ball element orbits around the bearing axis and the spinning speed ω_b at which the ball element rotates around its own axis as given in Equations (1) and (2), respectively.

$$\omega_c = \omega_i \frac{(1-\gamma_i)\cos(\alpha_e-\beta)}{(1+\gamma_e)\cos(\alpha_i-\beta)+(1-\gamma_i)\cos(\alpha_e-\beta)} \tag{1}$$

$$\omega_b = \omega_i \frac{d_m}{D} \frac{(1-\gamma_i)(1+\gamma_e)}{(1+\gamma_e)\cos(\alpha_i-\beta)+(1-\gamma_i)\cos(\alpha_e-\beta)} \tag{2}$$

where α_i and α_e are the contact angles between the ball and inner/outer raceway. γ_i and γ_e are equal to $D\cos\alpha_i/d_m$ and $D\cos\alpha_e/d_m$, respectively.β is the ball pitch angle.

Based on d'Alembert's inertia force principle, Ding [16] derived the following pitch angle formula as shown in Equation (3).

$$\tan\beta = \frac{\frac{M_{bi}^S}{M_{be}^S}\left(\frac{1+\gamma_e}{1-\gamma_i}+1\right)\sin\alpha_i + 2\sin\alpha_e}{\frac{M_{bi}^S}{M_{be}^S}\left(\frac{1+\gamma_e}{1-\gamma_i}+1\right)\cos\alpha_i + 2(\cos\alpha_e+\gamma') + \gamma'\cdot\frac{M_{bi}^S}{M_{be}^S}\left(\cos(\alpha_i-\alpha_e)-\frac{1+\gamma_e}{1-\gamma_i}\right)} \tag{3}$$

in which γ' is equal to D/d_m, M_{bi}^S and M_{be}^S are the friction moments for the ball and inner/outer raceways, respectively, and:

$$\frac{M_{bi}^S}{M_{be}^S} = \frac{Q_i a_i L_2(\kappa_i)}{Q_e a_e L_2(\kappa_e)} \tag{4}$$

where Q_i and Q_e are the contact forces between the ball and inner/outer raceway. a_i and a_e are the semi-major axes of the contact ellipses for the ball and inner/outer raceways, respectively. $L_2(\cdot)$ is the second kind of elliptic integral function. κ_i and κ_e are the ratios of the semi-major axis to the semi-minor axis of the contact ellipses at the ball inner and ball outer raceway contacts, respectively. The detailed derivation of the pitch angle is omitted here and can be found in Ding's work [16].

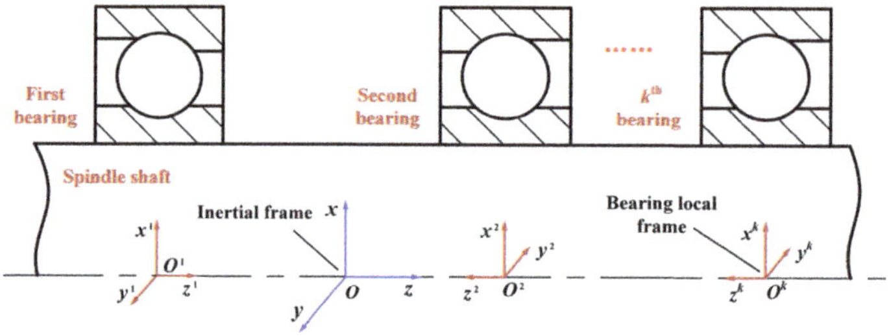

Figure 1. Geometrical configuration of a shaft-bearing system.

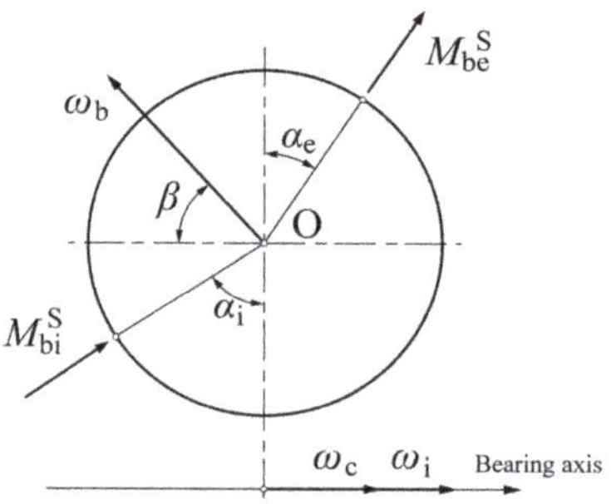

Figure 2. Kinematic relation of a ball element.

The contact angle and contact elastic deformation between the ball element and raceways are shown in Figure 3. When the external loads are applied on the static bearing, the ball center O and raceway groove curvature center O_e, O_i are collinear. These three points are no longer collinear due to the centrifugal force of the ball element when the rotating speed is imposed. The ball center will be shifted from O to O′, and the inner raceway groove curvature center moves from O_i to O'_i, while the groove curvature center of the outer raceway remains unchanged since the outer ring is fixed.

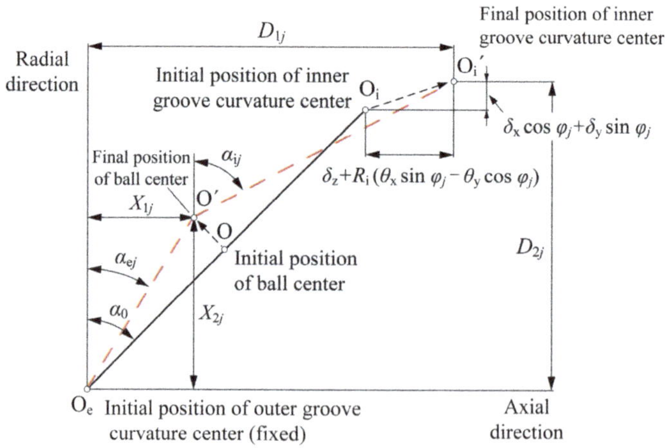

Figure 3. Positions of ball centers and raceway groove curvature centers.

In Figure 3, D_{1j}, D_{2j} denote the axial and radial distance between the inner and outer raceway groove curvature centers at the jth ball element position and can be written as:

$$D_{1j} = (f_e + f_i - 1)D\sin\alpha_0 + \delta_z + R_i(\theta_x \sin\varphi_j - \theta_y \cos\varphi_j) \tag{5}$$

$$D_{2j} = (f_e + f_i - 1)D\cos\alpha_0 + \delta_x \cos\varphi_j + \delta_y \sin\varphi_j \tag{6}$$

where δ_x, δ_y, and δ_z are the translational displacements of the inner ring along the x-, y-, and z-axis, respectively, θ_x and θ_y are the angular displacements around the x- and y-axis, respectively. α_0 is the bearing initial contact angle. f_i and f_e are the inner and outer raceway groove curvature coefficients, respectively. The radius of the locus of the inner raceway groove curvature centers R_i and the jth ball element azimuth angle φ_j are determined by:

$$R_i = 0.5d_m + (f_i - 0.5)D\cos\alpha_0 \tag{7}$$

$$\varphi_j = 2\pi(j-1)/Z \tag{8}$$

where Z is the ball element number in the bearing. The loaded contact angles α_i and α_e at ball-raceway contacts can be obtained:

$$\alpha_{ij} = \arctan\frac{D_{2j} - X_{2j}}{D_{1j} - X_{1j}} \tag{9}$$

$$\alpha_{ej} = \arctan\frac{X_{2j}}{X_{1j}} \tag{10}$$

The contact elastic deformations δ_{ij} and δ_{ej} can be given as:

$$\delta_{ij} = \sqrt{(D_{2j} - X_{2j})^2 + (D_{1j} - X_{1j})^2} - (f_i - 0.5)D \tag{11}$$

$$\delta_{ej} = \sqrt{X_{1j}^2 + X_{2j}^2} - (f_e - 0.5)D \tag{12}$$

where X_{1j} and X_{2j} are the axial and radial distances between ball center O' and outer raceway groove curvature center O_e, respectively.

As shown in Figure 4, considering the centrifugal force F_{cj}, gyroscopic moment M_{gj}, contact forces Q_{ij} and Q_{ej}, and friction forces F_{ij} and F_{ej}, the force and moment equilibrium equations for the jth ball element can be presented as:

$$\begin{cases} Q_{ij}\sin\alpha_{ij} - Q_{ej}\sin\alpha_{ej} + F_{ij}\cos\alpha_{ij} - F_{ej}\cos\alpha_{ej} = 0 \\ Q_{ij}\cos\alpha_{ij} - Q_{ej}\cos\alpha_{ej} - F_{ij}\sin\alpha_{ij} + F_{ej}\sin\alpha_{ej} + F_{cj} = 0 \\ F_{ej} + F_{ij} - 2M_{gj}/D = 0 \end{cases} \tag{13}$$

Here, an approximate relation between friction forces F_{ij}, F_{ej} and contact forces Q_{ij}, Q_{ej}, at ball-raceway contacts is applied, as follows:

$$F_{ij} = \mu_j Q_{ij} \tag{14}$$

$$F_{ej} = \mu_j Q_{ej} \tag{15}$$

where μ_j is the friction coefficients at ball-raceway contacts.

The centrifugal force and gyroscopic moment for the jth ball element can be written, respectively, as follows:

$$F_{cj} = 0.5d_m m\omega_{cj}^2 \tag{16}$$

$$M_{gj} = J\omega_{bj}\omega_{cj}\sin\beta_j \tag{17}$$

where m is the mass of the ball element and J is the mass moment of inertia:

$$m = \rho\pi D^3/6 \tag{18}$$

$$J = \rho\pi D^5/60 \tag{19}$$

Based on Hertz's work, the ball-raceway contact forces can be related to the elastic contact deformation presented in Equations (11) and (12).

$$Q_{ij/ej} = K_{i/e}\delta_{ij/ej}^{1.5} \tag{20}$$

in which K_i, K_e are the contact stiffness coefficients and can be calculated based on Harris and Kotzalas' work [17].

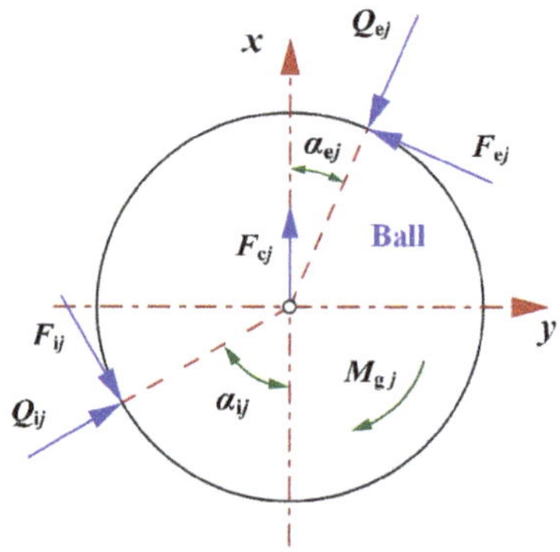

Figure 4. Forces on the ball element.

The bearing loads applied on the inner ring from the shaft include the translational forces F_x, F_y, and F_z and moments M_x, M_y. The total bearing loads should be balanced by contact forces and friction forces between ball elements and the inner raceway, and the equilibrium relations of the inner ring can be presented as:

$$\begin{cases} F_x = \sum_{j=1}^{Z}\left(Q_{ij}\cos\alpha_{ij} - F_{ij}\sin\alpha_{ij}\right)\cos\varphi_j \\ F_y = \sum_{j=1}^{Z}\left(Q_{ij}\cos\alpha_{ij} - F_{ij}\sin\alpha_{ij}\right)\sin\varphi_j \\ F_z = \sum_{j=1}^{Z}\left(Q_{ij}\sin\alpha_{ij} + F_{ij}\cos\alpha_{ij}\right) \\ M_x = \sum_{j=1}^{Z}\left[\left(Q_{ij}\sin\alpha_{ij} + F_{ij}\cos\alpha_{ij}\right)R_i - F_{ij}r_i\right]\sin\varphi_j \\ M_y = \sum_{j=1}^{Z}\left[\left(Q_{ij}\sin\alpha_{ij} + F_{ij}\cos\alpha_{ij}\right)R_i - F_{ij}r_i\right]\cos\varphi_j \end{cases} \tag{21}$$

in which r_i is the groove curvature radius of the inner raceway.

2.2. Shaft Model

In order to consider shaft flexibility, the finite element method is adopted here, and the shaft is discretized into N segments containing $N+1$ nodes using Timoshenko's beam element, as shown in Figure 5. *O-xyz* is the inertial coordinate system of the whole shaft-bearing system (mentioned in Section 2.1). The torsional deformation of the shaft is ignored; only the bending and axial deformations are considered. The ith node on beam element has three translational DOFs δ_x^i, δ_y^i, and δ_z^i along the

x-, y-, and z-axis, respectively, and two rotational DOFs θ_{x}^{i}, θ_{y}^{i} around the x- and y-axis, respectively. In the present shaft-bearing system, the bearing inner ring is close fit with the shaft; therefore, the inner ring of the bearing has the same displacements as the shaft node at the bearing mounting position. The displacement vector at the ith node location is:

$$\{\delta_i\} = \left\{\delta_{\text{x}}^{i}, \theta_{\text{y}}^{i}, \delta_{\text{z}}^{i}, \delta_{\text{y}}^{i}, \theta_{\text{x}}^{i}\right\}^{\text{T}} i = 1, 2, \cdots, N+1 \tag{22}$$

The corresponding load vector is:

$$\{F_i\} = \left\{F_{\text{x}}^{i}, M_{\text{y}}^{i}, F_{\text{z}}^{i}, F_{\text{y}}^{i}, M_{\text{x}}^{i}\right\}^{\text{T}} i = 1, 2, \cdots, N+1 \tag{23}$$

The stiffness matrix for each beam element can be obtained as follows:

$$[K^e] = \frac{EI}{L(L^2+12g)} \begin{bmatrix} 12 & 6L & 0 & 0 & 0 & -12 & 6L & 0 & 0 & 0 \\ & (4L^2+12g) & 0 & 0 & 0 & -6L & 2L^2-12g & 0 & 0 & 0 \\ & & \frac{A(L^2+12g)}{I} & 0 & 0 & 0 & 0 & -\frac{A(L^2+12g)}{I} & 0 & 0 \\ & & & 12 & -6L & 0 & 0 & 0 & -12 & -6L \\ & & & & (4L^2+12g) & 0 & 0 & 0 & 6L & 2L^2-12g \\ & & & & & 12 & 6L & 0 & 0 & 0 \\ & & symmetry & & & & (4L^2+12g) & 0 & 0 & 0 \\ & & & & & & & \frac{A(L^2+12g)}{I} & 0 & 0 \\ & & & & & & & & 12 & 6L \\ & & & & & & & & & (4L^2+12g) \end{bmatrix} \tag{24}$$

where g is a parameter including the transverse shear effect of beam, and its detailed expression can be found in [18].

The finite element equilibrium equation of the shaft can be written as:

$$K\delta = F \tag{25}$$

where K is the global stiffness matrix obtained by assembling all the beam element stiffness matrix K^{e}, δ is the node displacement vector of the shaft and can be expressed as $\{\delta\} = \left\{\delta_1^{\text{T}}, \delta_2^{\text{T}}, \cdots, \delta_i^{\text{T}}, \cdots \delta_{\text{N+1}}^{\text{T}}\right\}^{\text{T}}$, and F is the load vector applied on the shaft nodes given as $\{F\} = \left\{F_1^{\text{T}}, F_2^{\text{T}}, \cdots F_i^{\text{T}}, \cdots F_{\text{N+1}}^{\text{T}}\right\}^{\text{T}}$. The load vector F consists of two parts: one part is attributed to the external loads, and the other part is the supporting loads provided by the ball bearings and also the counter force of the bearing loads calculated in Equation (21). Based on the transformation relation between the bearing local coordinate system and the global inertial coordinate system, the displacements of the bearing inner ring in the bearing local coordinate system can be easily represented by the node displacements of the shaft in the inertial coordinate system at the location where the shaft is supported by the ball bearing. Therefore, the supporting loads applied on the shaft nodes provided by the ball bearings can be expressed as a nonlinear function of the node displacements of the shaft.

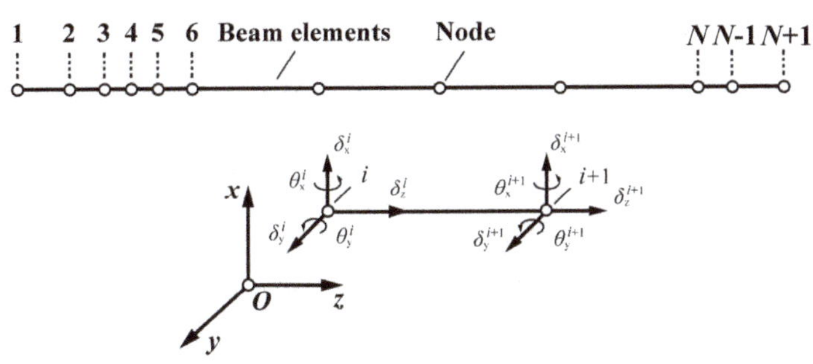

Figure 5. Finite element model for the shaft.

2.3. Preload and Angular Misalignment Factors

In the current analysis, the ball bearing is preloaded axially on the outer ring including constant-displacement preload and constant-force preload. The angular misalignment taken into account here is caused by mounting error, and the misaligned inner and outer rings due to mounting error can both lead to angular misalignment. In order to simplify the analysis, the inner ring is assumed to be mounted accurately. The outer ring is misaligned, and its misalignment angle keeps constant when the shaft-bearing is loaded.

Axial preload and angular misalignment can lead to additional deformation in each ball element and cause ball element-raceway contact forces to be redistributed; therefore, they should be included in the calculation of ball element deformation to reflect their effects on the fatigue life. Both the axial preload and angular misalignment affect mainly the axial distance D_{1j} between inner and outer raceway groove curvature centers at the jth ball element position. The translational displacement δ_z should be replaced by $\delta_z + \delta_a$ in Equation (5) to calculate the axial distance D_{1j} when the axial preload is taken into account. The axial displacement δ_a is caused by axial preload on the outer ring, which is known when constant-displacement preload is applied and unknown if constant-force preload is applied. When angular misalignment components of the outer ring of the ball bearing, θ_{x0} and θ_{y0}, are considered, the angular displacements θ_x and θ_y should be replaced with $\theta_x - \theta_{x0}$ and $\theta_y - \theta_{y0}$ in Equation (5), respectively. Then, the axial distance D_{1j} considering axial preload should be rewritten as:

$$D_{1j} = (f_e + f_i - 1)D\sin\alpha_0 + (\delta_z + \delta_a) + R_i(\theta_x \sin\varphi_j - \theta_y \cos\varphi_j) \tag{26}$$

The axial distance considering the angular misalignment is:

$$D_{1j} = (f_e + f_i - 1)D\sin\alpha_0 + \delta_z + R_i\left[(\theta_x - \theta_{x0})\sin\varphi_j - \left(\theta_y - \theta_{y0}\right)\cos\varphi_j\right] \tag{27}$$

3. Numerical Solution and Model Validity

3.1. Numerical Solution of the System Model

By combining the shaft model and quasi-static model of ACBB, the generic shaft-bearing system model is obtained. The unknowns in the system model include the ball orbital speed ω_c, spinning speed ω_b, pitch angle β, axial and radial distances (X_1, X_2) to imply the final position of the ball center, friction coefficients μ at ball-raceway contacts, inner ring displacements $\left(\delta_x, \delta_y, \delta_z, \theta_x, \theta_y\right)$ for ACBB, and the node displacements for the shaft. The unknowns are $6Z + 5$ for each ball bearing and $5(N+1)$ for the shaft. Considering that the inner ring displacements of ball bearings are the same as the displacements of shaft nodes where ball bearings are mounted, the total unknowns of the shaft-bearing system model are $k_b(6Z) + 5(N+1)$, and k_b is the number of ACBBs supporting the shaft. Obviously, the shaft-bearing system model is a huge system of equations provided that the shaft is supported by several ACBBs and the number of ball elements for each bearing and of the beam elements for the shaft is large.

It is computationally infeasible to solve all the equations of the shaft-bearing system model together. Hence, a numerical scheme is proposed to solve the system model. The numerical solution process is presented in Figure 6. The shaft-bearing system properties, including material properties, geometrical parameters, operation conditions, load conditions, axial preload, and/or angular misalignment of ball bearing, are taken as the input parameters. The initial guess for the node displacements of the shaft, the spinning speed ω_b, orbital speed ω_c, and pitch angle β of each ball element are also necessary to start the solution. Then, the distances parameters (X_1, X_2) and friction coefficients μ of all ball elements in the kth ball bearing are solved together by the global optimization method. The new spinning speed ω_b^*, orbital speed ω_c^*, and pitch angle β^* for ball elements are calculated based on Equations (1)–(3). The distances parameters (X_1, X_2), friction coefficients μ, speed parameters (ω_b,ω_c), and pitch angle β are solved alternately until the corresponding converge criterion is reached. Once the above calculations for all the ball bearings are finished, the equilibrium equations of the shaft are checked. Since the loads applied on the shaft nodes provided by the ball bearings are a nonlinear function of the node displacements of the shaft, the equilibrium equation of the shaft, Equation (25), is nonlinear as a function of the shaft nodes' displacements. Based on the first order Taylor expansion of nonlinear equilibrium equations of the shaft, the node displacements on the shaft are updated. The whole calculation process will be continued until the equilibrium equations of the shaft are satisfied. Then, the node displacements of the shaft, ball-raceway contact forces, spinning speed, orbital speed, and pitch angle for each ball element can be obtained. It should be noted that, in the innermost loop of the flowchart, the Newton–Raphson algorithm is not used to solve the equilibrium Equation (13) of all ball elements simultaneously to obtain the distance parameters X_1, X_2 and friction coefficients μ for the kth ball bearing, although this method was commonly used in the previously published works. The accuracy of the Newton–Raphson algorithm typically relies on the trial-and-error of initial estimates, and the numerical scheme will be very time-consuming and not be able to converge if the initial chosen solution is far away from the exact solution. In order to overcome this deficiency, the global optimization method is used to calculate the unknowns X_1, X_2 and μ of all ball elements simultaneously for the kth ball bearing. In the global optimization method, the sum of the square of $3Z$ mathematical formulas at the left of the equilibrium equations Equation (13) of all ball elements in the kth ball bearing is the objective function, and the aim is to find the correct X_1, X_2 and μ that make the objective function almost zero (global minimum). The scatter search [19] algorithm is used to generate trial points (initial estimates of independent variables X_1, X_2 and μ). The two filter conditions, the distance filter and merit filter [20], are used to examine trial points to ensure that the trial points that do not actually contribute to finding the local minimum of the objective function are excluded. After examination, the correct X_1, X_2 and μ of ball elements can be calculated, which make the objective function nearly zero (global minimum). The calculation shows that the global optimization method is very fast and reliable, and the computational efficiency of the system model is improved significantly.

3.2. Model Validity

In order to illustrate the validity of the ball bearing model, the contact angle results for the b218 bearing considering different load and speed conditions are compared in Figure 7 with that from Antoine's work [21]. It can be seen that the contact angle obtained by the present ball bearing model agreed well with the previously published results, and the maximum relative error between them was 16% as the preload was near 8000 N and the bearing speed was 10,000 rpm; However, the error was very small for other running conditions, which validated that the ball bearing model was effective.

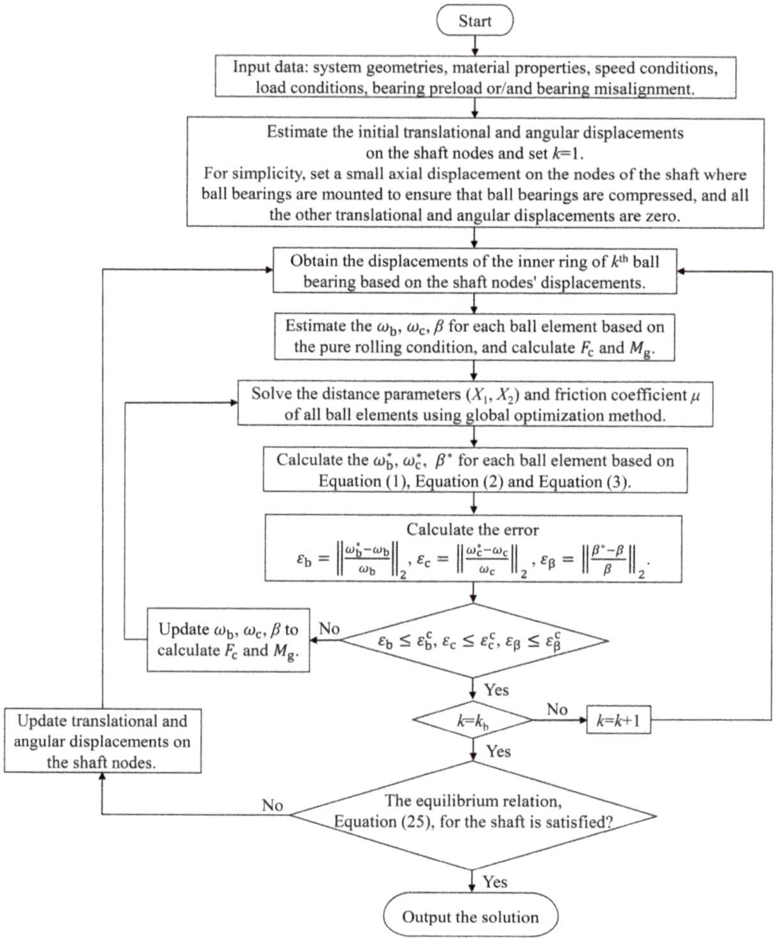

Figure 6. Flowchart of solving the shaft-bearing system.

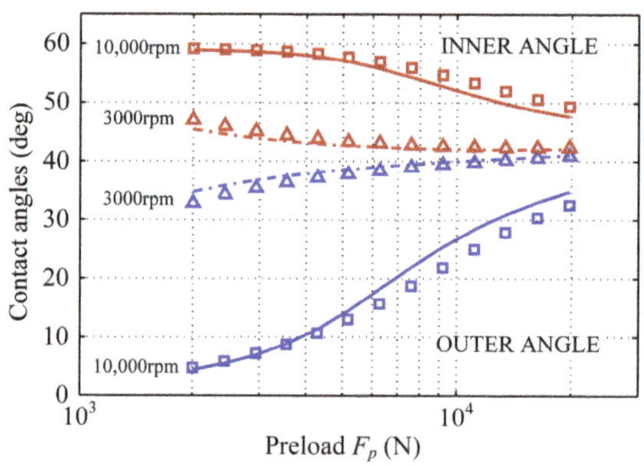

Figure 7. The comparison of contact angles from Antoine's model (markers) and from the present bearing model (lines).

Table 1 gives the stiffness comparison of ball bearing NSK 7014A5TYSULP4 between the NSK manual and the calculation results based on the present ball bearing model. The bearing stiffness formulas can be derived from the partial derivatives of the bearing load with respect to bearing displacement based on Equation (21). The detailed calculation process is omitted here, and a similar work can be found in [22–24]. It can be seen that the errors between the calculation results and manufacturer data were very small and less than 3%, which indicated that the ball bearing model was very valid.

Table 1. The bearing stiffness comparison between the numerical calculation results and the NSK manual.

Item	Super Light Preload		Light Preload		Medium Preload		Heavy Preload	
	Force (N)	Axial Stiffness (N/μm)	Force (N)	Axial Stiffness (N/μm)	Force (N)	Axial Stiffness (N/μm)	Force (N)	Axial Stiffness (N/μm)
Manual	245	170	490	218	1080	293	2160	390
Calculation	245	164.90	490	213.48	1080	291.27	2160	389.80
Error		3.00%		2.07%		0.59%		0.05%

To further validate the effectiveness of the system model, a gear shaft supported by a pair of angular contact ball bearings from Tong's work [25] was analyzed considering different shaft diameters and bearing arrangements. The bearing displacements and loads are presented in Figures 8 and 9. As shown in Figures 8 and 9, the results calculated by the present system model were very consistent with those from Tong's model [25], and the difference between the results from the two different models was less than 8%, which verified the system model and the calculation procedure.

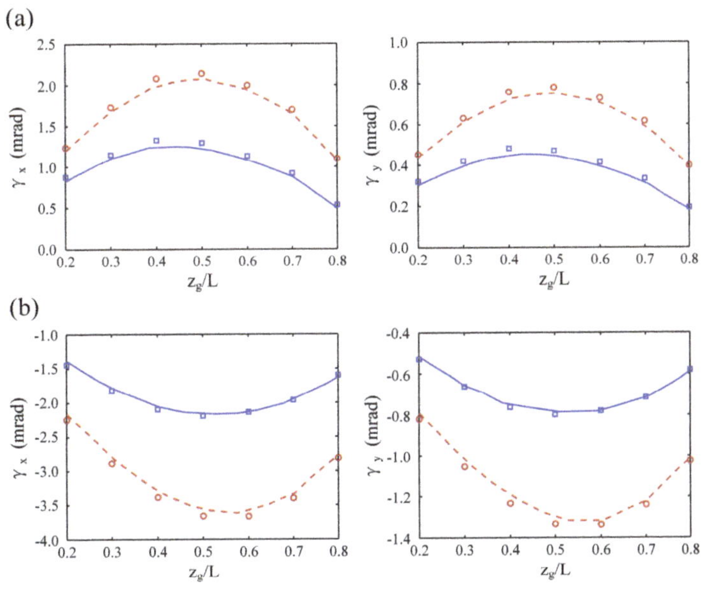

Figure 8. The comparison of the bearing angular displacements. ----- ds = 35 mm, ——— ds = 40 mm from Tong's model and ○ ds = 35 mm, □ ds = 40 mm from the present system model. (**a**) Face-to-Face arrangement; (**b**) Back-to-Back arrangement.

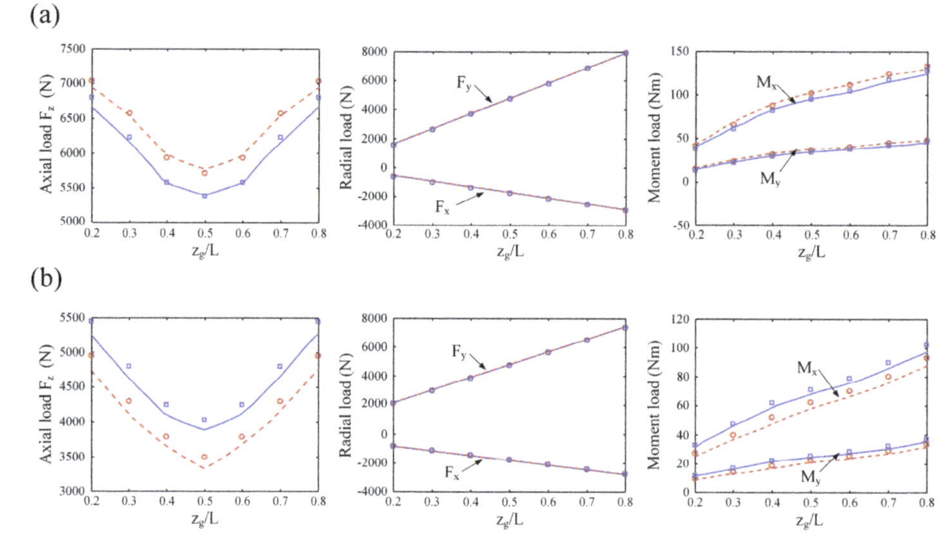

Figure 9. The comparison of the bearing loads. ----- ds = 35 mm, —— ds = 40 mm from Tong's model and ○ ds = 35 mm, □ ds = 40 mm from the present system model. (**a**) Face-to-Face arrangement; (**b**) Back-to-Back arrangement.

4. Fatigue Life Model

Based on the fatigue life theory proposed by Lunberg and Palmgren [26], the basic reference rating life model of the rolling element bearing, in million revolutions, can be calculated as:

$$L_{10r} = \left(L_i^{-10/9} + L_e^{-10/9}\right)^{-9/10} \tag{28}$$

In addition:

$$L_{i,e} = \left(\frac{Q_{ci/ce}}{Q_{ei/ee}}\right)^3 \tag{29}$$

in which the subscripts *i* and *e* refer to the inner and outer ring. The basic dynamic load rating of the inner and outer rings, $Q_{ci/ce}$, can be calculated as:

$$Q_{ci/ce} = 98.1\left(\frac{2f_{i,e}}{2f_{i,e}-1}\right)^{0.41}\frac{(1\mp\gamma)^{1.39}}{(1\pm\gamma)^{1/3}}\left(\frac{\gamma}{\cos\alpha_0}\right)^{0.3}D^{1.8}Z^{-1/3} \tag{30}$$

where γ is $D\cos\alpha_0/d_m$ and the upper sign and the lower sign denote the inner- and outer-raceway contact, respectively. For the current analysis, the equivalent dynamic loads for the rotating inner ring and the fixed outer ring are calculated, respectively, as:

$$Q_{ei} = \left(\frac{1}{Z}\sum_{j=1}^{Z} Q_{ij}^3\right)^{1/3} \tag{31}$$

$$Q_{ee} = \left(\frac{1}{Z}\sum_{j=1}^{Z} Q_{ej}^{10/3}\right)^{3/10} \tag{32}$$

Based on the geometry parameters of the ball bearing, the basic dynamic load rating of the inner and outer rings can be calculated by Equation (30). The contact force distribution in each bearing is solved by the presented system model considering bearing preload and angular misalignment, then the equivalent dynamic loads of the inner and outer rings can be obtained by Equations (31) and (32). The basic reference rating life of ball bearings taking into account different preload and angular misalignment conditions is calculated according to Equations (28) and (29).

5. Results and Discussion

A sample three-bearing shaft system was investigated as shown in Figure 10, and the shaft was supported by three identical 7008C ACBBs. All ball bearings' inner rings were perfectly fixed to the shaft. The outer ring of front bearing was free to move axially in order to apply preload, and the outer ring of each rear bearing was fixed either ideally or with a small misalignment angle caused by mounting error. The three ball bearings were assembled 32.5 mm, 187.5 mm, and 237.5 mm away from the left end of the shaft, respectively, and named as the front bearing, Rear Bearing 1 and Rear Bearing 2. The geometrical and material parameters for the bearing and shaft are given in Tables 2 and 3, respectively. The external loads acting on the shaft (radial loads F_x, F_y were 1500 N and 1000 N, respectively; axial load F_z was −1000 N; and moments M_x, M_y were 5000 N·mm and 6000 N·mm, respectively) were applied to the shaft node 110mm away from the left end of the shaft. The shaft rotation speed n_i was 10,000 r/min with a centrifugal force of around 11 N for each ball element. The two axial preload methods, including constant-displacement preload and constant-force preload, were considered, and the preload was applied on the outer ring of the front bearing. The outer ring of either Rear Bearing 1 or Rear Bearing 2 was subjected to angular misalignment caused by mounting error, and only the angular misalignment around the y-axis of the bearing local coordinate system was considered here. For brevity, Case I was used to indicate that Rear Bearing 1 was subjected to angular misalignment, and Case II implied that Rear Bearing 2 was subjected to angular misalignment.

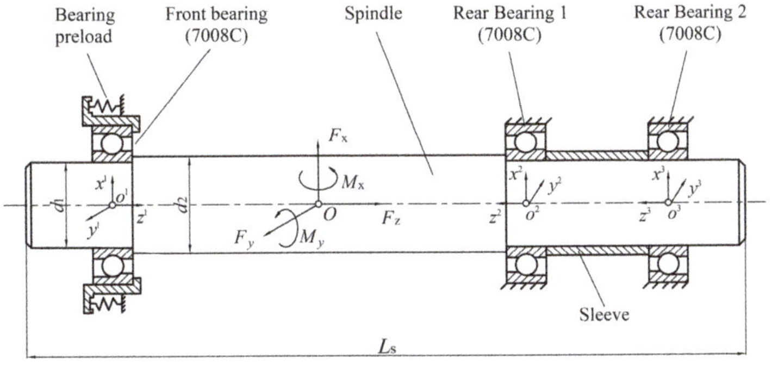

Figure 10. Shaft-bearing system with three angular contact ball bearings.

Table 2. The parameters of angular contact ball bearing 7008C.

Parameter	Value
Inner diameter of bearing d_i (mm)	40
Outer diameter of bearing d_e (mm)	68
Ball diameter D (mm)	7.9
Groove curvature coefficients of inner raceway f_i	0.52
Groove curvature coefficients of outer raceway f_e	0.52
Initial contact angle α_0 (deg)	15
Number of balls Z	16
Thickness B (mm)	15
Modulus of elasticity E (GPa)	206
Poisson's ratio ν	0.3
Density ρ (kg·m^{-3})	7890

Table 3. The parameters of the shaft.

Parameter	Value
d_1 (mm)	40
d_2 (mm)	44
Total length L_s (mm)	270
Modulus of elasticity E (GPa)	206
Poisson's ratio ν	0.3

5.1. The Constant-Displacement Preload Condition

Figure 11 shows the nonlinear dependence of the basic reference rating life of all three bearing on the angular misalignment in the constant-displacement preload condition. As can be seen, the fatigue life of each bearing varied significantly with the angular misalignment, and most of them were not the highest when the angular misalignment was zero, except for that of Rear Bearing 2 in the Case II condition. For each misaligned bearing, an optimal angular misalignment existed regardless of the preload value at which the misaligned bearing had the highest fatigue life. The optimal angular misalignment values were different for the two misaligned bearings, −0.8 mrad for Rear Bearing 1 and 0 mrad for Rear Bearing 2, which mainly depended on the contact force distribution of the misaligned bearings [27]. According to SKF [28], the permissible misalignment angle was 1.2 mrad. It could be seen that the fatigue life of all bearings was very low at either 1.2 mrad or −1.2 mrad compared with that within the permissible misalignment angle, as shown in Figure 11; the angular misalignment approaching the permissible misalignment angle could lead to a significant reduction of bearing life.

For a shaft-bearing system, the failure of any bearing will cause the system to fail. Due to the dispersion of material properties, the bearing with a high rated life may fail earlier than that with a low rated life, but the probability of this phenomenon decreases significantly as the gap between the high rated life and low rated life increases. In general, the bearing with a low rated life is more likely to fail first compared with the bearing with a high rated life. Without losing generality, it is supposed that the system life is mainly related to the bearing with the lowest rated life. The shaft fatigue life is not taken into account here. In the Case I condition, Rear Bearing 1 was misaligned. When the front bearing was preloaded by an axial displacement of 0.025 mm, the fatigue life of Rear Bearing 1 was always the lowest among the three bearings as the angular misalignment varied from −1.2 mrad to 1.2 mrad, therefore governing the system life. The shaft system had the same optimal angular misalignment, −0.8 mrad, and the same highest fatigue life with Rear Bearing 1. The highest system life was significantly greater than the system life at the angular misalignment of 0 mrad, reaching 1.55 times. As the preload displacement increased to 0.050 mm, the system life depended on the fatigue life of both Rear Bearing 1 and the preloaded front bearing, and the optimal angular misalignment turned to −0.21 mrad. The system optimal angular misalignment was no longer the same as the bearing optimal angular misalignment and approached 0 mrad, which was caused by the significant reduction of the fatigue life of the preloaded front bearing with increasing preload value. The system life at the optimal angular misalignment was only 1.12 times that at the angular misalignment of 0 mrad. The difference between the system life at the optimal angular misalignment and the system life at the angular misalignment of 0 mrad decreased with the increasing constant-displacement preload. In the Case II condition, the system life was mainly related to both the front bearing and Rear Bearing 1 due to the high fatigue life of the misaligned Rear Bearing 2. The optimal angular misalignments of the shaft system were 0.8 mrad, 0.8 mrad, and 0.21 mrad, respectively, as the preload displacement varied from 0.025 mm to 0.050 mm. The optimal angular misalignments of the shaft system for Case I and Case II were nearly opposite each other at the same preload displacement. Such a phenomenon was attributed to the negative optimal angular misalignment in the Case I condition and the positive optimal angular misalignment in the Case II condition causing a similar shaft elastic deformation, therefore producing a similar contact force distribution in Rear Bearing 1. The contact force distribution of Rear Bearing 1 at the preload displacements of 0.025 mm and 0.035 mm is given in Figure 12 to verify this point, and the contact force results at the preload displacement of 0.050 mm were omitted here.

In addition, the comparison of the highest system life showed that the Case I condition could lead to a reduction of system life compared with the Case II condition, the reduction amount being from 15% to 1% with the constant-displacement preload varying from 0.025 mm to 0.050 mm. The reduction amount was significant at the low preload displacement and could be ignored at the high preload displacement.

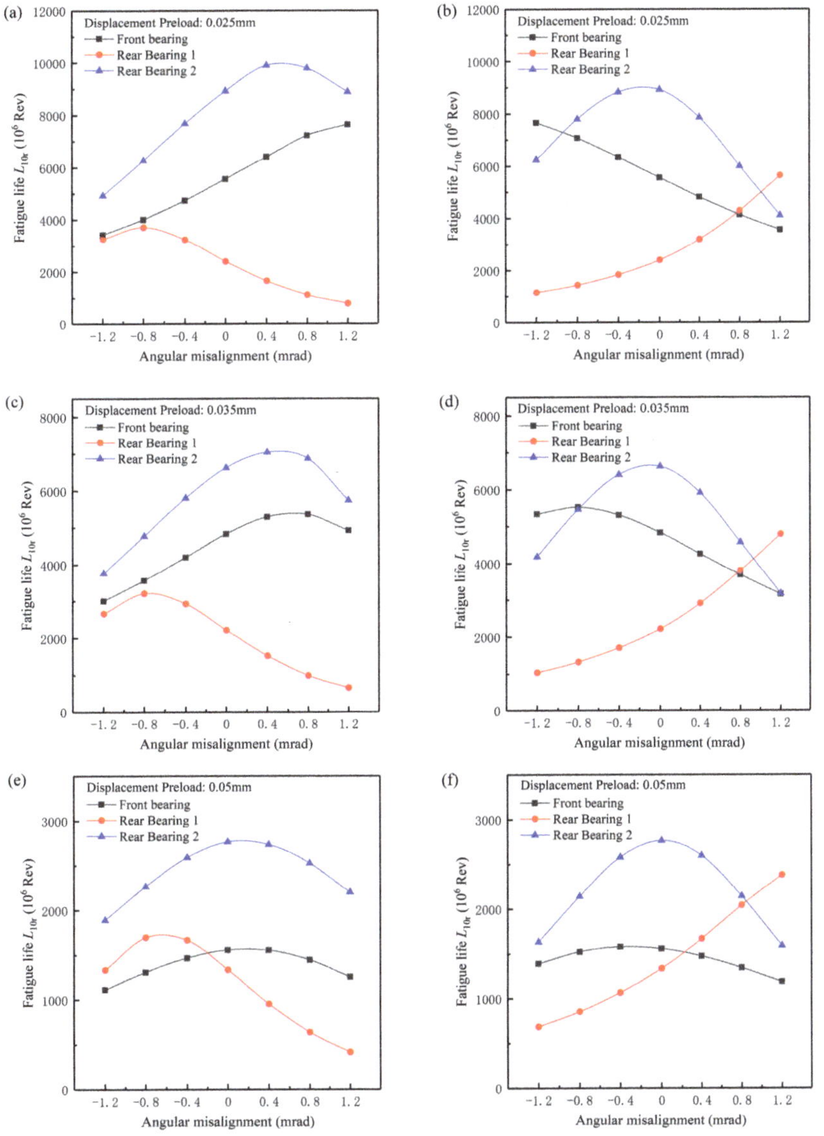

Figure 11. The variation of bearing life with the angular misalignment in the Case I condition ((**a**), (**c**) and (**e**)) and the Case II condition ((**b**), (**d**) and (**f**)) for constant-displacement preload.

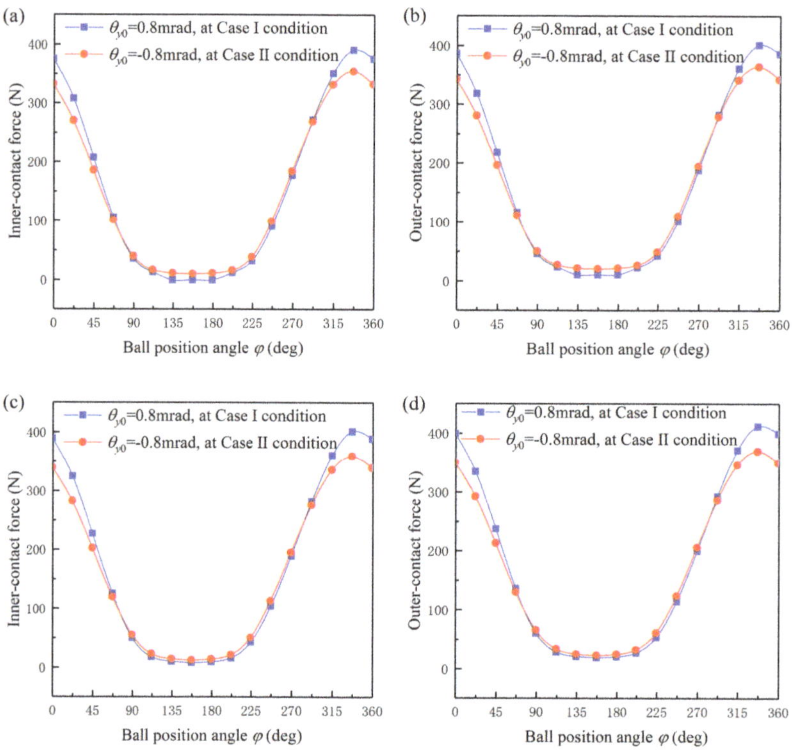

Figure 12. The contact force distribution in Rear Bearing 1 at the optimal angular misalignment of the shaft-bearing system. The preload displacements are 0.025 mm ((**a**) and (**b**)) and 0.035 mm ((**c**) and (**d**)), respectively.

5.2. The Constant-Force Preload Condition

Figure 13 shows the variation of bearing life with the angular misalignment in the constant-force preload condition. In order to analyze the effects of the preload method on the bearing life, the preload force was taken as 500 N, 1000 N, and 1500 N. When the shaft was running without external loads, the preload force of 500 N could cause 0.0355 mm axial displacement of the outer ring of the front bearing, which was very close to 0.035 mm, and the preload force of 1000 N could cause 0.050 mm axial displacement. Therefore, the preload force of 500 N and 1000 N at the constant-force preload could be assumed to be equivalent to the preload displacement of 0.035 mm and 0.050 mm in the constant-displacement preload, respectively. The life comparisons can be made between Figures 11c–f and 13a–d. As can be seen, the two preload methods had a certain effect on the variation of the fatigue life of the preloaded front bearing with the angular misalignment. For the constant-force preload condition, the fatigue life of front bearings was basically linear with angular misalignment, and for the constant-displacement condition, a nonlinear relation could be found. The above difference was mainly due to the outer ring of the front bearing moving slightly with the angular misalignment for the constant-force preload condition, while the outer ring of the front bearing was immovable for the constant-displacement preload condition, which affected the contact force distribution in the front bearing, therefore resulting in different trends of the front bearing life with angular misalignment. With the increasing of the constant-displacement preload, the nonlinear relation between the fatigue life of the front bearing and the angular misalignment became more obvious, as shown in Figure 11. For the constant-force preload condition, the increasing preload force reduced the dependence of the front bearing life on the angular misalignment, and the relation between the front bearing life and the angular misalignment approached a horizontal line at the preload force of 1500 N, as shown in Figure 13. Among the three bearings in the shaft system, the effect of the preload method on the fatigue life of Rear Bearing 1 was the weakest, and the two preload methods only caused a small change in the fatigue life of Rear Bearing 1. From Figures 11c–f and 13a–d, it can be seen that the fatigue life of Rear Bearing 1 was lowest over a wide angular misalignment range, therefore playing a significant role in the system life. Due to the effect of the preload method on the fatigue life of Rear Bearing 1 being very limited, the optimal angular misalignment and highest fatigue life for the shaft system were the same for the two preload methods.

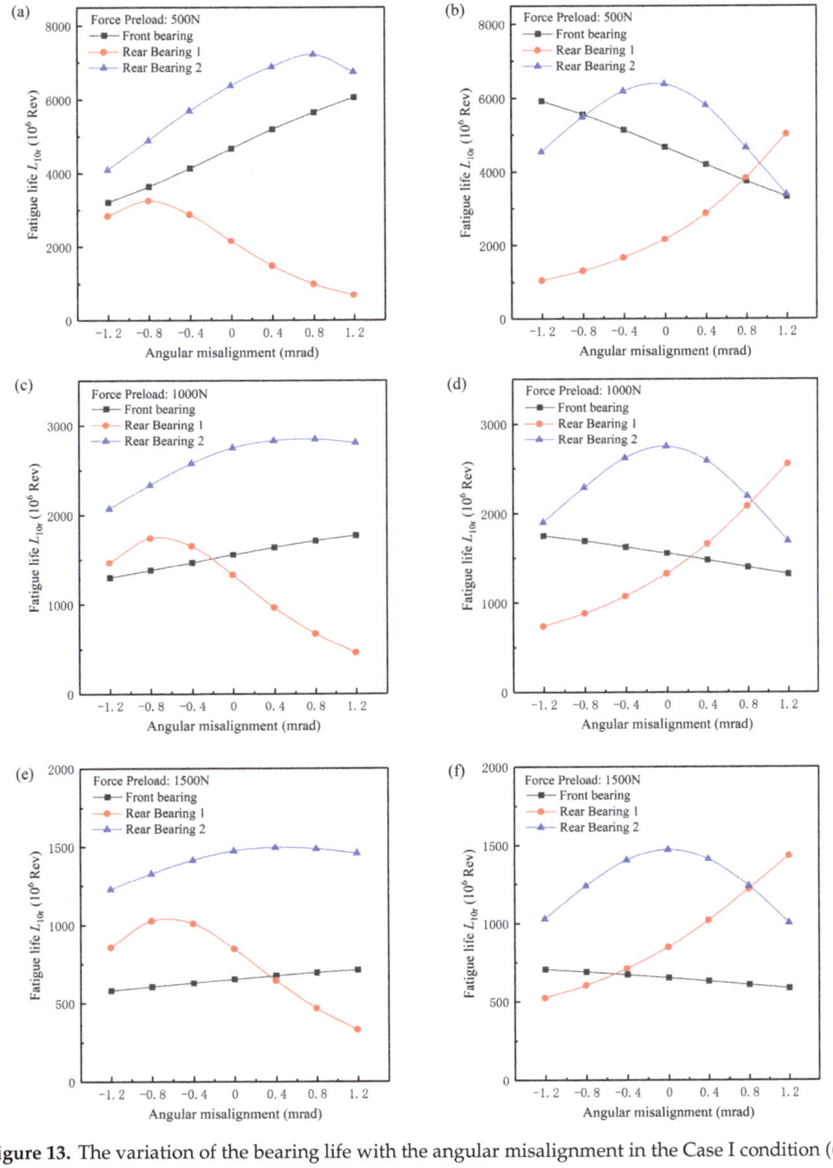

Figure 13. The variation of the bearing life with the angular misalignment in the Case I condition ((**a**), (**c**) and (**e**)) and in the Case II condition ((**b**), (**d**) and (**f**)) for constant-force preload.

As the preload force reached 1500 N, the system life depended on the fatigue life of both the front bearing and Rear Bearing 1. The optimal angular misalignment for the shaft system was close to 0.4 mrad in the Case I condition and −0.4 mrad in the Case II condition. As mentioned above, the fatigue life of the front bearing varied slowly with angular misalignment, which resulted in very small variation of the system life near the optimal angular misalignment. Therefore, a large range, for example from −0.4 mrad to 0.4 mrad for the Case I condition and from −0.8 mrad to 0.4 mrad for the Case II condition, could be regarded as the optimal angular misalignment range of the shaft system. Such a phenomenon implied that the increasing constant-force preload could weaken the effects of angular misalignment on the system life.

6. Conclusions

In this paper, to analyze the effects on the fatigue life of bearings and the shaft system resulting from the combined preload and angular misalignment induced by inaccurate mounting, a shaft-bearing system with bearing preload and angular misalignment was investigated. By improving the computational efficiency of ball element force analysis, the system model was solved fast and reliably. Comparisons were also made to verify the correctness of the system model and calculation program. Based on the contact force distribution of ball bearings, the fatigue life of each ball bearing was obtained. The results showed that both the preload and angular misalignment had significant effects on the fatigue life of ball bearings and the shaft-bearing system.

An optimal angular misalignment existed for the shaft-bearing system and could prolong the system life. At the low preload value, the system life at the optimal angular misalignment was much higher than that at 0mrad, and such a difference decreased with the increasing preload value. The optimal angular misalignment of the shaft system was not always the same as that of the misaligned bearing, which depended on the preload value and the fatigue life of each bearing. The two preload methods, constant-displacement preload and constant-force preload, had significant effects on the variation of the preloaded bearing life with the angular misalignment, but weak effects on the highest system life. At high constant-force preload, the system life varied very slowly with angular misalignment, and an optimal angular misalignment range existed. The angular misalignment occurring on different bearings could lead to a significant difference of the highest system life when the preload value was low, but the difference could be ignored when the preload value was high.

Author Contributions: Conceptualization, Y.Z. and L.X.; Methodology, Y.Z. and M.Z.; Software, M.Z.; Validation, Y.Z. and M.Z.; Analysis, Y.Z. and Y.W.; Writing—Original Draft Preparation Y.Z., M.Z. and Y.W.; Writing—Review and Editing, Y.Z. and M.Z.; Supervision, Y.Z. and L.X. All authors have read and agreed to the published version of the manuscript.

Funding: This research received no external funding.

Acknowledgments: This work was supported by the Liaoning Provincial Natural Science Foundation of China (Grant Number 20170540739), the Open Fund of Key Laboratory of Vibration and Control of Aero-Propulsion System Ministry of Education, Northeastern University (Grant Number VCAME201802), and the Natural Science Foundation of China (Grant Number U1708255).

Conflicts of Interest: The authors declared no potential conflict of interest with respect to the publication of this article.

Nomenclature

a	major semi-axis for ball-raceway contact ellipse, mm
b	minor semi-axis for ball-raceway contact ellipse, mm
D	ball diameter, mm
d_{m}	bearing pitch diameter, mm
d_{i}	inner diameter of bearing, mm
d_{e}	outer diameter of bearing, mm
B	thickness of bearing, mm
f_{e}	groove curvature coefficients of outer raceway

f_i	groove curvature coefficients of inner raceway
J	mass moment of inertia, kg·mm^2
ρ	density, kg/m^3
m	ball mass, kg
N	number of beam elements
Z	number of balls
γ_e	$D\cos\alpha_e/d_m$
γ_i	$D\cos\alpha_i/d_m$
γ'	D/d_m
γ	$D\cos\alpha_0/d_m$
α_0	initial contact angle, rad
α_e	contact angle at ball-outer raceway contact, rad
α_i	contact angle at ball-inner raceway contact, rad
β	ball pitch angle, rad
ω_i	angular speed of inner ring/shaft around the bearing axis, rad/s
ω_c	ball orbital speed around the bearing axis, rad/s
ω_b	ball spinning speed around its own axis, rad/s
M_{be}^S	friction moment at ball-outer raceway contact due to spinning, N·mm
M_{bi}^S	friction moment at ball-inner raceway contact due to spinning, N·mm
$L_2(\cdot)$	second kind of elliptic integral function
Q_i	contact force at ball-inner raceway contact, N
Q_e	contact force at ball-outer raceway contact, N
F_c	ball centrifugal force, N
M_g	gyroscopic moment, N·mm
F_i	friction force at ball-inner raceway contact, N
F_e	friction force at ball-outer raceway contact, N
F_x	bearing force applied on the inner ring of ball bearing along x-axis of bearing local coordinate system, N
F_y	bearing force applied on the inner ring of ball bearing along y-axis of bearing local coordinate system, N
F_z	bearing force applied on the inner ring of ball bearing along z-axis of bearing local coordinate system, N
M_x	moment applied on the inner ring of ball bearing around x-axis of bearing local coordinate system, N·mm
M_y	moment applied on the inner ring of ball bearing around y-axis of bearing local coordinate system, N·mm
F	external load vector applied on the shaft nodes, N
μ	ball-raceway friction coefficient
O	initial position of ball center
O′	final position of ball center
O_e	outer raceway groove curvature center
O_i	initial position of inner raceway groove curvature center
O'_i	final position of inner raceway groove curvature center
X_1	axial distance between ball center O′ and outer raceway groove curvature center O_e, mm
X_2	radial distance between ball center O′ and outer raceway groove curvature center O_e, mm
δ_x	translational displacement of inner ring along x-axis of bearing local coordinate system, mm
δ_y	translational displacement of inner ring along y-axis of bearing local coordinate system, mm
δ_z	translational displacement of inner ring along z-axis of bearing local coordinate system, mm
θ_x	angular displacement of inner ring around x-axis of bearing local coordinate system, rad
θ_y	angular displacement of inner ring around y-axis of bearing local coordinate system, rad
δ_e	ball-outer raceway contact deformation, mm
δ_i	ball-inner raceway contact deformation, mm
n_i	rotational speed of inner ring/shaft, rpm
R_i	radius of locus of inner raceway groove curvature centers, mm

φ	ball azimuth angle, rad
κ	a/b
ν	Poisson's ratio
K_e	deflection coefficient at ball and outer raceway contact, N/mm$^{1.5}$
K_i	deflection coefficient at ball and inner raceway contact, N/mm$^{1.5}$
r_i	groove curvature radius of inner raceway, mm
K	global stiffness matrix of the shaft
K^e	stiffness matrix of the beam element
δ	displacement vector of the shaft nodes, mm
E	modulus of elasticity, GPa
I	moment of inertia of the shaft section, mm^4
L	length of beam element, mm
L_s	total length of the shaft, mm
k_b	number of angular contact ball bearings supporting the shaft
ε_b	calculation error of ball spinning speed
ε_c	calculation error of ball orbit speed
ε_β	calculation error of ball pitch angle
δ_a	axial displacement caused by constant-displacement preload or constant-force preload, mm
θ_{x0}	angular misalignment of outer ring of ball bearing around x-axis of bearing local coordinate system, mrad
θ_{y0}	angular misalignment of outer ring of ball bearing around y-axis of bearing local coordinate system, mrad
L_{10r}	basic reference rating life of bearing, in million revolutions
L_i	basic reference rating life of inner ring, in million revolutions
L_e	basic reference rating life of outer ring, in million revolutions
Q_{ci}	basic dynamic load rating of inner ring, N
Q_{ce}	basic dynamic load rating of outer ring, N
Q_{ei}	equivalent dynamic load for rotating inner ring, N
Q_{ee}	equivalent dynamic load for fixed outer ring, N

References

1. Jin, Y.; Yang, R.; Lei, H.; Chen, Y.; Zhang, Z. Experiments and Numerical Results for Varying Compliance Contact Resonance in a Rigid Rotor–Ball Bearing System. *J. Tribol.* **2017**, *139*, 041103. [CrossRef]
2. Yang, Y.; Yang, W.; Jiang, D. Simulation and experimental analysis of rolling element bearing fault in rotor-bearing-casing system. *Eng. Fail. Anal.* **2018**, *92*, 205–221. [CrossRef]
3. Wang, H.; Han, Q.; Zhou, D. Nonlinear dynamic modeling of rotor system supported by angular contact ball bearings. *Mech. Syst. Signal Process.* **2017**, *85*, 16–40. [CrossRef]
4. Cao, H.; Niu, L.; Xi, S.; Chen, X. Mechanical model development of rolling bearing-rotor systems: A review. *Mech. Syst. Signal Process.* **2018**, *102*, 37–58. [CrossRef]
5. Harris, T.A. The Effect of Misalignment on the Fatigue Life of Cylindrical Roller Bearings Having Crowned Rolling Members. *J. Tribol.* **1969**, *91*, 294–300. [CrossRef]
6. Hagiu, G.; Gafitanu, M. Preload-service life correlation for ball bearings on machine tool main spindles. *Wear* **1994**, *172*, 79–83. [CrossRef]
7. Hwang, Y.-K.; Lee, C.M. A review on the preload technology of the rolling bearing for the spindle of machine tools. *Int. J. Precis. Eng. Manuf.* **2010**, *11*, 491–498. [CrossRef]
8. Tong, V.-C.; Hong, S.-W. Fatigue life of tapered roller bearing subject to angular misalignment. *Proc. Inst. Mech. Eng. Part C J. Mech. Eng. Sci.* **2015**, *230*, 147–158. [CrossRef]
9. Yang, L.; Xu, T.; Xu, H.; Wu, Y. Mechanical behavior of double-row tapered roller bearing under combined external loads and angular misalignment. *Int. J. Mech. Sci.* **2018**, *142*, 561–574. [CrossRef]
10. Warda, B.; Chudzik, A. Fatigue life prediction of the radial roller bearing with the correction of roller generators. *Int. J. Mech. Sci.* **2014**, *89*, 299–310. [CrossRef]
11. Warda, B.; Chudzik, A. Effect of ring misalignment on the fatigue life of the radial cylindrical roller bearing. *Int. J. Mech. Sci.* **2016**, *111*, 1–11. [CrossRef]

12. Jiang, S.; Mao, H. Investigation of variable optimum preload for a machine tool spindle. *Int. J. Mach. Tools Manuf.* **2010**, *50*, 19–28. [CrossRef]
13. Xu, T.; Xu, G.; Zhang, Q.; Hua, C.; Tan, H.; Zhang, S.; Luo, A. A preload analytical method for ball bearings utilising bearing skidding criterion. *Tribol. Int.* **2013**, *67*, 44–50. [CrossRef]
14. Than, V.-T.; Huang, J.H. Nonlinear thermal effects on high-speed spindle bearings subjected to preload. *Tribol. Int.* **2015**, *96*, 361–372. [CrossRef]
15. Zhang, J.; Fang, B.; Hong, J.; Zhu, Y. Effect of preload on ball-raceway contact state and fatigue life of angular contact ball bearing. *Tribol. Int.* **2017**, *114*, 363–372. [CrossRef]
16. Ding, C. Raceway control assumption and the determination of rolling element attitude angle. *Chin. J. Mech. Eng.* **2001**, *37*, 58–61. (In Chinese) [CrossRef]
17. Harris, T.A.; Kotzalas, M.N. *Rolling Bearing Analysis*; Taylor & Francis: Boca Raton, FL, USA, 2007.
18. Logan, D.L. *A First Course in the Finite Element Method*; Cengage: Boston, MA, USA, 2012.
19. Glover, F. *A Template for Scatter Search and Path Relinking, Lecture Notes in Computer Science 1363*; Springer: Berlin/Heidelberg, Germany, 1998.
20. Ugray, Z.; Lasdon, L.; Plummer, J.; Glover, F.; Kelly, J.; Martí, R. Scatter Search and Local NLP Solvers: A Multistart Framework for Global Optimization. *INFORMS J. Comput.* **2007**, *19*, 328–340. [CrossRef]
21. Antoine, J.-F.; Abba, G.; Molinari, A. A New Proposal for Explicit Angle Calculation in Angular Contact Ball Bearing. *J. Mech. Des.* **2006**, *128*, 468–478. [CrossRef]
22. Zhang, X.; Han, Q.; Peng, Z.-K.; Chu, F. Stability analysis of a rotor–bearing system with time-varying bearing stiffness due to finite number of balls and unbalanced force. *J. Sound Vib.* **2013**, *332*, 6768–6784. [CrossRef]
23. Yang, Z.; Chen, H.; Yu, T. Effects of rolling bearing configuration on stiffness of machine tool spindle. *Proc. Inst. Mech. Eng. Part C J. Mech. Eng. Sci.* **2017**, *232*, 775–785. [CrossRef]
24. Zhang, J.; Fang, B.; Zhu, Y.; Hong, J. A comparative study and stiffness analysis of angular contact ball bearings under different preload mechanisms. *Mech. Mach. Theory* **2017**, *115*, 1–17. [CrossRef]
25. Tong, V.-C.; Hong, S.-W. Study on the running torque of angular contact ball bearings subjected to angular misalignment. *Proc. Inst. Mech. Eng. Part J J. Eng. Tribol.* **2017**, *232*, 890–909. [CrossRef]
26. Lundberg, G.; Palmgren, A. Dynamic capacity of rolling bearings. In *Acta Polytechnica, Mechanical Engineering Series I*; Royal Swedish Academy of Engineering Sciences: Stockholm, Sweden, 1947; pp. 1–50.
27. Li, X.; Li, H.; Hong, J.; Zhang, Y. Heat analysis of ball bearing under nonuniform preload based on five degrees of freedom quasi-static model. *Proc. Inst. Mech. Eng. Part J J. Eng. Tribol.* **2015**, *230*, 709–728. [CrossRef]
28. SKF Group. *SKF General Catalogue 6000/I EN*; SKF Group: Gothenburg, Sweden, 2008.

applied sciences

MDPI

Article

Planetary Gearbox Fault Diagnosis Based on ICEEMD-Time-Frequency Information Entropy and VPMCD

Yihan Wang [1,2], Zhonghui Fan [1,2], Hongmei Liu [3,4,*] and Xin Gao [1,2]

1 CEPREI, Guangzhou 510610, China; zy1714208@buaa.edu.cn (Y.W.); 18620809907@163.com (Z.F.); gaoxin@ceprei.com (X.G.)
2 Key laboratory of Industrial Equipment Quality Big Data, MIIT, Guangzhou 510610, China
3 School of Reliability and Systems Engineering, Beihang University, Beijing 100191, China
4 Science & Technology Laboratory on Reliability & Environmental Engineering, Beihang University, Beijing 100191, China
* Correspondence: liuhongmei@buaa.edu.cn

Received: 5 August 2020; Accepted: 8 September 2020; Published: 13 September 2020

Abstract: Planetary gearboxes are more and more widely used in large and complex construction machinery such as those used in aviation, aerospace fields, and so on. However, the movement of the gear is a typical complex motion and is often under variable conditions in real environments, which may make vibration signals of planetary gearboxes nonlinear and nonstationary. It is more difficult and complex to achieve fault diagnosis than to fix the axis gearboxes effectively. A fault diagnosis method for planetary gearboxes based on improved complementary ensemble empirical mode decomposition (ICEEMD)-time-frequency information entropy and variable predictive model-based class discriminate (VPMCD) is proposed in this paper. First, the vibration signal of planetary gearboxes is decomposed into several intrinsic mode functions (IMFs) by using the ICEEMD algorithm, which is used to determine the noise component by using the magnitude of the entropy and to remove the noise components. Then, the time-frequency information entropy of intrinsic modal function under the new decomposition is calculated and regarded as the characteristic matrix. Finally, the fault mode is classified by the VPMCD method. The experimental results demonstrate that the method proposed in this paper can not only solve the fault diagnosis of planetary gearboxes under different operation conditions, but can also be used for fault diagnosis under variable operation conditions. Simultaneously, the proposed method is superior to the wavelet entropy method and variational mode decomposition (VMD)-time-frequency information entropy.

Keywords: planetary gearbox; ICEEMD; time-frequency information entropy; VPMCD; fault diagnosis

1. Introduction

As the key unit of a transmission device, planetary gearboxes have the advantages of a compact structure, large transmission moment, and accurate transmission ratio, and so on [1,2]. Once the planetary gearbox is damaged, it will cause machine breakdowns. In severe cases, it will lead to huge economic losses [3,4]. What is more, planetary gearboxes are often operated under heavy loads, which are very prone to failure [5]. Therefore, fault diagnosis of planetary gearboxes is significant to production safety and cost efficiency [6]. Vibration signal analysis method is one of the key tools for planetary gearbox fault diagnosis [7]. The vibration of planetary gearboxes are complex and very easy to be drowned out by noise, which makes the weak features of incipient faults difficult to detect [8]. Traditional vibration signal analysis methods such as time domain feature analysis and frequency domain analysis do not reflect the fault information of planetary gearboxes and the diagnostic accuracy of fault diagnosis is unsatisfactory.

For this reason, time-frequency analysis has emerged as an important signal processing method and has a wide range of applications in the field of vibration signals. Many time-frequency methods, such as wavelet transform and empirical mode decomposition (EMD), are applied to analyze nonlinear and nonstationary signals of planetary gearboxes [9]. However, the basis function of wavelet transform has no self-adaptability. It needs to be a preselected fixed-basis function according to the characteristics of the different signals. The empirical mode decomposition algorithm can adaptively decompose the vibration signal into a series of intrinsic mode functions (IMFs) from high frequency to low frequency [10], but usually generates problem of mode mixing. Later, a series of derived methods came into being such as ensemble empirical mode decomposition (EEMD), local characteristic-scale decomposition (LCD), and complementary ensemble empirical mode decomposition (CEEMD), and so on. However, low-speed and heavy-duty operation conditions with high noise environments frequently cause faults in the planetary gearboxes. It is more difficult to diagnose a fault than to fix the axis gearboxes, in which accurate diagnosis is very important. Due to the interference of noise, the problem of mode mixing is more severe, which leads to difficulty in completing the subsequent feature extraction. The above methods do not solve the problem caused by high noise. It is necessary to find a new method of signal features analysis with noise reduction to suppress the influence of noise on the vibration signal. The permutation entropy algorithm, a useful tool in amplifying slight changes of signals, has advantages of generality and easy computing in nonlinear and nonstationary signals [11]. Based on the permutation entropy, Aziz et al. proposed multi-scale permutation entropy (MPE) to depict tiny changes of signals from different local scales rather than a single scale [12]. Therefore, the MPE algorithm was utilized to effectively distinguish components between noise and original vibration signals of planetary gearboxes. For the reasons given above, in this paper, a method of improved complementary ensemble empirical mode decomposition (ICEEMD) is proposed for the first time.

In terms of feature extraction, time-frequency information entropy is utilized to reveal the complexity hidden in signals and can effectively reflect fault information of time-frequency characteristics. Thus, time-frequency information entropy is regarded as a feature characteristic parameter and is applied to realize the robust feature extraction [13].

For mechanical fault diagnosis, different fault classification methods directly affect the efficiency and accuracy of fault recognition. The algorithm of variable predictive model-based class discriminate (VPMCD) makes use of the inherent correlation of sample eigenvalues to establish a feature learning model [14], which can be effectively applied to multi-classification mechanical fault diagnosis with small samples [15]. In addition, the VPMCD algorithm avoids the problem of over-fitting of neural network classification and kernel function selection of SVM (Support Vector Machine) classification. At the same time, the computing time is reduced and the robustness of the algorithm is improved [16].

In the study presented in this paper, a fault diagnosis method of planetary gearboxes based on ICEEMD-time-frequency information entropy and VPMCD is first proposed. First, vibration signals of planetary gearboxes are collected by the accelerometer. Second, several IMFs are obtained by using the method of ICEEMD. The method proposed can eliminate noise on the basis of each IMF, which avoids the disadvantages of loss of feature information caused by overall noise reduction. Afterwards, each IMF is calculated by time-frequency information entropy as feature vectors. Then, by the method of principal component analysis (PCA), the dimension of feature vectors is reduced. In order to verify the effectiveness of the proposed method, the diagnostic performance of other two methods under various working conditions is also investigated in fault feature extraction. Compared with the other two methods, the ICEEMD-time-frequency information entropy method proposed in this paper has an advantage for feature extraction and allows the fault features to be effectively distinguished in the feature space. Finally, the VPMCD model is utilized to classify different fault modes.

This paper is organized as follows: Section 2 introduces the relevant feature extraction methodology, which includes the ICEEMD, time-frequency information entropy, and principal component analysis (PCA); Section 3 presents the principle of classification about VPMCD. In Section 4, the scheme of the planetary gearbox based on ICEEMD-time-frequency information entropy and VPMCD are described. Section 5 describes the case study to validate the entire method; Conclusions are drawn in Section 6.

2. Feature Extraction Method Based on ICEEMD-Time-Frequency Information Entropy

Vibration signals of planetary gearboxes are complex, with multiple components and nonlinear and nonstationary characteristics that are obviously interfered with by noise [17]. How to obtain fault characteristic information effectively is an important part of fault diagnosis of planetary gearboxes. In this study, the vibration signal feature is extracted by ICEEMD-time-frequency information entropy to characterize the operating state of the planetary gearbox.

2.1. A Description of Improved Complementary Ensemble Empirical Mode Decomposition (ICEEMD)

The vibration signal of planetary gearboxes, with multiple components and nonlinear, nonstationary, and strong time-varying characteristics, is complex and easily buried by noise. However, traditional fault diagnosis methods fail to effectively diagnose the faults of planetary gearboxes. It is essential to find out a suitable method to suppress interference components in order to obtain intrinsic features of the vibration signal.

Comparing complementary ensemble empirical mode decomposition (CEEMD) with ensemble empirical mode decomposition (EEMD) and empirical mode decomposition (EMD) [18], CEEMD is an improved algorithm that reduces the mode mixing problem [19]. However, the reconstructed components of CEEMD contain noise components that cannot be eliminated. In order to eliminate the influence of noise on feature extraction, we used multi-scale permutation entropy with signal noise reduction for the first time. Multi-scale permutation entropy combines multi-scale entropy with permutation entropy to analyze sequence information more effectively [20–22], which can measure noise characteristics from multiple scales rather than a single scale. Using this method, the noise components can be accurately identified. In this paper, the improved complementary ensemble empirical mode decomposition algorithm (ICEEMD) was used to analyze the vibration signal of planetary gearboxes for the first time. The specific ICEEMD algorithm steps are shown in Figure 1.

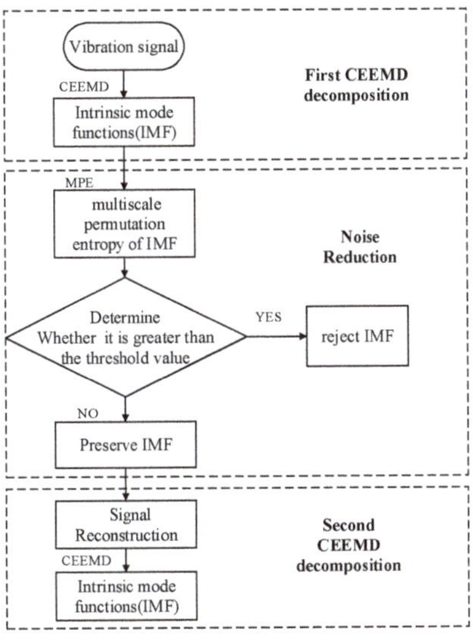

Figure 1. Algorithm flowchart of improved complementary ensemble empirical mode decomposition (ICEEMD).

Step 1: Use CEEMD to decompose the original signal.

(1) The original signal is $S(t)$. Add white noise signals $n_i(t)$ and $-n_i(t)$ with the mean value of zero onto the original signal. Then, the signal after adding white noise is obtained:

$$S_i^+(t) = S(t) + a_i n_i(t) \tag{1}$$

$$S_i^-(t) = S(t) - a_i n_i(t) \tag{2}$$

where a_i is the amplitude of the white noise, $i = 1, 2, \ldots, Ne$, and Ne is the logarithm of the added white noise.

(2) Signal $S_i^+(t)$ and signal $S_i^-(t)$ are decomposed by EMD to obtain IMF_{i1}^+ and IMF_{i1}^-. Then, the integrated average value of IMF_{i1}^+ and IMF_{i1}^- are obtained by Equation (3).

$$IMF_1 = \frac{1}{2Ne}\sum_{i=1}^{Ne}\left[IMF_{i1}^+ + IMF_{i1}^-\right] \tag{3}$$

(3) Repeat the above steps until the termination condition of EMD is met.

Step 2: Reduce noise by using the MPE algorithm.

The basic idea of multi-scale permutation entropy is the coarse graining of the original signal [23]. The specific steps for calculating the multi-scale arrangement entropy of IMF are as follows:

(1) Coarse graining of IMF.

$IMF_p = \{x_i, i = 1, 2, \ldots, N\}, 1 \le p \le M$, p is the number of IMFs.

Coarse graining sequence:

$$y_j^s = \frac{1}{s}\sum_{i=(j-1)s+1}^{js} x_i,\ j = 1, 2, \ldots \frac{N}{s} \text{ and } j \text{ is an integer.} \tag{4}$$

where s is the scale factor.

(2) Time reconstruction of coarse graining sequence y_j^s.

$$Y_l^s = \left\{y_l^s, y_{l+\tau}^s, \ldots, y_{l+(m-1)\tau}^s\right\}$$

where m is embedding dimension, τ is delay time, and l is a reconstructed component, where the expression for l is $l = 1, 2, \ldots, N-(m-1)\tau$.

Arrange the Y_l^s in ascending order and calculate the probability P_r of each sequence, where $r = 1, 2, \ldots, R, R \le m!$

Calculate the multi-scale permutation entropy of each coarse graining sequence.

$$IMF_P(m) = -\sum_{r=1}^{R\Sigma_r} P_r ln \tag{5}$$

(3) Eliminate noise components.

The multi-scale permutation entropy of each IMF component is calculated and compared with the set threshold value. Then, eliminate noise components larger than the threshold value.

Step 3: Second CEEMD decomposition.

Reconstruct signal. The reconstructed signal $X(t)$ is decomposed by CEEMD to obtain new intrinsic modal components (IMFs). Refer to step 1 for detailed steps.

In order to test the advantages of the method proposed in this paper, the ICEEMD method and CEEMD method were used to decompose the simulation signal, where Y is the simulation signal, x_n is the noise signal, and S is the simulation signal with noise.

$$Y = 2.5sin(2\pi 8t + \pi/6) + 3tsin(2\pi 4t + \pi/2) \quad (6)$$

The simulation signal, noise signal, and simulation signal with noise are shown in Figure 2.

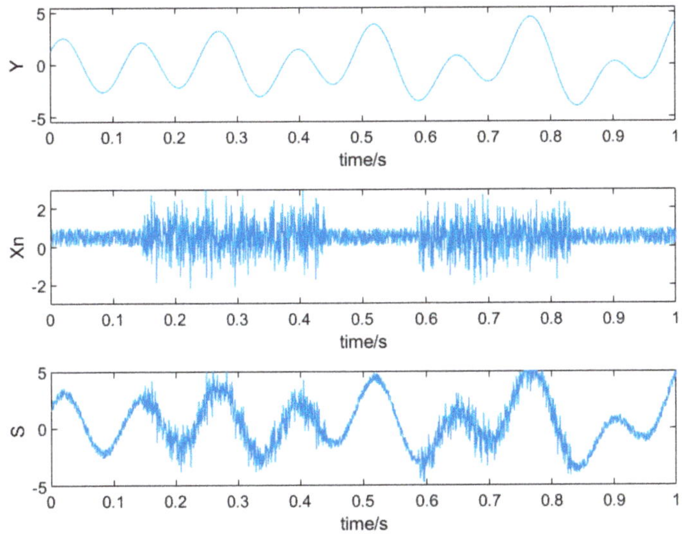

Figure 2. Simulation signal, noise signal, and simulation signal with noise.

The seven sub-signals decomposed by ICEEMD are shown in Figure 3 and the seven sub-signals decomposed by CEEMD are shown in Figure 4.

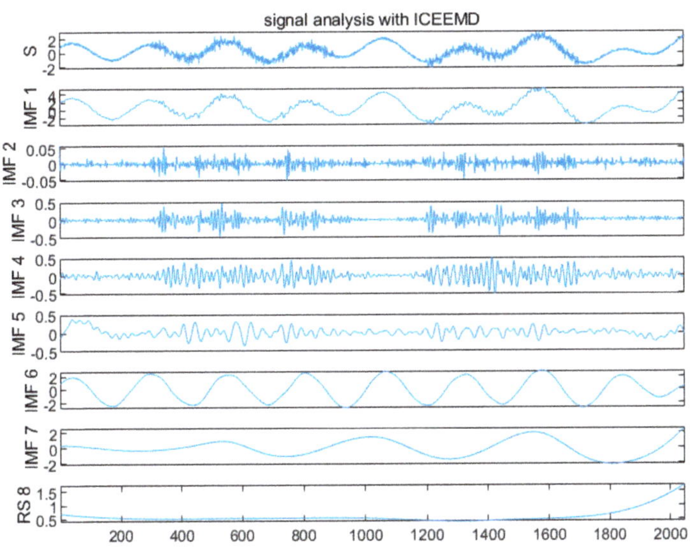

Figure 3. Seven sub-signals decomposed by ICEEMD.

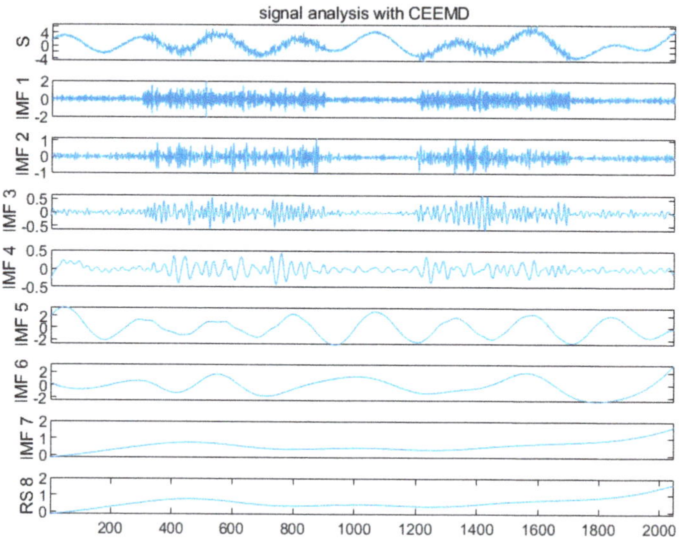

Figure 4. Seven sub-signals decomposed by complementary ensemble empirical mode decomposition (CEEMD).

As shown in Figures 3 and 4, seven IMFs are sequentially decomposed in order of their frequency using the two methods. We can see that the decomposed components of CEEMD are more severely affected by noise than those of ICEEMD. In order to further test the performance of the two methods in eliminating noise, the simulation signal was compared with the reconstructed signals of ICEEMD and CEEMD, respectively, and the results were shown in Table 1.

Table 1. The comparison of ICEEMD and CEEMD.

Method	MSE (Mean Squared Error)	IO (Index of Orthogonality)	R (Coefficient of Determination)
ICEEMD	4.0286	0.9073	0.9925
CEEMD	6.2469	1.3127	0.9489

By comparison, the correlation between the simulation signal and the reconstructed signal of ICEEMD is higher and the mean squared error is smaller than that of CEEMD. In addition, compared with CEEMD, the components of the ICEEMD method have better orthogonality. Therefore, the method proposed in this paper is better at removing noise and restoring the signal characteristics.

2.2. A Description of Time-Frequency Information Entropy

Time-frequency distribution of the signal reflects its energy variation at each frequency. It can quantitatively describe the different degrees of planetary gearboxes under different operation conditions. Time-frequency information entropy can describe the time-frequency distribution complexity of signals, so it is especially suitable for fault feature extraction. In this paper, time-frequency information entropy is the feature vector for fault diagnosis. It can represent the fault information hidden in the signal, with high diagnostic accuracy and good robustness.

The steps of time-frequency information entropy can be described as follows [24,25]:

(1) Short-time Fourier transform for each IMF component.

$$F(t,\omega) = \int IMF(i)w(i-t)e^{-j\omega t}d\iota \tag{7}$$

where $w(\tau)$ is the window function.

(2) Calculate the time-frequency energy spectrum of each IMF.

The time-frequency energy spectrum is used to describe the time-frequency distribution of the signal. The expression is as follows [26]:

$$S(t,\omega) = |F(t,\omega)|^2 \tag{8}$$

(3) Energy normalization.

The difference of energy distribution of time-frequency block on time-frequency plane reflects the difference in signal time-frequency distribution. The time-frequency energy W_i of IMF_i is obtained by the Equations (6) and (7). The time-frequency energy of all IMFs is A.

Energy normalization of IMF_i:

$$q_i = W_i/A(i = 1, \ldots, M) \tag{9}$$

(4) Calculate the entropy value of each time-frequency block:

$$s(q) = -\sum_{i=1}^{M} q_i \ln q_i \tag{10}$$

3. Fault Classification Based on VPMCD

In mechanical fault diagnosis, there is an intrinsic relationship between the characteristic values of signals. Their intrinsic relationships will change differently under different operation conditions [27–29].

The variable predictive model based class discriminate (VPMCD) method is a pattern recognition method based on the variable predictive model (VPM). According to the relationship between characteristic values, a prediction model is established to classify the test samples [30–33]:

Linear model (L):

$$X_i = b_0 + \sum_{j=1}^{r} b_j X_j \tag{11}$$

Linear interaction model (LI):

$$X_i = b_0 + \sum_{j=1}^{r} b_j X_j + \sum_{j=1}^{r} \sum_{k=j+1}^{r} b_{jk} X_j X_k \tag{12}$$

Quadratic interaction model (QI):

$$X_i = b_0 + \sum_{j=1}^{r} b_j X_j + \sum_{j=1}^{r} b_{jj} X_j^2 + \sum_{j=1}^{r-1} \sum_{k=j+1}^{r} b_{jk} X_j X_k \tag{13}$$

Quadratic model (Q):

$$X_i = b_0 + \sum_{j=1}^{r} b_j X_j + \sum_{j=1}^{r} b_{jj} X_j^2 \tag{14}$$

where r is model order, $r \le k-1$, k is number of feature values for each fault type, and j is number of fault types. $X_j (j \ne i)$ is the predictive variable, X_i is the predicted variable, and b_0, b_j, b_{jj} and b_{jk} are model parameters.

Select one of the four models as a predictive model, and its expression is as follows [34,35]:

$$X_i = f\left(X_j, b_0, b_j, b_{jj}, b_{jk}\right) + e \tag{15}$$

where e is the prediction error.

Suppose there are n fault types, and k prediction models are built for each fault type to obtain $n \times k$ prediction models. Calculate the sum of squared prediction errors of k eigenvalues under the same fault type. The fault category corresponding to the model with the smallest sum of squares of the prediction error is the final output fault mode. Therefore, the VPMCD classification algorithm can be applied not only into linear classification, but also into nonlinear classification.

4. Fault Diagnosis Scheme for a Planetary Gearbox

This paper proposes a fault diagnosis method based on ICEEMD-time-frequency information entropy and VPMCD. The specific procedure of the diagnosis scheme is shown in Figure 5.

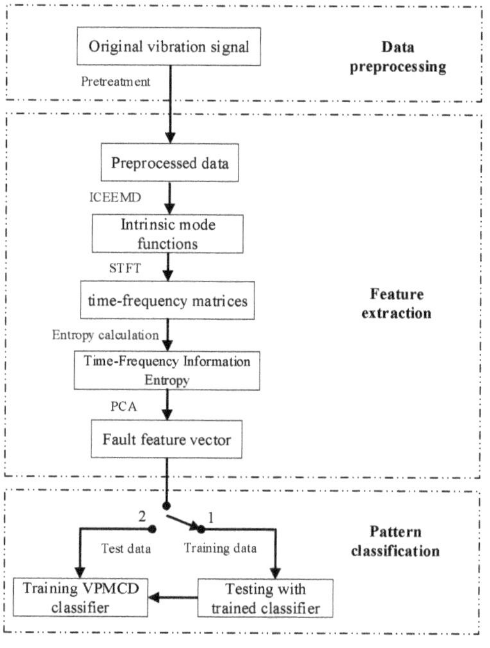

Figure 5. The fault diagnosis procedure for a planetary gearbox.

(1) Preprocess the data of the original vibration signal.
(2) Decompose the vibration signals of the planetary gearbox into a series of IMFs by utilizing ICEEMD.
(3) Calculate the time-frequency information entropy of the IMFs as feature vectors by using short-time Fourier transform (STFT) and information entropy.
(4) Reduce the dimension of the feature vectors by utilizing principal components analysis (PCA) to improve the accuracy and robustness of pattern recognition.
(5) Classify the fault modes by using VPMCD.

5. Experimental Verification

Since the gear motion in the planetary gearbox is a typical composite motion, the vibration signal is more complex than that of a fixed shaft gearbox. Moreover, the gear is the most fault-prone part of the planetary gearbox. Therefore, this paper takes the gear of the planetary gearbox as the diagnostic object to verify the effectiveness of the proposed method. ICEEMD-time-frequency information entropy method was utilized to extract the fault feature of the gear vibration data from a fault-prediction test bed. We used the power transmission fault prediction test bed (DPS) manufactured by Spectra Quest,

USA. The test bed is shown in Figure 6, which includes a centrally fixed sun gear, a planetary carrier and ring gear, and four planetary gears that change with the center of rotation of the sun gear. In addition, the planetary gearbox of this experimental bench was small in volume, and the vibration caused by the fault was weak, which increases the difficulty of diagnosis. In this experiment, a total of 12 operation conditions were collected. The specific operation condition information is shown in Table 2. The signal sampling frequency was 12,800 Hz and the signal sampling point was 524,288. The experimental fault data includes five states: gear tooth crack fault, gear wear fault, tooth breaking fault, gearing missing fault, and normal. In this study, experimental data were divided into 40 segments, and each segment had 12,800 points.

Figure 6. Test bed of planetary gearbox.

Table 2. The description of signal operation conditions.

Operation Condition Number	Speed (Hz)	Load (Nm)
Condition 1	20	0
Condition 2	40	0
Condition 3	60	0
Condition 4	20	0.6
Condition 5	40	0.6
Condition 6	60	0.6
Condition 7	10	1.2
Condition 8	20	1.2
Condition 9	30	1.2
Condition 10	40	1.2
Condition 11	50	1.2
Condition 12	60	1.2

5.1. Fault Feature Extraction Based on ICEEMD-Time-Frequency Information Entropy

First of all, the vibration signal collected by the sensor was preprocessed.

Secondly, the ICEEMD algorithm was applied to decompose each fault mode. The threshold of multi-scale permutation entropy θ_0 was 0.6, embedded dimension m was 6, time delay factor τ was 1, and the scale factor s was 5. The high frequency part of the vibration signal of the planetary gearbox contained the main fault signal. Therefore, the first six intrinsic modal components (IMFs) of the second CEEMD decomposition were taken.

Thirdly, the time-frequency distribution of IMF was computed by means of STFT. The time-frequency entropy of each failure mode was taken as the fault feature vector. The length of the time-frequency block was 64 and width was 64. The number of the lateral and longitudinal slip steps was equal to 32. Then, principal component analysis (PCA) was used to reduce the fault characteristics and obtain the 3-dimensional fault feature vector. The clustering results are shown in Figures 7–10 for the first four operation conditions.

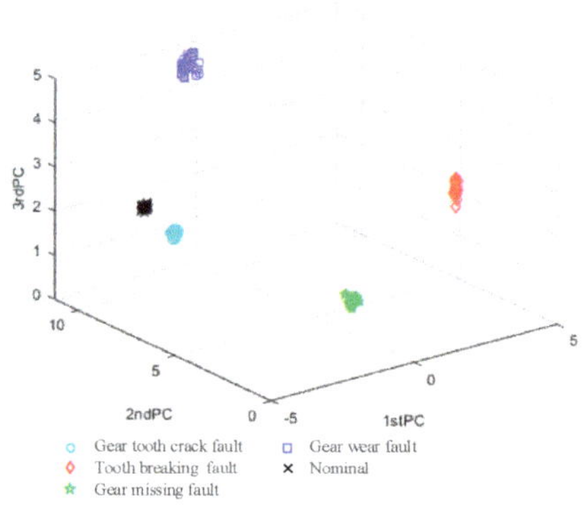

Figure 7. Clustering result of gear fault features under condition 1.

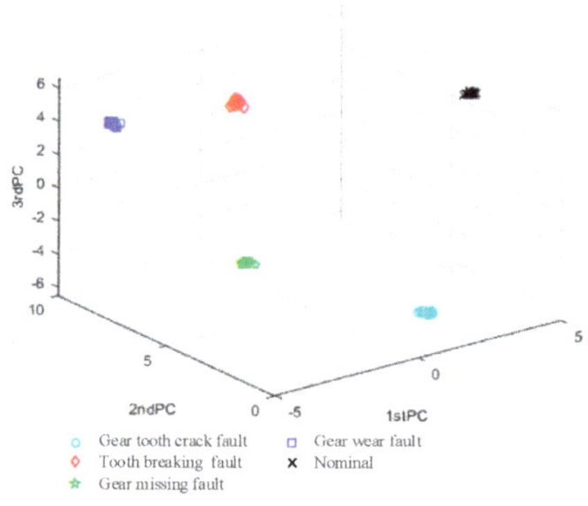

Figure 8. Clustering result of gear fault features under condition 2.

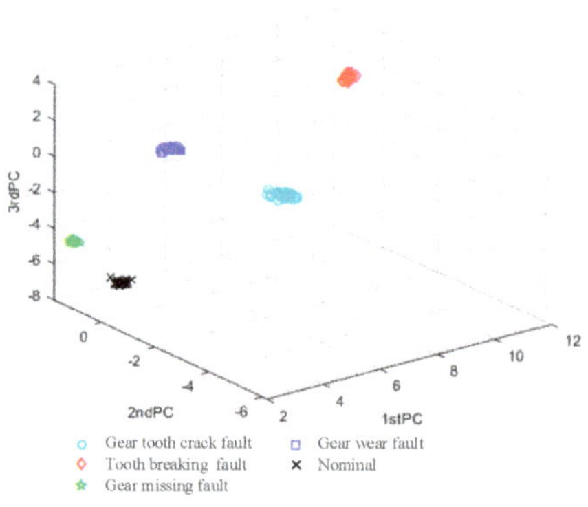

Figure 9. Clustering result of gear fault features under condition 3.

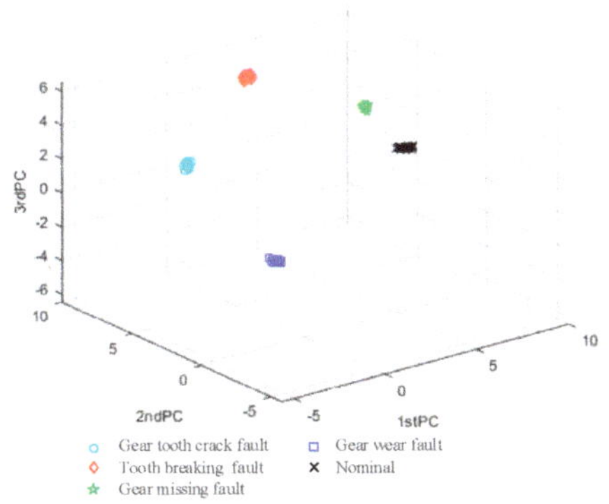

Figure 10. Clustering result of gear fault features under condition 4.

From Figures 7–10, it can be seen that different fault types can be well separated under the same operation condition. There is no feature vector mixing under different fault states. Moreover, the clustering effect about the eigenvalues of the same fault type shows satisfactory performance and no divergence.

5.2. Method Comparison

In order to reflect upon the effectiveness of the proposed algorithm, two feature extraction methods were used to compare with the proposed method in this paper.

5.2.1. Comparison between the Proposed Method with the Wavelet Entropy Method

Wavelet entropy, as a traditional time-frequency analysis method, is widely used in weak signal fault diagnosis. The specific algorithm steps are as follows:

First, the vibration signal is preprocessed.

Second, the planetary gearbox vibration signal is decomposed into three layers to obtain eight sub-band signals by using the sym6 wavelet base.

The third step is to calculate the total energy of each frequency band signal and normalize the energy. The wavelet entropy of each sub-band is calculated to obtain an 8-dimensional feature vector. The feature vector is dimension-reduced by principal component analysis (PCA). The clustering results are shown in Figures 11–14.

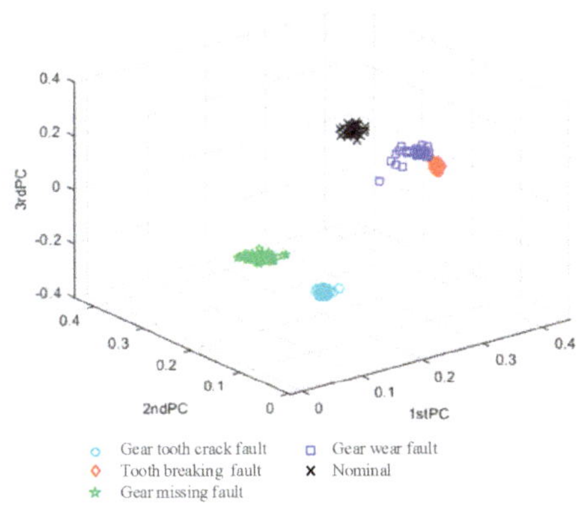

Figure 11. Clustering result of wavelet entropy under condition 1.

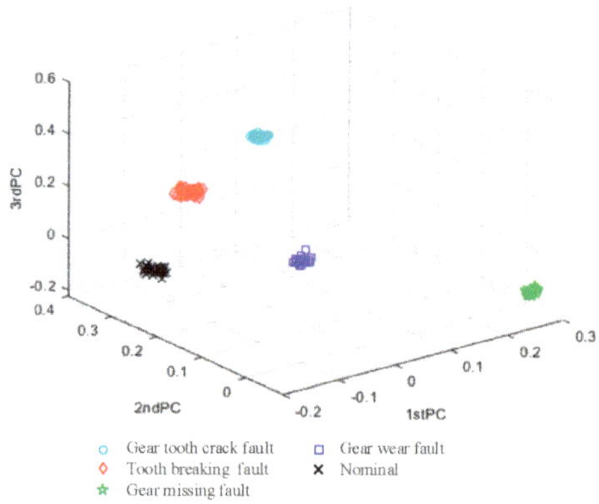

Figure 12. Clustering result of wavelet entropy under condition 2.

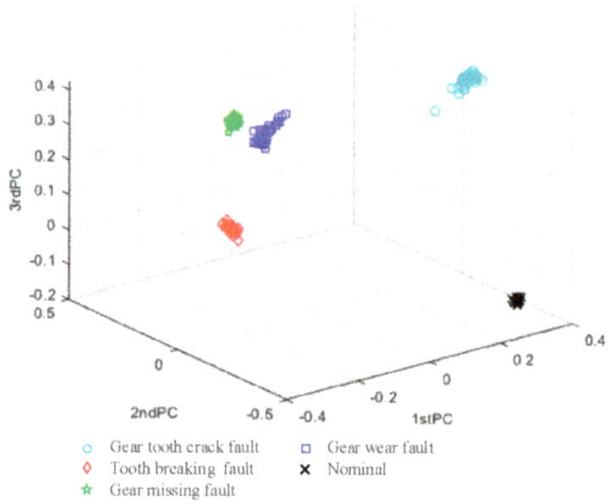

Figure 13. Clustering result of wavelet entropy under condition 3.

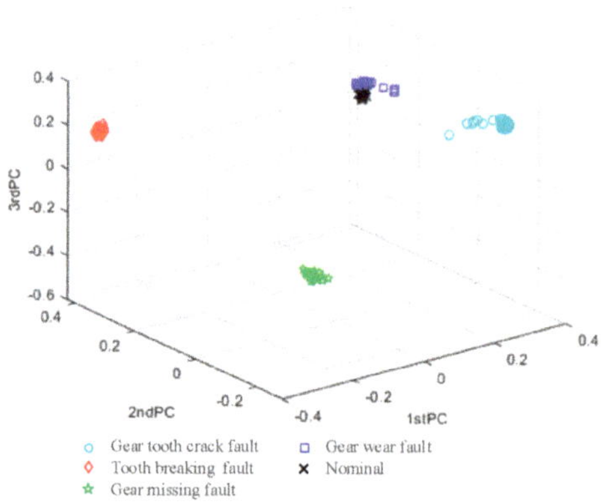

Figure 14. Clustering result of wavelet entropy under condition 4.

As shown in Figures 11–14, when the feature vector extracted by the wavelet entropy method was used to identify the degree of fault type, a certain degree of aliasing appeared under the operation condition 1 and operation condition 4. Therefore, the method proposed in this paper is desirable and has better clustering results than does the wavelet entropy method in fault feature extraction.

5.2.2. Comparison between the Proposed Method with VMD-Time-Frequency Information Entropy

Variational mode decomposition (VMD) is a completely non-recursive, adaptive signal processing method proposed by Dragomiretskiy, K. [36]. The method has higher decomposition accuracy and can effectively filter out the white noise in the signal. The specific algorithm steps are as follows:

Step 1: Preprocess the vibration signal.

Step 2: Decompose the signal into six IMFs using the VMD method.

Step 3: Calculate the time-frequency information entropy of each IMF and obtain the 6-dimensional feature vector. Similarly, the feature vector is reduced by PCA. The clustering results are shown in Figures 15–18.

As can be seen from Figures 15–18, the VMD-time-frequency information entropy method also has the phenomenon of fault mixing under the operation condition 1 and the operation condition 4. By contrast, it can be proved that the method proposed in this paper has better performance than the VMD-time-frequency information entropy method in fault feature extraction.

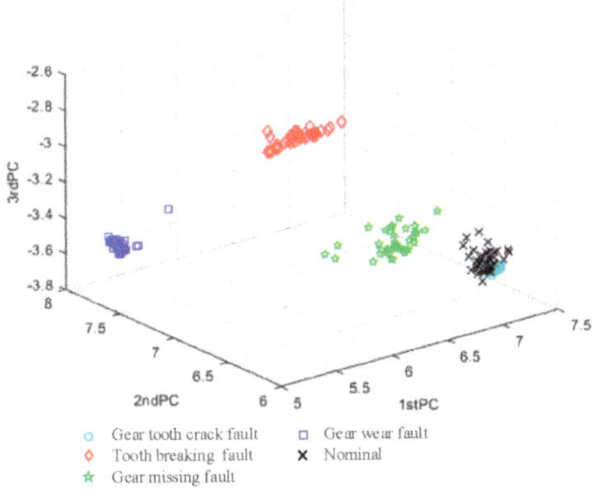

Figure 15. Clustering result of variational mode decomposition (VMD)-time-frequency information entropy under condition 1.

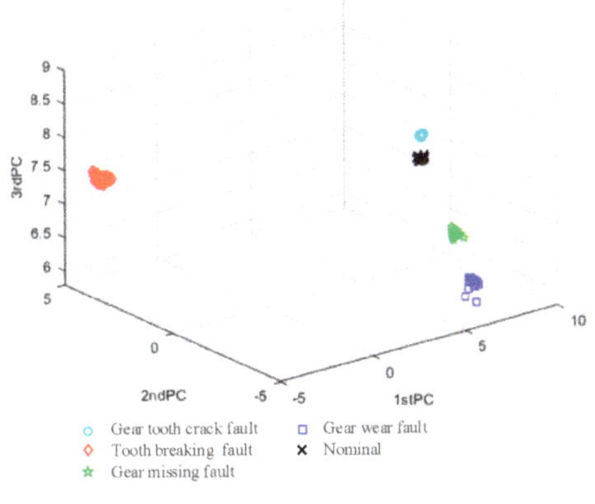

Figure 16. Clustering result of VMD-time-frequency information entropy under condition 2.

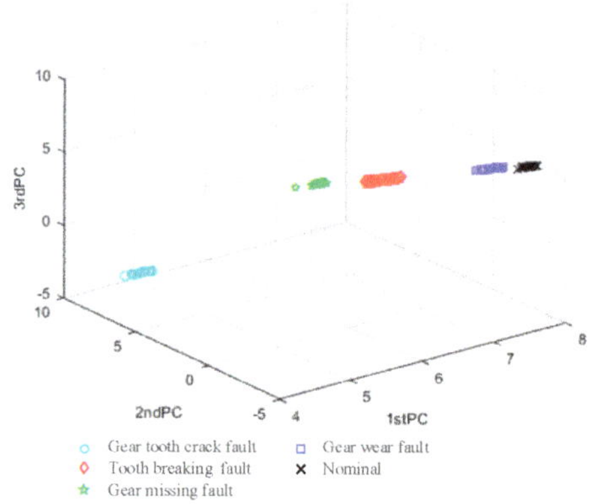

Figure 17. Clustering result of VMD-time-frequency information entropy under condition 3.

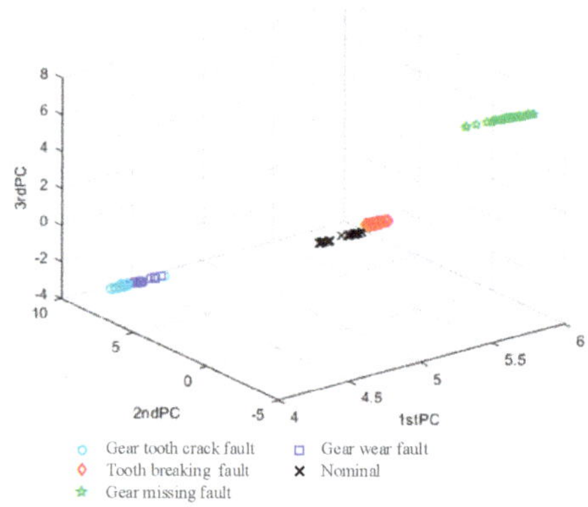

Figure 18. Clustering result of VMD-time-frequency information entropy under condition 4.

5.2.3. Comparison under Variable Operation Conditions

In order to further demonstrate the advantages of the proposed algorithm, the diagnostic performance of the three methods were compared under variable operation conditions. In this paper, the data under variable operation conditions were spliced together. The specific information is shown in Table 3. We compared the three algorithms mentioned in this paper. The results are shown in Figures 19–21. In total, 128,000 sampling points were extracted from each operation condition, and each fault consisted of a total of 384,000 sampling points under three operation conditions.

Table 3. Multi-operation condition data.

Failure Mode	Variable Operation Conditions (Arranged According to the Order of Operation Conditions)
Gear tooth crack fault	condition 1–condition 5–condition 9
Tooth breaking fault	condition 2–condition 4–condition 7
Gear breaking fault	condition 3–condition 5–condition 9
Gear wear fault	condition 12–condition 11–condition 10
Normal	condition 1–condition 6–condition 10

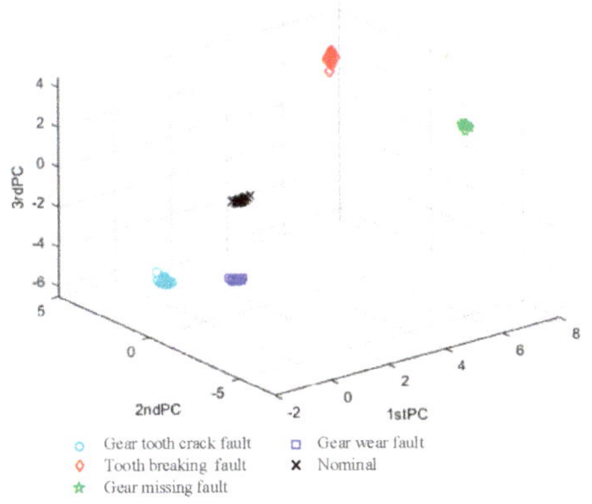

Figure 19. Clustering result of ICEEMD-time-frequency information entropy under variable conditions.

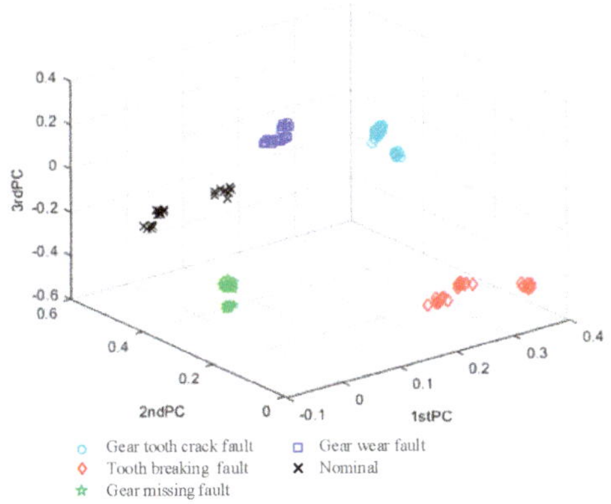

Figure 20. Clustering result of wavelet entropy under variable conditions.

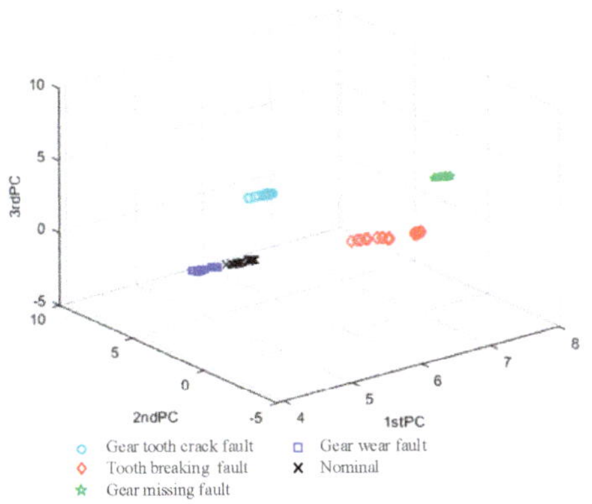

Figure 21. Clustering result of VMD-time-frequency information entropy under variable conditions.

Under the variable operation conditions, the algorithm proposed in this paper had obvious clustering of fault features and no fault mixing. However, the wavelet entropy algorithm and VMD-time-frequency information entropy algorithm had a certain degree of mixing for the same fault under different operation conditions, and the degree of polymerization was not ideal. Therefore, neither of these methods can be applied to the fault diagnosis of variable operation conditions.

5.3. Fault Classification of Planetary Gearbox Based on VPMCD

Under a single operation condition, 100 groups of data were selected as the VPMCD training set, and the remaining 100 groups of data were used as the VPMCD test set. Similarly, under variable operation conditions, there were 75 groups of data used as the training set and the remaining 75 groups of data as the test set. The feature vector after PCA dimension reduction was used as the input of VPMCD, and the fault type was used as the output of VPMCD. The corresponding relationship between the number and the fault type is shown in Table 4.

Table 4. Corresponding information between number and fault type.

Number	Fault Type
1	Gear tooth crack fault
2	Tooth breaking fault
3	Gear breaking fault
4	Gear wear fault
5	Normal

The classification results are shown in Figures 22–26. It can be seen from the classification results that the diagnostic accuracy rate reached 100% under the single operation condition and variable operation conditions.

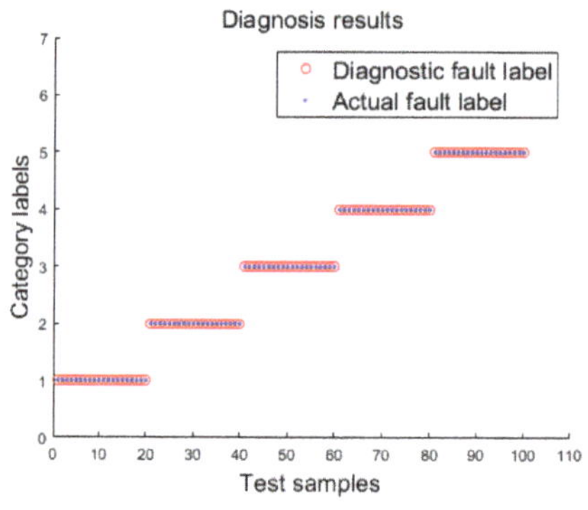

Figure 22. Classification result under condition 1.

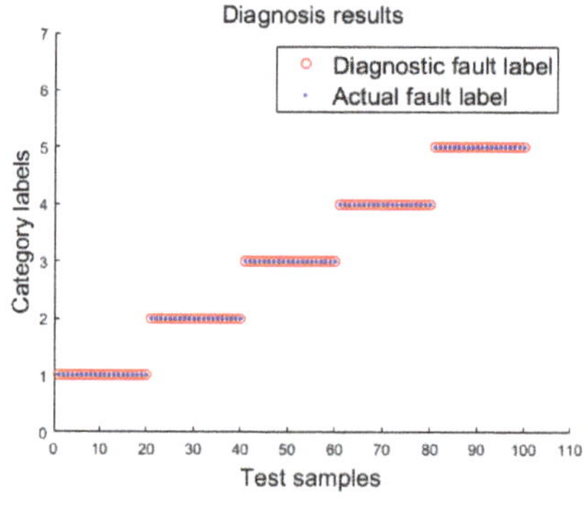

Figure 23. Classification result under condition 2.

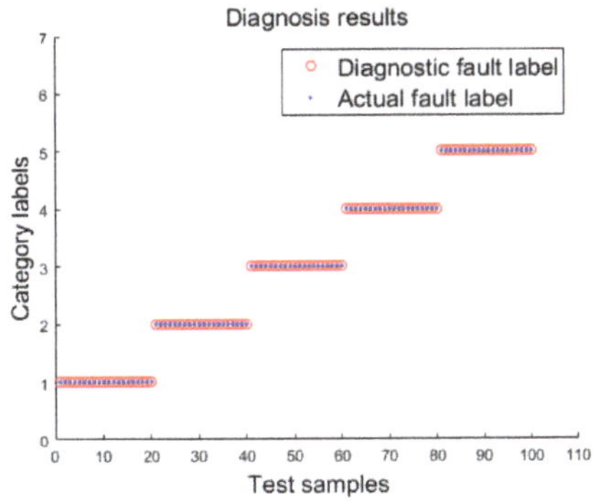

Figure 24. Classification result under condition 3.

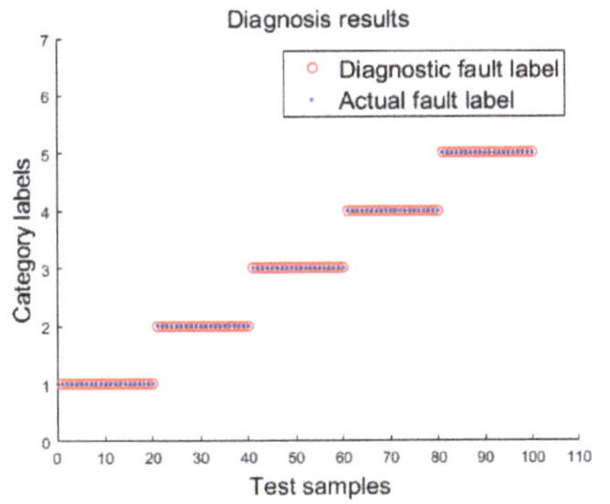

Figure 25. Classification result under condition 4.

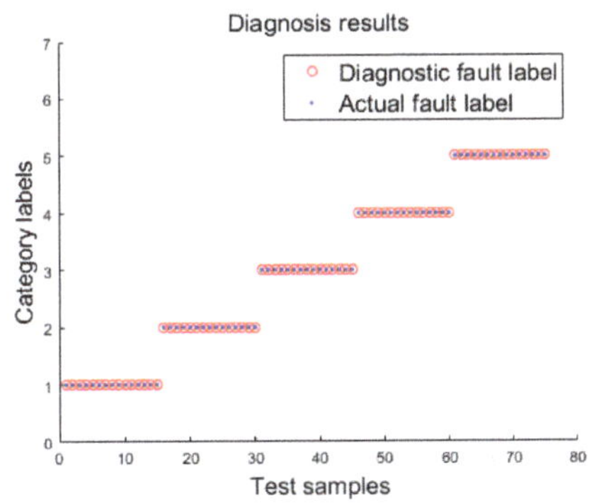

Figure 26. Classification result under variable conditions.

In order to reduce the randomness of the diagnosis, we selected 75 groups of data randomly as training sets and the remaining 75 groups of data as test sets. After 100 random classifications, the average classification accuracy of VPMCD was 100% and no testing sample wase misclassified.

6. Conclusions

The effective fault diagnosis of planetary gearboxes is indeed difficult and has always attracted the attention of researchers. The weak fault features are difficult to extract from a planetary gearbox because the signal components are nonlinear, nonstationary, and easily drowned out by noise. In this paper, we propose a fault diagnosis method based on ICEEMD-time-frequency information entropy. The effectiveness of the method is verified by a case study where the proposed method outperforms the wavelet entropy method and the VMD-time-frequency information entropy method in extracting critical fault features. In regard to fault classification, the VPMCD algorithm is adopted to comprehensively consider the correlation about eigenvalues, which can effectively identify the characteristic information of various types of small samples. Moreover, experimental results show that the proposed method can not only accurately diagnose faults under multiple operating conditions, but also produces satisfactory diagnostic performance under variable operating conditions. Therefore, the ICEEMD-time-frequency information entropy and VPMCD method, with favorable robustness and diagnostic performance, has wide applicability in other similar fault diagnoses of rotating machinery. However, to some extent, the performance of the proposed method is limited by computer resources. Future work will focus on optimization of the algorithm to improve its performance.

Author Contributions: Conceptualization, Y.W.; Data curation, Y.W.; Methodology, Y.W., H.L., and Z.F.; Writing original draft, Y.W.; Writing review and editing, X.G. All authors have read and agreed to the published version of the manuscript.

Funding: This research was funded by Key R & D projects of Shanxi province (No. 2019ZDLGY17-06), the Key Projects for Quality Reliability Design, and Analysis Technology Breakthrough of MIIT under Tender (No. TC190A4DA/6).

Conflicts of Interest: The authors declare no conflict of interest.

References

1. Khater, F.; Omar, A. A Review of Direct Driven PMSG for Wind Energy Systems. *J. Energy Power Eng.* **2013**, *7*, 1592–1603.
2. Villecco, F. On the Evaluation of Errors in the Virtual Design of Mechanical Systems. *Machines* **2018**, *6*, 36. [CrossRef]
3. Sun, H.; Wu, C.; Liang, X.; Zeng, Q. Identification of Multiple Faults in Gearbox Based on Multipoint Optional Minimum Entropy Deconvolution Adjusted and Permutation Entropy. *Entropy* **2018**, *20*, 850. [CrossRef]
4. Formato, A.; Guida, D.; Ianniello, D.; Villecco, F.; Lenza, T.L.; Pellegrino, A. Design of Delivery Valve for Hydraulic Pumps. *Machines* **2018**, *6*, 44. [CrossRef]
5. Shen, Z.; Shi, Z.; Zhen, D.; Zhang, H.; Gu, F. Fault Diagnosis of Planetary Gearbox Based on Adaptive Order Bispectrum Slice and Fault Characteristics Energy Ratio Analysis. *Sensors* **2020**, *20*, 2433. [CrossRef] [PubMed]
6. Tempe, D.K.; Satyanarayana, L. Time-frequency analysis of nonlinear and non-stationary weak signals of corona discharge. *J. Phys.* **2013**, *92*, 1–3.
7. Zhang, X.; Liang, Y.; Zhou, J. A novel bearing fault diagnosis model integrated permutation entropy, ensemble empirical mode decomposition and optimized SVM. *Measurement* **2015**, *69*, 164–179. [CrossRef]
8. Wang, D.F.; Guo, Y.; Wu, X.; Na, J. Planetary-Gearbox Fault Classification by Convolutional Neural Network and Recurrence Plot. *Appl. Sci.* **2020**, *10*, 932. [CrossRef]
9. Thirumala, K.; Umarikar, A.C.; Jain, T. Estimation of Single-Phase and Three-Phase Power-Quality Indices Using Empirical Wavelet Transform. *IEEE Trans. Power Deliv.* **2015**, *30*, 445–454. [CrossRef]
10. Wang, L.; Liu, S.; Wei, M. The time-frequency analysis of nonlinear and non-stationary weak signals. In Proceedings of the 7th International Conference on Applied Electrostatics, Dalian, China, 17–19 September 2012.
11. Feng, Z.; Liang, M. Fault diagnosis of wind turbine planetary gearbox under nonstationary conditions via adaptive optimal kernel time-frequency analysis. *Renew. Energy* **2014**, *66*, 468–477. [CrossRef]
12. Liu, H.; Zhang, J.; Cheng, Y.; Lu, C. Fault diagnosis of gearbox using empirical mode decomposition and multi-fractal detrended cross-correlation analysis. *J. Sound Vib.* **2016**, *385*, 350–371. [CrossRef]
13. Li, Y.; Chen, X.; Yu, J. A Novel Feature Extraction Method for Ship-Radiated Noise Based on Variational Mode Decomposition and Multi-Scale Permutation Entropy. *Entropy* **2017**, *19*, 342–349.
14. Li, G.; Deng, C.; Wu, J.; Chen, Z.; Xu, X. Rolling Bearing Fault Diagnosis Based on Wavelet Packet Transform and Convolutional Neural Network. *Appl. Sci.* **2020**, *10*, 770. [CrossRef]

15. Su, X.; Liu, H.; Tao, L. TF Entropy and RFE Based Diagnosis for Centrifugal Pumps Subject to the Limitation of Failure Samples. *Appl. Sci.* **2020**, *10*, 2932. [CrossRef]
16. Luo, S.; Cheng, J.; Zheng, J. UA-VPMCD method and its application in machinery fault intelligent diagnosis. *J. Vib. Eng.* **2014**, *27*, 289–294.
17. Lin, B.; Zeng, K.; Xu, G.; Lu, C. Fault Recognition Method for Rolling Bearing Bearings Based on RQA and V-VPMCD. *J. Vib. Meas. Diagn.* **2018**, *38*, 314–318.
18. Wang, F.; Fang, L. Fault Diagnosis Method for Automata Machine Based on Multiple Scale Sample Entropy and VPMCD. *J. Vib. Meas. Diagn.* **2018**, *38*, 564–569.
19. Zheng, J.; Cheng, J.; Yang, Y. Modified EEMD algorithm and its applications. *J. Vib. Shock* **2013**, *32*, 21–26.
20. Yeh, J.R.; Shieh, J.S.; Huang, N.E. Complementary ensemble empirical mode decomposition: A noise enhanced data analysis method. *Adv. Adapt. Data Anal.* **2010**, *2*, 135–156. [CrossRef]
21. Yangde, G.; Francesco, V.; Ming, L. Multi-Scale Permutation Entropy Based on Improved LMD and HMM for Rolling Bearing Diagnosis. *Entropy* **2017**, *19*, 176.
22. Liu, T.; Yao, W.; Wu, M. Multiscale permutation entropy analysis of electrocardiogram. *Acta Phys.* **2014**, *471*, 978. [CrossRef]
23. Liu, B.; Hu, W.-P. Recognition of denatured biological tissue based on variational mode decomposition and multiscale permutation entropy. *Chin. Phys.* **2019**, *68*, 028702.
24. Mu, Z.; Hu, J.; Min, J.-L. Driver Fatigue Detection System Using Electroencephalography Signals Based on Combined Entropy Features. *Appl. Sci.* **2017**, *7*, 150. [CrossRef]
25. Le, D.H.; Cheng, J.; Yang, Y. Gear Fault Diagnosis Method Based on Local Characteristic-Scale Decomposition Multi-Scale Permutation Entropy and Radial Basis Function Network. *J. Comput. Theor. Nanosci.* **2017**, *14*, 5054–5063. [CrossRef]
26. Costa, M.; Goldberger, A.L.; Peng, C.K. Multiscale Entropy Analysis of Complex Physiologic Time Series. *Phys. Rev. Lett.* **2007**, *89*, 705–708. [CrossRef] [PubMed]
27. Jing, J.; Liu, H.; Lu, C. Fault diagnosis of electro-mechanical actuator based on WPD-STFT time-frequency entropy and PNN. *Vibroengineering PROCEDIA* **2017**, *14*, 130–135.
28. Wang, X.; Benesty, J.; Chen, J. A single-channel noise cancelation filter in the short-time-Fourier-transform domain. In Proceedings of the 2016 IEEE International Conference on Acoustics, Speech and Signal Processing (ICASSP), Shanghai, China, 20–25 March 2016.
29. Cheng, J.S.; Ma, X.W. Gear fault diagnosis method based on VPMCD and EMD. *J. Vib. Shock* **2013**, *32*, 9–13.
30. Tao, R.; Xu, Y.C.; Li, X.S. The Research of Fault Diagnosis Method of Roller Bearing Based on EMD and VPMCD. *Adv. Mater. Res.* **2014**, *1014*, 505–509. [CrossRef]
31. Wang, X.; Makis, V.; Yang, M. A wavelet approach to fault diagnosis of a gearbox under varying load conditions. *J. Sound Vib.* **2010**, *329*, 1570–1585. [CrossRef]
32. Wang, U.; Duo, G. A Novel Approach for the Pattern recognition of hand movements based on EMG and VPMCD. *J. Mech. Med. Biol.* **2018**, *18*, 1750115. [CrossRef]
33. Tang, T.; Bo, L.; Liu, X. Variable predictive model class discrimination using novel predictive models and adaptive feature selection for bearing fault identification. *J. Sound Vib.* **2018**, *425*, 137–148. [CrossRef]
34. Rao, R.; Lakshminarayanan, S. Variable predictive models—A new multivariate classification approach for pattern recognition applications. *Pattern Recognit.* **2009**, *42*, 7–16.
35. Kankar, P.K.; Sharma, S.C.; Harsha, S.P. Fault diagnosis of ball bearing using continuous wavelet transform. *Appl. Soft Comput.* **2011**, *11*, 2300–2312. [CrossRef]
36. Dragomiretskiy, K.; Zosso, D. Variational Mode Decomposition. *IEEE Trans. Signal Process.* **2014**, *62*, 531–544. [CrossRef]

applied sciences

MDPI

Article

Analysis of the Frequency Interaction of the Turbine Block in the Stand for the Magnitude of the Error in Measuring the Turbine's Power

Anton Petrochenkov [1,*], Aleksey Sal'nikov [1], Sergey Bochkarev [1] and Pavel Ilyushin [2]

[1] Electrical Engineering Faculty, Perm National Research Polytechnic University, 614990 Perm, Russia; afsalnikov_1@mail.ru (A.S.); bochkarev@msa.pstu.ru (S.B.)
[2] Research and Educational Center for Geology, Oil and Gaz Fields Development, 614990 Perm, Russia; ilushin-pavel@yandex.ru
* Correspondence: pab@msa.pstu.ru; Tel.: +7-3422391821

Abstract: An algorithm for constructing a dynamic analysis during the formation of a wave field of stand for testing turbines and the effect of the frequency interaction of the stand's elements on the measurement of its magnitude is described. The research algorithm involves the use of theoretical solutions of nonlinear wave processes using linear oscillations, refined by experiments. The diagnostic model can determine the technical condition of the stand's elements and also determine the causes of the discrepancies between the calculated and measured turbine power values. To clarify the stiffness coefficients between the stand's elements, a modal analysis was used to obtain the range of their changes depending on the external dynamic load, which made it possible to assess the impact of changes in the frequency interaction conditions on the turbine power measurement at different test modes. The conditions for amplifying the amplitude of oscillations at their eigenfrequencies are obtained, and the value of the possible deviation of the expected power value at its measurement for specific modes of the turbine is calculated. The algorithm allows to estimate the dynamic state of the stand-in different research modes of turbines and give recommendations for reducing the level of frequency interaction.

Keywords: non-destructive testing; technical diagnostics; eigenfrequency; amplitude-frequency characteristic; wave field; frequency analysis

Citation: Petrochenkov, A.; Sal'nikov, A.; Bochkarev, S.; Ilyushin, P. Analysis of the Frequency Interaction of the Turbine Block in the Stand for the Magnitude of the Error in Measuring the Turbine's Power. *Appl. Sci.* **2021**, *11*, 4149. https://doi.org/10.3390/app11094149

Academic Editor: Toomas Vaimann

Received: 31 March 2021
Accepted: 21 April 2021
Published: 1 May 2021

1. Introduction

The stringent requirements that are imposed on the quality of generated energy necessitate making significant fundamental modifications to existing turbine plant prototypes. The conditions of a test stand offer the possibility of combining the stages of computer-aided tests, i.e., simulation of the automatic control system, and hardware-in-the loop-tests, i.e., tests of mockups and experimental and pilot prototypes, also using a computer model of the power system [1]. During the experimental testing of new and modernized products, it is important to take into account the influence of external factors, in particular, frequency interactions with the stand elements caused by vibrations and their influence on the receipt of measurement information from measuring devices built into the stand equipment. In addition, vibration processes can cause damage to the stand and measuring devices.

To control vibration parameters, various methods and devices are offered. For example, in work [2], the prediction of bearing failure based on the results of vibration monitoring of bearing assemblies based on the damage accumulation model is considered. In work [3], a model of optimal frequency control of vibration by a formative active excitation of indefinite flexible mechanical systems with constant excitation is presented. In work [4], a model of studies of nonlinear analysis of the frequency response of hydrodynamic plain bearings with external disturbances according to the results of monitoring of bearing assemblies is presented.

In work [5], the results of an experimental determination of the dynamic force coefficients of a flexible shaft in bearing assemblies are presented. In work [6], the results of experimental vibration tests of a turbo compressor of a high-speed diesel engine are presented. In work [7], questions of the influence of nonlinearity in the dynamic analysis of the machines' operations are considered; also, the examples of the behavior of their structures are shown. Work [8] described a micro turbocharger with airfoil bearings for a 100 W micro-electric power station, and dynamic analysis to improve the dynamic characteristics of the rotor is given. In work [9], the results of a study of the effect of rotor imbalance in response (vibration) on its stability in hybrid airfoil bearings are presented.

In paper [10], the technical problems of using the Markov process for managing and planning the operation of equipment using a servicing strategy relevant to the technical state are considered. In paper [11], the authors considered the problem of torque pulsation and its stabilization by means of control for switched reluctance drives. The authors introduced the method of model predictive control with reduced integration step size in order to implement precise torque stabilization using pulse width modulation.

The paper [12] presents the experimental results of the combustion diagnosis in the cylinders of heavy diesel engines with direct injection. In work [13], devices are considered to improve the accuracy of parameter measurement: For the vibrational movement of elements of electronic equipment, a three-component measurement method is used. Additionally, modal impedance curves calculated using the linear perturbation method are used to predict rotor-dynamic characteristics of the rotor.

The scheme of the information model of provision of multi-aspect diagnostics of the condition of the electrotechnical complex elements in the standard IDEFlx is given in paper [14].

In this regard, attention should be paid to transients and the restructuring of the "product-stand" self-oscillating system; for example, when the power (number of rotations and other parameters) of the product in the measuring devices of the stand's modes. Receiving and studying the correct measurement data is the actual task.

To understand the physical aspects of the dynamic behavior of products, experimental complexes are developed based on which issues of the product and stand's dynamic interaction are investigated. In addition, it is necessary to simultaneously solve the problem of not only obtaining actual results but also to assess the impact of transients during the receiving of a new self-oscillating "product-stand" mode. This is a multi-parameter problem wherein various methods are used to solve, including reducing the dynamic effect of rotation elements (rotors) [15,16] or due to the think selection of turbine block blades [17] or integrated approaches [18,19].

However, insufficient attention is being paid to the influence of frequency interactions with measurement elements embedded in the stand's equipment since frequency interaction can lead to incorrect results of measuring the output parameters of the equipment under test at the stand. In this case, it is necessary to study the eigenfrequencies and forced frequencies of the elements of the stand's equipment, both with the tested product and without it. To control the vibration parameters, various means and devices and various methods for processing vibration signals are offered.

Equally important is the identification of the causes leading to a change in the error of the attorney measuring devices during stand tests of various mechanical systems [9,20]. This is especially important for units that provide a minimum dispersion of traction characteristics of various power plants; therefore, strict requirements are imposed on them to meet the specified power characteristics. Therefore, measuring the actual power of the turbine of a turbopump unit depending on the speed (number of rotations) of the turbine allows us to build processes for regulating the operation of the power plant; thereby, minimizing losses (an increase in the error in measuring the power of the turbine is the additional mass of fuel for the gas generator to work).

There are quite a lot of works on the actual change in the error when changing the number of rotations [21], which allows us to take into account when developing the power unit operation algorithm both when entering the mode and in control problems during stand tests of the turbine.

Stand tests of turbines usually have an electric drive part, which provides loading of the turbine during its tests as part of the stand. Typically, this is a generator with a speed reduction system (multiplier), which is mounted on the same frame with elements for measuring the speed of the turbine and its torque. Consequently, the presence of certain structural elements can be based on the works [5,7]. It can be assumed that these elements of the stand will affect the dynamic behavior of the stand during turbine tests. Figure 1 shows a part of the installation of the stand with a generator and a multiplier, indicating the points of measurement of the level of vibration during testing and the study of the modal characteristics of these elements mounted on a common frame of the stand.

Figure 1. General view of the installation generator and frame with measuring points.

When testing the elements of mechanical systems, there is a significant impact on their operation. It is important to understand the influence of the stand's elements on the measurement results.

In this paper, we do not evaluate the errors of the measuring system, but we pose the problem of identifying the discrepancy between the expected value of the turbine power at certain numbers of its rotations (calculated value) and its measured value during testing in the stand.

Nowadays, there are practically no studies aimed at obtaining the influence of the stand's equipment on the stand's measuring systems. For example, when a turbine block at the stand is working, a significant spread in the magnitude of the realized torque with its calculated value was revealed. This is especially observed during transients associated with a change in the turbine operating modes. In addition, during the steady-state process, at certain rotations of the turbine, oscillations of the measured torque in time relative to its average value are recorded. It can be assumed that these oscillations of the measured torque are associated with the occurrence of self-oscillating processes of the stand's wave field with the turbine. Therefore, the analysis of the reasons leading to the discrepancy between the "expected" characteristics of the turbine power and the actually measured values during the turbine working at the stand is an urgent research task. This task allows us to understand and evaluate the impact of dynamic interactions of the structural elements of the stand and the turbine on the reliability of the power plant.

2. Research Methods

To study the modal frequencies of the test stand and measure the level of vibration, we used the tuning measuring and vibration analyzer Kamerton [22] with a set of VK 310A vibration sensors that comply with ISO 8042:1988 [23]. The characteristics of the measuring device are shown in Table 1.

Table 1. Technical characteristics of the measuring device Kamerton.

No	Description	Measurement Limits
1	Number of channels	8
2	Operating frequency ranges (Hz) - vibration acceleration - vibration speed - vibration displacement	5–5000 5–1000 5–200
3	Measuring ranges - vibration acceleration (peak) (mm/s^2) - vibration velocity (SKZ) (mm/s) - vibration displacement (mm)	0.5–80 0.8–100 5.0–500
4	Eigen noise: - vibration acceleration (peak) (mm/s^2) - vibration velocity (SKZ) (mm/s) - vibration displacement (mm)	0.1 0.1 5.0
5	The limit of permissible basic relative error when measuring the general level of vibrations and for spectral analysis, not more than, %	±5
6	Frequency response in the frequency range 10–4000 Hz, not more than, %	±10
7	Frequency response at frequencies of 3 and 5000 Hz, not more than, %	±25

The eigenfrequencies (modal frequencies) and vibrations of the stand were measured at the same points (bearing units, mounting units of the stand elements, on the frame and foundation), which made it possible to conduct a frequency analysis and build a model of the frequency interaction of the stand elements of the measuring elements (springs) with the excitation frequencies initiated by the operation of the turbine installed for stand testing. As an example, to estimate the eigenfrequencies, Table 2 shows the values of the excitation and response frequencies by which the computational algorithm below was tested (the selected points No. 10–No. 14 are the key points at which the amplitude-frequency values of the oscillations at the turbine power measurement unit were obtained during the experiment).

In the framework of this task, a mathematical model and an automated computational algorithm for studying the dynamic behavior of a turbine as a part of a stand during turbine testing were developed, taking into account the frequency interactions of elements and nodes modal vibrations with the turbine's rotational frequencies when changing its operating modes when working on the stand [24,25].

Based on the analysis of the stand's elements, a range of initial and boundary conditions was selected for the process of mathematical modeling of dynamic changes in the turbine power during its operation, taking into account the interaction with the stand's elements [1,26]. Changes in the dynamics of the stand's structural elements' behaviors are also taken into consideration. The conditions for the occurrence of increased dynamic loads acting on the stand's measurement elements are modeled and determined, taking into account the frequency interactions when changing the turbine speed and measuring power under the conditions of wave interactions of the stand's elements and power measurement unit. Thus, the general task of research is to determine the modal frequencies of the stand, their changes when exposed to a load (power load, temperature) within the range of the measured power, and temperature changes of the gas medium used during the operation of the turbine (compressed air is used on the stand).

Table 2. Studies of the modal frequencies of the measuring base of the stand "spacers" (points No. 10–No. 14 on the stand).

No. of Measuring Points	Vibration Velocity			Vibration Acceleration		
	Vertical Direction	Horizontal Direction	Axial Direction	Vertical Direction	Horizontal Direction	Axial Direction
10 (hit)	98 342 537 732 1025	98 293 488 732 928	146 342 537 732 1221 1318 1660	146 342 537 732 1025 1318 1514 1660 1855	293 488 635 732 781 977 1172 1367 1611 1807 2490 2637 3711	146 342 537 732 1221 1318 1514 1660 1807 2002 2393 2637 2783 3027 3711
10 (response)	391 879 1270	293 977 1270	293	391 879 1270 1758 2441 2930	391 977 1270 2734	293 1074 1270 1758 2148 4004
11 (hit)	146 391 830	146 391 830 1270	98 391 830 1221	146 191 830	146 439 830 1270	98 391 830 1270
11 (response)	293 586	195 586	293 586	293 586	195 586	293 586
12 (hit)	195 488	195 586 781	293 781 1270 1563	586 781 977 1367 1660 1953	195 586 781 1270 1465 2441	488 781 1270 1563 1758 2539
12 (response)	98 488 879	146 293 488	147 342 488 781 1318	98 293 188 879	293 488 830 1123	146 342 488 781 879 1318
13 (hit)	195 391 879	195 781 1270	195 391 879	195 391 879 1270	195 488 781 1270	879 1367
13 (response)	293 586 879 1270	195 781 1270	293 684 1172 1563	293 586 879 1270 2246 2441	195 481 1270 1465 2930 4004	293 781 1270 1465 2930

Table 2. *Cont.*

No. of Measuring Points	Vibration Velocity			Vibration Acceleration		
	Vertical Direction	Horizontal Direction	Axial Direction	Vertical Direction	Horizontal Direction	Axial Direction
14 (hit)	195 488 781 1172 2148 3125	195 391 879 1660 2148	293 684 1270 1465	195 488 879 1172 2148 3223 3418	195 391 977 1660 2148 2441 2930 3223 3418	293 781 1270 1465 1458 2148 2930 3223
14 (response)	195 391 586 879 1172	391 684 1465 2441	195 488 879 1270	195 391 586 879 1172	391 684 1465 2441 2832	195 488 879 1270

In addition, the rotation of the turbine and the power measurement system creates a certain dynamic effect in the stand's elements and the stand's measuring system with a certain frequency and amplitude, i.e., a number of frequency harmonics, which lead to the formation of the wave field of the stand when testing the turbine.

The algorithm for determining the eigenfrequencies of the stand's elements is based on a mathematical model as a system of discrete masses with elastic-dissipative bonds.

A generalization of the design scheme is the introduction of elastic-dissipative bonds between the non-adjacent masses of the chain [27,28]. Figure 2 shows the stand system in the form of discrete masses with elastic-dissipative bonds. The following values are indicated in Figure 2: $x_1, \ldots, x_6$—directions of the possible movement (3 directions are considered); $R_1, \ldots, R_6$—viscosity forces between bodies; $G_1, \ldots, G_6$—inertia forces of bodies; $C_1, \ldots, C_6$—body stiffness forces; ζ (t)—exposure time.

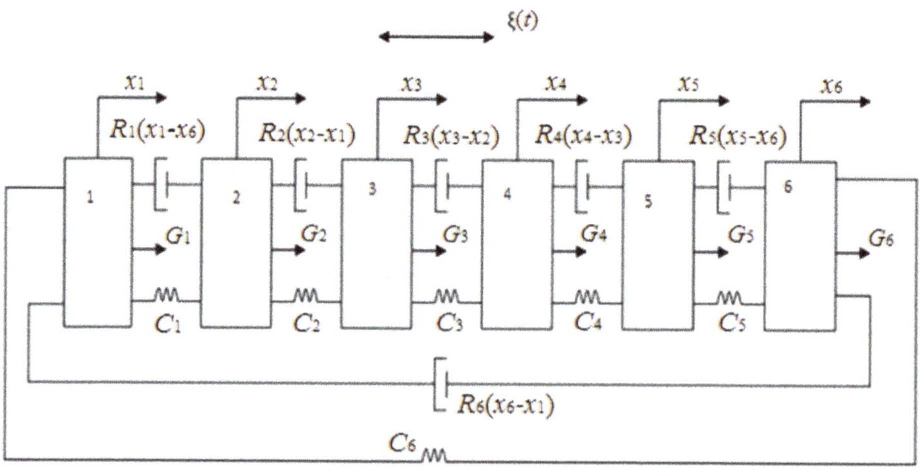

Figure 2. Stand model in the form of six discrete masses with elastic-dissipative bonds: 1—Turbine; 2—Spacer; 3—Measurement system; 4—Multiplier; 5—Generator; 6—Frame with foundation.

To determine the eigenfrequencies of the oscillations of the stand's elements with the turbine, as well as the eigenfrequency of the system itself, it is necessary to obtain matrix forms for the following parameters: generalized coordinates of displacements q, inertial forces F0(q) with inertial parameter A, and elastic forces with quasielastic coefficient C, without taking into account dissipative forces (no losses), which will reduce the amount of computation and obtain modal frequencies of structural elements.

The condition of free oscillations of the stand's elements system is determined from the equality of the kinetic and potential energies. For free oscillations, the kinetic and potential energies of the system are written accordingly:

$$\begin{aligned} T &= \tfrac{1}{2}\mathbf{A}\dot{\mathbf{q}}^2 \\ U &= \tfrac{1}{2}\mathbf{C}\mathbf{q}^2 \end{aligned} \tag{1}$$

where T and U are the kinetic and potential vibrational energies in the system; A is the matrix of inertial coefficients; **C** is the matrix of quasielastic coefficients; $\mathbf{q}$ = (ξ, η, ζ, α, β, and γ) is the vector characterizing the movement of the body during oscillations in the directions: ξ and η, which are the indices of longitudinal and torsional vibrations along the axis of rotation of the turbine; ζ, α are the indices of longitudinal and torsional vibrations in the vertical plane to the axis of rotation of the turbine; β, γ, and α are the indices of longitudinal and torsional vibrations in the horizontal plane to the axis of rotation of the turbine.

The differential equation of free oscillations in matrix form will have the form:

$$\mathbf{A}\ddot{\mathbf{q}} + \mathbf{C}\mathbf{q} = 0, \tag{2}$$

The modal (free) frequency f of longitudinal vibrations of any element of the stand, in any direction, as an elastic system is written:

$$\omega = \left(\frac{c}{a}\right)^{1/2} = \frac{1}{(af)^{1/2}}, \tag{3}$$

where f = s^{-1} is the unit compliance corresponding to the quasi-elastic coefficient c; a is the size of the element in the direction of propagation of the elastic wave.

For the stand considered as a system of discrete masses of its elements connected in a series with a common frame with a foundation (shown in Figure 2), the equations of eigenfrequencies will be determined by the following matrix:

$$\begin{pmatrix} 1-\omega^2\sum\limits_{k=1}^{n} f_{1k}A_{k1} & -\omega^2\sum\limits_{k=1}^{n} f_{1k}A_{k2} & \cdots & -\omega^2\sum\limits_{k=1}^{n} f_{1k}A_{k6} \\ -\omega^2\sum\limits_{k=1}^{n} f_{2k}A_{k1} & 1-\omega^2\sum\limits_{k=1}^{n} f_{2k}A_{k2} & \cdots & -\omega^2\sum\limits_{k=1}^{n} f_{2k}A_{k6} \\ \cdots & \cdots & \cdots & \cdots \\ -\omega^2\sum\limits_{k=1}^{n} f_{6k}A_{k1} & -\omega^2\sum\limits_{k=1}^{n} f_{6k}A_{k2} & \cdots & 1-\omega^2\sum\limits_{k=1}^{n} f_{6k}A_{k6} \end{pmatrix} \tag{4}$$

where k is the number of the body in the stand scheme (according to Figure 2); n is the number of stand elements.

The general solution of the equation of free (modal) oscillations with the eigenfrequency ω_a and the natural form v_a can be written in the form:

$$q(t) = \sum_{a=1}^{n} v_a(C_a \cos\omega_a t + D_a \sin\omega_a t), \tag{5}$$

where C_a and D_a are the oscillation amplitudes upon expansion in a Fourier series.

The solution to the differential equation of free oscillations is the expression:

$$q(t) = v\sin(\omega t + \chi), \tag{6}$$

where v is a vector (column matrix) that characterizes the relationship between the generalized coordinates, and these are also eigenmodes. χ is the phase angle of the shift between them, while ω and v must satisfy the matrix expression:

$$(\mathbf{C} - \omega^2\mathbf{A})v = 0. \tag{7}$$

The solution of the differential equation of free vibrations has the form:

$$\begin{pmatrix} C_{11}-\omega^2A_{11} & C_{12}-\omega^2A_{12} & C_{13}-\omega^2A_{13} & C_{14}-\omega^2A_{14} & C_{15}-\omega^2A_{15} & C_{16}-\omega^2A_{16} \\ C_{21}-\omega^2A_{21} & C_{22}-\omega^2A_{22} & C_{23}-\omega^2A_{23} & C_{24}-\omega^2A_{24} & C_{25}-\omega^2A_{251} & C_{26}-\omega^2A_{26} \\ C_{31}-\omega^2A_{31} & C_{32}-\omega^2A_{32} & C_{33}-\omega^2A_{33} & C_{34}-\omega^2A_{34} & C_{35}-\omega^2A_{35} & C_{36}-\omega^2A_{36} \\ C_{41}-\omega^2A_{41} & C_{42}-\omega^2A_{42} & C_{43}-\omega^2A_{43} & C_{44}-\omega^2A_{44} & C_{45}-\omega^2A_{45} & C_{46}-\omega^2A_{46} \\ C_{51}-\omega^2A_{51} & C_{52}-\omega^2A_{52} & C_{53}-\omega^2A_{53} & C_{54}-\omega^2A_{54} & C_{55}-\omega^2A_{55} & C_{56}-\omega^2A_{56} \\ C_{61}-\omega^2A_{61} & C_{62}-\omega^2A_{62} & C_{63}-\omega^2A_{63} & C_{64}-\omega^2A_{64} & C_{65}-\omega^2A_{65} & C_{66}-\omega^2A_{66} \end{pmatrix} \tag{8}$$

where the serial number of matrix elements corresponds to the number of stand elements, according to Figure 2.

To simplify the solution of this complex system of equations, it is necessary to divide it into six subsystems (partial systems). For example, Formula (9) considers only one coordinate of the stand's wave processes formation and makes it possible to determine the eigenfrequencies of the stand's elements in a given direction. For this, it is necessary to find the partial frequencies of each element of the system. These frequencies will coincide with the eigenfrequencies of the stand's elements system at: A12 = A21 = 0; C12 = C21 = 0:

$$\begin{cases} \omega_I = \left(\frac{C_{11}}{A_{11}}\right)^{1/2} \\ \omega_{II} = \left(\frac{C_{22}}{A_{22}}\right)^{1/2} \\ \omega_{III} = \left(\frac{C_{33}}{A_{33}}\right)^{1/2} \\ \omega_{IV} = \left(\frac{C_{44}}{A_{44}}\right)^{1/2} \\ \omega_{V} = \left(\frac{C_{55}}{A_{55}}\right)^{1/2} \\ \omega_{VI} = \left(\frac{C_{66}}{A_{66}}\right)^{1/2} \end{cases} \tag{9}$$

where ω_I, ω_{II}, ω_{III}, ω_{IV}, ω_V, and ω_{VI} are the partial frequencies of the components of the stand's element base, respectively. Using this method, an algorithm was developed to find the eigenfrequencies of the stand's elements by analyzing the four subsystems and approximating the change in stiffness from the action of the twist torque of the measuring spring. That allows us to relate the obtained results of the vibration level and frequency implementations with the conditions of the dynamic interaction of the element base of the stand with the results of measuring the turbine power when changing the range of its rotations.

The Equations (1)–(8) show an example of a specific implementation of the theory of linear vibrations as applied to a real design of the stand for a one-dimensional model in order to determine the eigenfrequencies of the elements of a coupled system. The whole algorithm for calculating the eigenfrequencies assumes their determination in two more directions, which form their own model of the stand for a specific direction with discrete masses, as in the example, but with different elastic-dissipative bonds. The calculation algorithm is formed based on generalized coordinates of the three-dimensional model of the stand, which requires taking the nonlinear spatial-correlation relationships into account, therefore, a new calculation method.

The conditions for the dynamic interaction of the turbine block with the elements of the measuring system (the spacer where the spring is installed to measure the realized torque of the turbine) were identified. The main emphasis in calculating the eigenfrequencies according to this algorithm was associated with the analysis of their changes when the external load changes. Therefore, the stress changes in the connecting nodes of the stand's elements, which leads to a change in the stiffness of their joints and the eigenfrequency (modal frequency) of the structure under consideration and partial frequencies.

With an increase in the external load from the turbine, the partial frequencies also change previously linearly with increasing stresses in the connecting elements. In this case, the algorithm for calculating step supply (s^{-1}) will be determined from Hooke's linear law, connecting strains and stresses through the elastic modulus of the material of the connecting elements, which can significantly simplify the algorithm for calculating the modal frequencies of each of the stand elements, according to Figure 2, for any range of changes in the number of rotations of the turbine, and, consequently, its power.

3. Results

The results of the frequency analysis of the modal characteristics of the stand were verified according to the results of the modal analysis performed on the stand. For example, Figure 3 shows a general view of the multiplier with measuring points; Figures 4 and 5 show the spacer of the stand (element No. 2, according to Figure 2) with measuring points (structural elements for installing the turbine power measurement unit).

Figure 3. General view of the multiplier with measurement points.

Figure 4. General view of the spacer with measuring points.

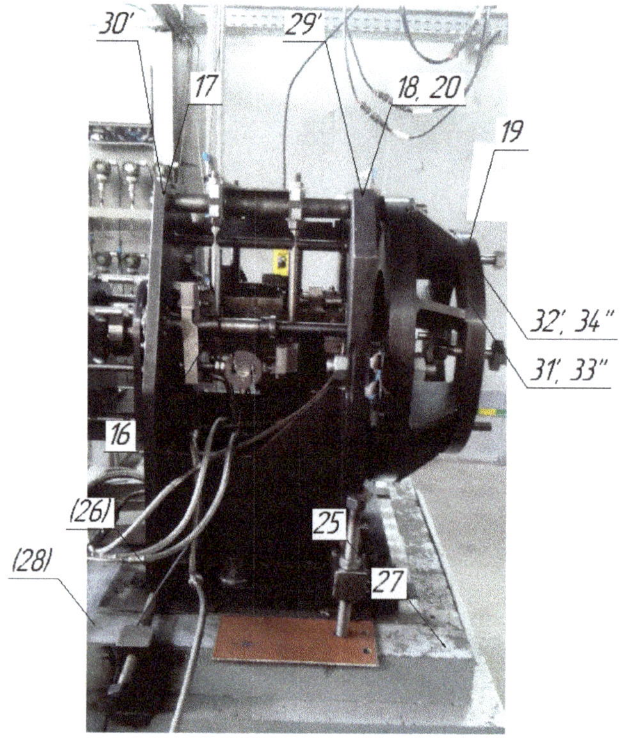

Figure 5. General view of the spacer with frame.

A comparison of the results of the modal analysis of the unloaded stand conducted on its element base (Figure 2) and fitting the above algorithm to the experimental frequencies made it possible to refine the matrix of quasielastic coefficients C at zero load (unit compliance was determined from the tightening conditions of the mounting bolts of the spacer with the turbine block and unit of the turbine torque measurement system). This made it possible to predict a change in the matrix of quasielastic elements depending on the load realized on the stand's loading device from the generator through the multiplier, directly on the measuring spring used as a turbine power sensor.

Knowing the range of the percentage change in the spring stiffness coefficient (the eigenfrequencies of the spring grow almost linearly from the value of the applied torque since the stress, and the value of the spring angle of the spring increase linearly because the measuring element operates in the elastic zone of deformation—twisting, and, therefore, its torsional eigenfrequencies, i.e., oscillations, occur in the elastic strain zone), it can be assumed that the eigenfrequencies of the stand's elements change with the same linear dependence, according to the realized turbine power (Hooke's linear law).

Then, this algorithm allows us to connect the changing eigenfrequencies of the spring with the harmonic frequencies when the turbine rotates, and its speed and torque are realized by the turbine through the loading device, taking into account measurements of the vibration level on non-rotating stand elements (stator) through the transfer function from the spring to the vibration measurement point on the housing of the stand's elements.

Consequently, it becomes possible to clarify the actual power of the turbine due to the frequency interaction of the changing modal frequencies of the spring with the rotational frequencies of the turbine, which forms distinctive indications of the power of the turbine during the turbine research with the calculated ones. The magnitude of the discrepancy in the measurement of the torque when determining the power of the turbine will be determined through the coefficient of dynamism in the frequency interaction of eigenfrequencies and forced frequencies (frequencies associated with the number of rotations of the turbine):

$$\eta = \frac{1}{\sqrt{\left[1 - \left(\frac{\omega_k}{\omega_h}\right)^2\right]^2 + 4\xi_k \frac{\omega_k^2}{\omega_h^2}}}, \tag{10}$$

where ω_k is the spring eigenfrequency; ω_h is the reverse (harmonic) frequency of the gas of rotation of the turbine; ξ_k is the coefficient determining losses in the spring material. In this case, it is necessary to find frequencies close to each other in the frequency domain:

$$0.85 \geq \frac{\omega_k n}{\omega_h m} \leq 1.15, \tag{11}$$

where n and m are integers from 1 to 8 (eighth mode). Sorting frequencies above 8 harmonics does not make sense since the energy of the frequency interaction at high frequencies is insignificant; however, even if the frequency difference is more than 15%, the effect on the error of the frequency interaction also gives a slight error.

4. Analysis of the Results

Figures 6 and 7 show the vibration acceleration spectrum at point 16 (the bearing assembly of the spring of the turbine torque measuring system) at different turbine rotations. An analysis of the results of vibration acceleration makes it possible to evaluate the effect of the inertial force on the oscillatory system, which is realized at the stand's measurement point, and to track its change when the speed of the turbine changes, and, consequently, its power. In addition to torsional vibrations, bending and longitudinal arises in the spring that certain instantaneous precession values are formed on the bearing of the spacer (measuring unit), which form a vibration field on the fixed part of the spacer in the bearing installation area. Measurement of these vibrations is carried out in three directions. Bending and torsional vibrations create a displacement of the axis of rotation (precession) relative to the geometric axis of rotation of the spring. The spatial addition of eigen vibrations and forced vibrations allows us to build the trajectory of the instantaneous center of the axis of rotation of the spring. The analysis, in this case, can be carried out by comparing one of the three types of occurrence of vibrations of an element (spring) in the bearing assembly.

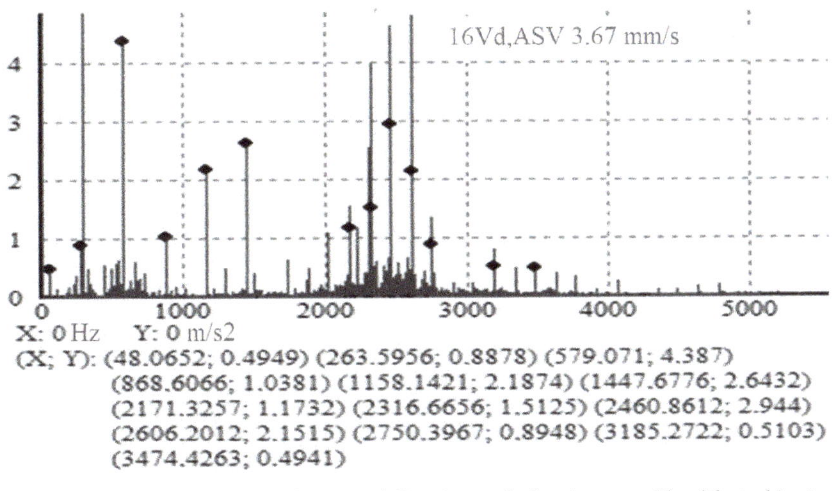

Figure 6. Vibration accelerations spectrum in the vertical direction on the bearing assembly of the turbine torque measurement system's spring at a speed of 8500 rpm.

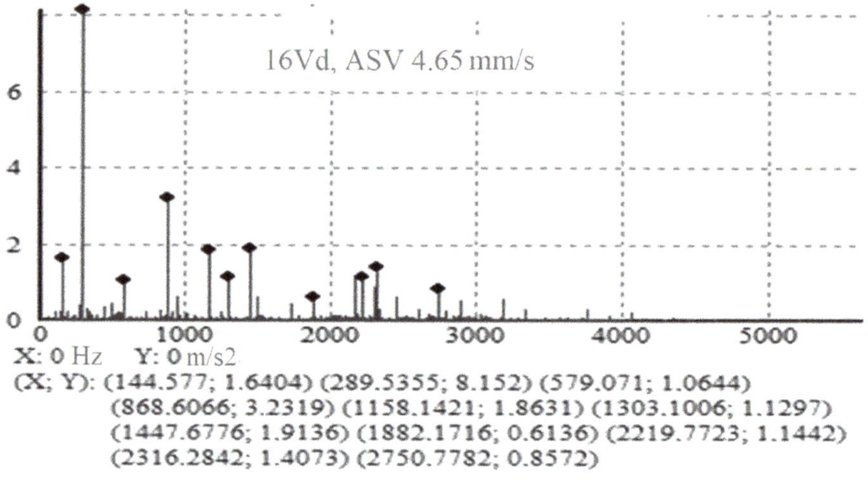

Figure 7. Vibration accelerations spectrum in the vertical direction on the bearing assembly of the turbine torque measurement system's spring at a speed of 12,000 rpm.

For example, the eigenfrequency of spring-bending vibrations is 579 Hz, and the harmonic frequency is 141.6 Hz (at turbine speed of 8500 rpm), which is close to the fourth harmonic of the external influence on the spring (4th harmonic is 566.5 Hz). Therefore, taking into account the possibility of frequency interaction between the torsion eigenfrequencies of the spring and the rotating frequencies, the amplitude of the partial frequency at the measuring point (bearing) on the spacer increases to 4.378 $\mu m/s^2$. The amplitude square value (ASV) is 3.67 mm/s.

Increasing the turbine speed to 12,000 rpm (Figure 7), which corresponds to a frequency of dynamic action of 200 Hz, the conditions for the frequency interaction of spring vibrations change. It leads to a decrease in the amplitude of vibration acceleration on the spacer bearing at a frequency that is a multiple of the eigenfrequency of the spring bending vibrations up to 1.054 $\mu m/s^2$. If the amplitude of oscillations in vibration velocity at a frequency of 141.6 Hz at rotations of 8500 rpm was 0.163 mm/s. With an increase in the number of rotations to 12,000 rpm, causing an increase in turbine power at a frequency of 200 Hz, the amplitude in vibration velocity increased to 0.323 mm/s. The ASV is 4.65 mm/s.

Consequently, the energy of oscillatory processes in the stand has grown, but the frequency palette of its wave field has changed. This confirms the wave restructuring of the oscillatory processes of the stand's elements, i.e., the amplitude-frequency characteristics change in the process of changing the turbine operation parameters (visual proof when comparing the vibration acceleration spectra shown in Figures 6 and 7). Such changes in the frequency-amplitude field of the stand are characteristic for all directions of vibration measurement (measurement was carried out in the horizontal and vertical planes and the axial direction). Simultaneous processing of the vibration signal in three directions in time allows us to get the trajectory of the precession of the axis of rotation of the spring when measuring the torque and power of the turbine when changing its speed. Therefore, by analyzing one of the components of the vibration field, it is possible to qualitatively assess the change in other parameters of the manifestation of vibrations, including the change in torsional vibrations of the spring, which form the torque measurement at this moment in time.

A comparison of the frequency spectra when the turbine rotational speed changes on the spacer's bearing units shows that the stand's structural elements change the energy of the spacer vibrations quite significantly; therefore, the dynamic interaction between the measuring system of the stand and the spring vibrations, which can significantly affect the accuracy of measuring the turbine power, not only during transients but also in the steady-state during the formation of a new level of self-oscillation mode of the turbine's torsional vibrations and changing the spring's modal vibrations.

If the discrepancy between the results of measuring the turbine power at 8500 rpm was 18%, then taking into account the frequency interaction according to Formula (10) for this mode, it is 4.3%, respectively. The discrepancy between the measured turbine power at 12,000 rpm and its calculated value is 15% for these speeds, respectively, and taking into account the conversion of 5.6%. Thus, the influence of frequency interaction as applied to a particular stand design allows one to take into account its influence on the results of the actual value of the measured turbine power.

When using the algorithm for calculating the eigenvibrations of an element of this stand with a change in the mass-rigid characteristics of its elements (in particular, changes in the design of the measuring unit (spacers) cause an increase in the mass of the bearing units by 10%), the calculated value of the measured torque, taking into account the influence of frequency interaction, decreased to 3%.

The results of evaluating the frequency indicators of the installation elements based on the measurement data at the corresponding points are shown in Figures 8–11.

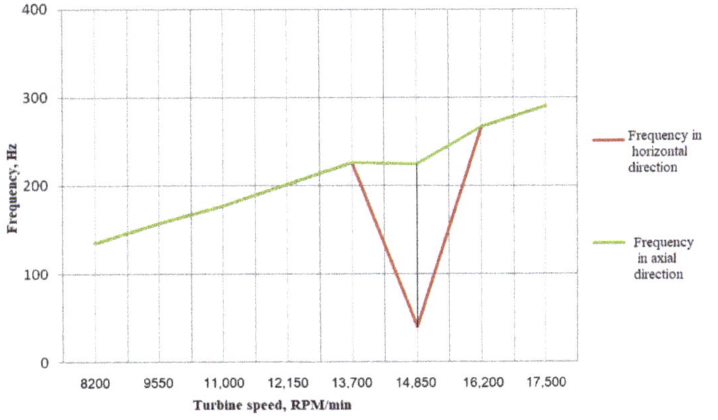

Figure 8. Change in the forced frequency on the electric motor, depending on the frequency of rotation of the turbine (measuring point No. 1).

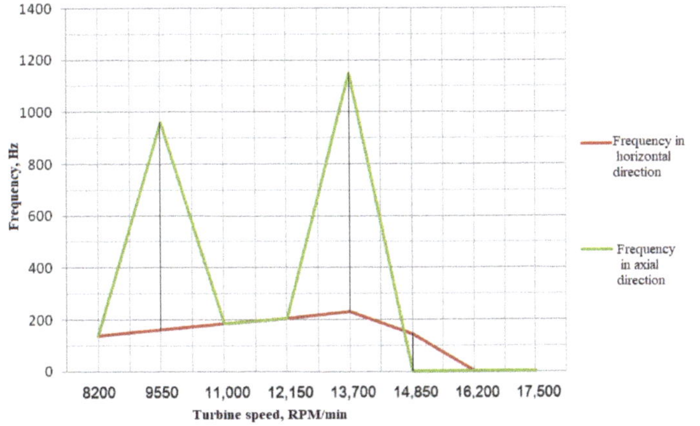

Figure 9. Change in the forced frequency on the spacer depending on the speed of the turbine (measuring points No. 11 and No. 13).

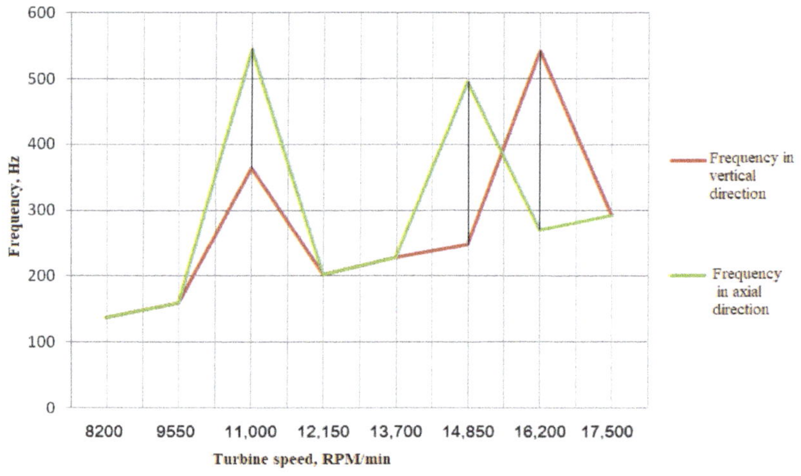

Figure 10. Change in the forced frequency on the spacer depending on the speed of the turbine (measuring point No. 17).

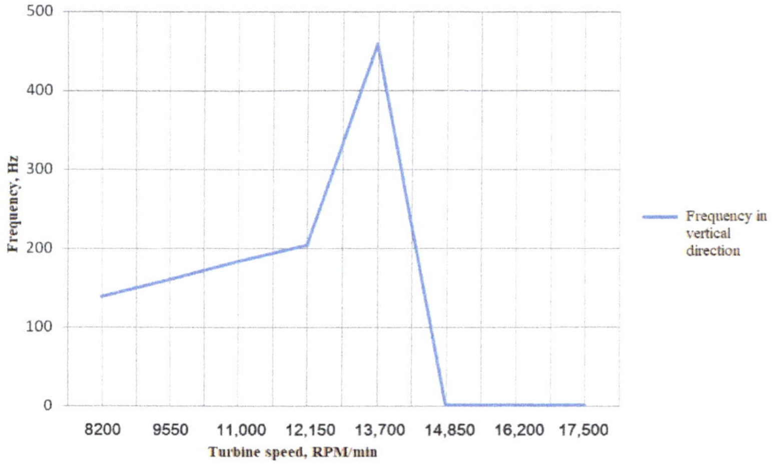

Figure 11. Change in the forced frequency on the spacer frame depending on the speed of the turbine (measuring point No. 28).

An analysis of the data in Figure 8 shows that the axial component in frequency increases with the increasing speed of the turbine. Significant damping by the stand system is observed in the range of shell frequencies from 14,000 to 16,000 rpm.

At a frequency of 8000 to 11,000 rpm, the frame structure does not provide a rigid fixation of the multiplier. A significant increase in frequency at rotations of 12,000 to 15,000 rpm is associated with the conditions for the dynamic interaction of the coupling joints of the multiplier with the spacer measuring element (shaft eigenfrequencies lie in the range of rotating frequencies).

On the spacer at measuring points No. 11, No. 13, and No. 17, the same dynamic character of the shaft with the measuring system is noted (Figures 9 and 10).

The spacer frame in the vertical direction slightly reduces the level of vibration. At 13,700 rpm, the measurement cuts off due to the high level of vibration at the measurement point (Figure 11).

5. Discussion

The conducted frequency analysis, taking into account the change in the eigenfrequencies of the structural elements of the stand, evaluates the actual value when measuring the power of the tested turbine in a certain range of working frequencies. Additionally, the analysis evaluates the structure of the dynamic interaction of the stand's elements and, having determined the possible frequency ranges of the interaction of eigenfrequencies and working frequencies with the help of certain upgrades, significantly reduces the discrepancy between the measured power and the actual (calculated) one; thereby, solving the problem of controlling the operation of the turbine as part of a power plant. In addition, the frequency analysis allows us to solve the problem of increasing the reliability and increasing the resource of using stand equipment.

6. Conclusions

A mathematical apparatus based on point masses has been developed for a stand to verify the operability of turbines of various powers in power plants. Using the procedure for studying the modal characteristics of stand equipment, we obtained the correct quasielastic coefficients for calculating the eigenfrequencies of the nodes and stand's elements and determined their variation in the load value acting on the structure when testing the turbine in various modes without taking into account the influence of temperature changes during testing.

The conditions for amplifying the amplitude of oscillations at eigenfrequencies are obtained, and the value of the possible discrepancy between the measured power results and the calculated (expected) results when measuring power for specific modes (cycles) of turbine operation is calculated. It is shown that taking into account the frequency interaction between the eigenfrequencies and rotating frequencies of the power measurement elements reduces the difference in the power measurement from 15% in the entire range of the studied turbine rotations to an acceptable maximum value of 3%.

The developed mathematical algorithm makes it possible to assess the dynamic state of the stand at different turbine research modes and give recommendations for reducing the level of frequency interaction; therefore, reducing dynamic loads on the stand base, increasing the reliability of the stand and, as a result, increasing the duration of its use without attracting additional material costs.

Author Contributions: Conceptualization, A.P. and A.S.; methodology, A.S.; software, A.S.; validation, S.B. and P.I.; formal analysis, A.P., S.B. and A.S.; writing—original draft preparation, A.P.; writing—review and editing, A.P. and A.S.; visualization, S.B.; supervision, A.P. and A.S. All authors have read and agreed to the published version of the manuscript.

Funding: This research was carried out with financial support from the Russian Ministry of Education and Science in the framework of scientific projects carried out by teams of research centers and scientific laboratories of educational institutions of higher education, subordinate to the Ministry of Education and Science of Russia (project 8.4157.2017/4.6/3305 "Development, design and manufacture of a prototype universal stand for model tests of a turbine of promising engines"). This research also was carried out with the financial support of the Ministry of Science and Higher Education of the Russian Federation in the framework of the program of activities of the Perm Scientific and Educational Center "Rational Subsoil Use."

Institutional Review Board Statement: Not applicable.

Informed Consent Statement: Not applicable.

Conflicts of Interest: The authors declare no conflict of interest.

References

1. Tarasov, V.A.; Petrochenkov, A.B.; Kavalerov, B.V. A computer-aided facility foreigen testing gas-turbine power stations. *Russ. Electr. Eng.* **2017**, *88*, 750–754. [CrossRef]
2. Cheng, Z.; Steven, Y.L.; Jing, Q. Bearing Failure Prognostic Model Based on Damage Mechanics and Vibration Monitoring. *Tribol. Trans.* **2001**, *44*, 603–608.

3. Zheng, L.-A.; Chen, S.-H.; Chou, J.-H. Robust Frequency-Shaping Optimal Active Vibration Control of Uncertain Flexible Mechanical Systems with Persistent Excitation. *Chin. J. Mech.* **2002**, *18*, 163–171. [CrossRef]
4. Byoung-Hoo, R.; Kyung-Woong, A.K. Study on Nonlinear Frequency Response Analysis of Hydrodynamic Journal Bearings with External Disturbances. *Tribol. Trans.* **2002**, *45*, 117–121.
5. Santiago, O.D.; Andrés, L.S. Experimental Identification of Bearing Dynamic Force Coefficients in a Flexible Rotor—Further Developments. *Tribol. Trans.* **2007**, *50*, 114–126. [CrossRef]
6. Kirk, R.G.; Alsaeed, A.; Liptrap, J.; Lindsey, C.; Sutherland, D.; Dillon, B.; Saunders, E.; Chappell, M.; Nawshin, S.; Christian, E.; et al. Experimental Test Results for Vibration of a High Speed Diesel Engine Turbocharger. *Tribol. Trans.* **2008**, *51*, 422–427. [CrossRef]
7. Moon, F.C. Reflections on Nonlinear Dynamics of Machines and Structures. *J. Mech.* **2000**, *16*, 79–83. [CrossRef]
8. Lee, Y.-B.; Kwak, Y.-S.; Chung, J.T.; Sim, K. Microturbocharger with Air Foil Bearings for a 100-W Class Micro Power System and Improvement of Rotordynamic Performance. *Tribol. Trans.* **2011**, *54*, 939–948. [CrossRef]
9. Kim, D.; Varrey, M.K. Imbalance Response and Stability Characteristics of a Rotor Supported by Hybrid Air Foil Bearings. *Tribol. Trans.* **2012**, *55*, 529–538. [CrossRef]
10. Petrochenkov, A.B.; Bochkarev, S.V.; Romodin, A.V.; Eltyshev, D.K. The Planning Operation Process of Electrotechnical Equipment Using the Markov Process. *Russ. Electr. Eng.* **2011**, *82*, 592–595. [CrossRef]
11. Anuchin, A.; Podzorova, V.; Hirz, M.; Datlinger, C.; Aarniovuori, L.; Popova, V.; Demidova, G.; Hao, C. Model predictive control with reduced integration step size of a switched reluctance drive. In Proceedings of the 2020 11th International Conference on Electrical Power Drive Systems, ICEPDS 2020, Saint Petersburg, Russia, 4 October 2020; p. 9249077. [CrossRef]
12. Vattulainen, J.; Nummela, V.; Hernberg, R.; Kytl, J. A system for quantitative imaging diagnostics and its application to pyrometric in-cylinder flame-temperature measurements in large diesel engines. *Meas. Sci. Technol.* **2000**, *11*, 103–119. [CrossRef]
13. Paimushin, V.N.; Firsov, V.A.; Shishkin, V.M. Complex eigenfrequencies and damping properties of an elongated plate with an integral damping coating. *J. Appl. Mech. Tech. Phys.* **2020**, *61*, 599–610. [CrossRef]
14. Petrochenkov, A.B.; Bochkarev, S.V.; Ovsyannikov, M.V.; Bukhanov, S.A. Construction of an ontological model of the life cycle of electrotechnical equipment. *Russ. Electr. Eng.* **2015**, *86*, 320–325. [CrossRef]
15. Kordjazi, A.; Coe, J.T.; Afanasiev, M. Nondestructive Evaluation of Drilled Shaft Construction Anomalies Using Full Waveform Tomography of Simulated Crosshole Measurements. *J. Nondestruct. Evaluation* **2021**, *40*, 1–13. [CrossRef]
16. Markelov, M.K.; Churikov, P.P. Ustrojstvo s vihretokovym preobrazovatelem dlya kontrolya parametrov vibracii. *Izmer. Tekhnika* **2013**, *12*, 40–43. (In Russian)
17. Yoshioka, T.; Shimizu, S. Monitoring of Ball Bearing Operation under Grease Lubrication Using a New Compound Diagnostic System Detecting Vibration and Acoustic Emission. *Tribol. Trans.* **2009**, *52*, 725–730. [CrossRef]
18. Khalyutkin, V.A.; Alekseenko, V.A.; Ulyanov, A.P.; Krasov, P.S. Device for Measurement of Torque, Shaft Speed and Power on Shaft. Patent Russian Federation No. 2585482, 20 March 2015.
19. He, J.; Fu, Z.-F. *Modal Analysis*; Butterworth-Heinemann: Oxford, UK, 2001; p. 291.
20. Coolen, F.P.A.; Utkin, L.V. Imprecise reliability: A concise overview. In Proceedings of the European Safety and Reliability Conference 2007, ESREL 2007 Risk, Reliability and Societal Safety, Stavanger, Norway, 25 June 2007; pp. 1959–1966.
21. Crystal Instruments. *Basics of Structural Vibration Testing and Analysis*; Crystal Instruments: Santa Clara, CA, USA, 2016. Available online: https://www.crystalinstruments.com/basics-of-structural-vibration-testing-and-analysis (accessed on 31 March 2021).
22. Vibration Analyzer Kamerton. Available online: http://ros-diagnostics.ru/component/content/article/85.html (accessed on 31 March 2021).
23. ISO 8042:1988. *Shock and Vibration Measurements—Characteristics to be Specified for Seismic Pick-Ups*; International Organization for Standardization (ISO): Geneva, Switzerland, 1988.
24. Petrova, E.N.; Salnikov, A.F. Gas-dynamic modeling of gas flow in semi-closed space including channel surface fluctuation. *AIP Conf. Proc.* **2016**, *1770*, 30071.
25. Bochkarev, S.V.; Ovsyannikov, M.V.; Petrochenkov, A.B.; Bukhanov, S.A. Structural synthesis of complex electrotechnical equipment on the basis of the constraint satisfaction method. *Russ. Electr. Eng.* **2015**, *86*, 362–366. [CrossRef]
26. Permyakov, K.V.; Sal'nikov, A.F. Exhaust-line performance in a gas-pumping system. *Russ. Eng. Res.* **2014**, *34*, 557–559. [CrossRef]
27. Mormul, R.V.; Pavlov, D.A.; Sal'nikov, A.F. Vibration-Based Diagnostics and Structural Analysis of Research on Nonlinear Dynamic Processes in Maglev-Suspension Rotor Turbomachines. *Russ. J. Nondestruct. Test.* **2018**, *54*, 487–494. [CrossRef]
28. Sal'nikov, A.F. Diagnostika tekhnicheskogo sostoyaniya krupnogabaritnyh detalej gazoturbinnyh dvigatelej po modal'nym harakteristikam. *Vestn. Permsk. Nation. Issledovatel'skogo Politekh. Universiteta. Aerokosmicheskaya Tekhnika* **2018**, *55*, 61–69. (In Russian)

applied sciences

MDPI

Article

Degradation State Recognition of Piston Pump Based on ICEEMDAN and XGBoost

Rui Guo [1,2,*], Zhiqian Zhao [1,3], Tao Wang [1,4], Guangheng Liu [1,4], Jingyi Zhao [2,4] and Dianrong Gao [3]

1 Hebei Provincial Key Laboratory of Heavy Machinery Fluid Power Transmission and Control, Yanshan University, Qinhuangdao 066004, China; zzq623310996@163.com (Z.Z.); m13237195956@163.com (T.W.); lgh18733407119@163.com (G.L.)
2 The State Key Laboratory of Fluid Power and Mechatronic Systems, Zhejiang University, Hangzhou 310027, China; zjy@ysu.edu.cn
3 Key Laboratory of Advanced Forging and Stamping Technology and Science, Yanshan University, Qinhuangdao 066004, China; gaodr@ysu.edu.cn
4 Hebei Key Laboratory of Special Delivery Equipment Yanshan University, Qinhuangdao 066004, China
* Correspondence: guorui@ysu.edu.cn

Received: 15 August 2020; Accepted: 17 September 2020; Published: 21 September 2020

Abstract: Under different degradation conditions, the complexity of natural oscillation of the piston pump will change. Given the difference of the characteristic values of the vibration signal under different degradation states, this paper presents a degradation state recognition method based on improved complete ensemble empirical mode decomposition with adaptive noise (ICEEMDAN) and eXtreme gradient boosting (XGBoost) to improve the accuracy of state recognition. Firstly, ICEEMDAN is proposed to alleviate the mode mixing phenomenon, which decomposes the vibration signal and obtain the intrinsic mode functions (IMFs) with less noise and more physical meaning, and subsequently the optimal IMF is found by using the correlation coefficient method. Then, the time domain, frequency domain, and entropy of the effective IMF are calculated, and the new characteristic values which can represent the degradation state are selected by principal component analysis (PCA) that it realizes dimension reduction. Finally, the above-mentioned characteristic indexes are used as the input of the XGBoost algorithm to achieve the recognition of the degradation state. In this paper, the vibration signals of four different degradation states are generated and analyzed through the piston pump slipper degradation experiment. By comparing the proposed method with different state recognition algorithms, it can be seen that the method based on ICEEMDAN and XGBoost is accurate and efficient, the average accuracy rate can reach more than 99%. Therefore, this method can more accurately describe the degradation state of the piston pump and has a highly practical application value.

Keywords: piston pump; degraded state recognition; slipper; improved complete ensemble empirical mode decomposition with adaptive noise; principal component analysis; eXtreme gradient boosting

1. Introduction

As the power source, the piston pump affects the function realization of the whole hydraulic system. According to the performance degradation of the piston pump so far, the degradation state recognition is carried out to determine the performance of the piston pump and the whole hydraulic system [1], which can provide decision-making information for condition-based maintenance (CBM), prognostics, and health management (PHM) [2]. Due to the influence of oil, temperature, load, and other factors [3,4], the components of piston pump will deteriorate because of wear [5], and this degradation will eventually lead to changes in the main performance indicators of the hydraulic pump.

The sealing belt on the bottom surface of the slipper will deform under the working condition, which will cause the inner edge to bulge. At the same time, the oil film thickness between the slipper and the swash plate will become thinner [6]. If there are pollution particles in the hydraulic oil, the inner edge of the sealing belt is easy to be worn [7]. Therefore, in actual production, the inner edge of the slipper sealing belt is most seriously worn, which will directly lead to the performance degradation of the pump, as shown in Figure 1. Therefore, the wear of the inner edge of slipper should be considered as the main degradation form when evaluating the degradation state of piston pump. According to the different wear states of slipper, the vibration signals of piston pump shell are different in multi-directional and multi-dimensional features, so how to extract effective degradation index from them is the key to state recognition.

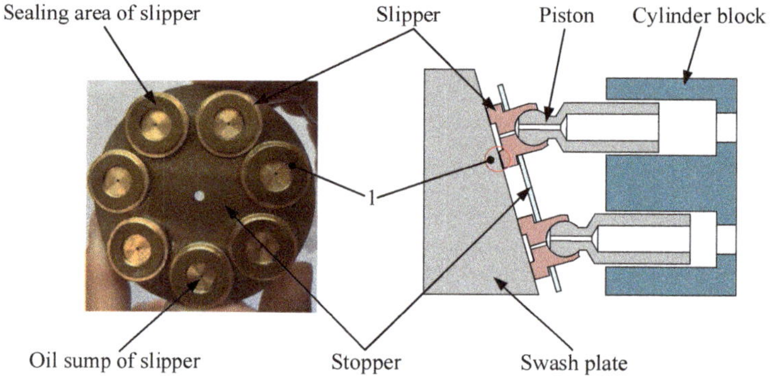

Figure 1. Structure of slipper pair. 1 is the inner edge of the slipper.

The vibration signal is usually transient and periodic pulse behavior, which contains critical information about the status of the mechanical equipment. Scholars have carried out a lot of research and analysis of the vibration signal of mechanical equipment to extract effective information. Tian et al. [8] used and improved the multi-fractal detrended fluctuation analysis (MF-DFA) to extract the performance degradation characteristics of the hydraulic pump vibration signal, which improved the accuracy of the degradation state recognition, but did not process the vibration signal noise of the pump. Xiao et al. [9] used wavelet packet transform to analyze the bearing vibration signal, and extracted the node energy and its total energy as features. The method has a certain effect on white noise suppression, but the ability to suppress signal pulse interference is not strong. Singh et al. [10] applies pseudo-fault signal (PFS) assisted empirical mode decomposition (EMD) on the envelope, which solves the problem that different fault frequencies are not obvious due to mode aliasing, but the decomposition efficiency of the algorithm is low. Lei et al. [11] proposed an improved Hilbert–Huang transform based on ensemble empirical mode decomposition (EEMD) and sensitive mode IMFs, and achieved good results to a certain extent. For EEMD [12,13], the method reduces the effect of mode aliasing by adding auxiliary white noise; however, different implementation of signal plus noise may produce a different number of modes, making the final averaging difficult [14]. To solve this problem, Torres et al. [15] proposed complete ensemble empirical mode decomposition with adaptive noise (CEEMDAN), which added adaptive white noise at each stage of decomposition, and then obtained the modal components of each layer by calculating the unique residual signal.

For the degradation state recognition, Tian et al. [16] judged the degradation state of hydraulic pump by calculating the Jensen–Renyi Divergence (JRD) distance between different characteristic variables. For the fuzziness of the boundary of different degradation states, Wang et al. [17] used the fuzzy c-means (FCM) method to cluster and identified the degradation state of bearing performance according to the maximum membership degree law, the overall recognition rate reached 96%. Zhang et al. [18] combined neighborhood preserving embedding (NPE) with Self-organizing Map

(SOM) in bearing degradation state recognition, and achieved good results. As classical classifiers, Support Vector Machine (SVM) [19,20] and Artificial Neural Network (ANN) [21,22] have been widely used in the field of mechanical fault diagnosis and condition recognition. For instance, an incomplete wavelet packet analysis (WPA) model composed of five-level discrete wavelet transform (DWT) and four-level WPA was established in the literature [23], and it was applied to multi-layered ANN engine failure classification. In Ref. [24], Laplace scoring algorithm is introduced to automatically select sensitive features according to the the importance of each feature, and the multi-state recognition of rolling bearing is realized based on particle swarm optimization-based support vector machine (PSO-SVM). However, SVM also has some shortcomings, that is, after selecting the appropriate kernel function, the quadratic programming of the function is required when dealing with the classification problem [25]. In this process, a lot of storage space is needed. Tree-based ensemble learning is an extensive and efficient method [26–28]. However, the traditional method is the model based on serial structure, which will inevitably increase the computational complexity and time [29]. XGBoost [30], as a new tree-based ensemble learning algorithm, can process data in parallel. It has minimal specification requirements for characteristic values, which can intelligently handle missing data and avoid overfitting [31]. Compared with SVM, the prediction accuracy is higher under relatively less parameter adjustment time [32]. Compared with deep learning, XGBoost is easier to classify small data sets [33].

In this paper, a new method for degradation state recognition of piston pump based on a further optimized decomposition method–improved complete ensemble empirical mode decomposition with adaptive noise (ICEEMDAN) [34–36] and XGBoost is proposed. Firstly, ICEEMDAN is used to decompose the vibration signal to generate a series of IMFs. This method greatly suppresses the illusive components and mode mixing caused by the initial decomposition process, and has better decomposition results. According to the correlation coefficient method, the IMFs which have a great correlation with the original signal is extracted, and the different characteristic values of IMFs are analyzed and selected. Finally, the new characteristic values with low coincidence degree are selected by PCA as the input, and the XGboost model is used for training to complete the degradation state identification. The superiority of this method is verified by comparing with different algorithms. An overview of the proposed methods and analysis steps in the paper are shown in Figure 2.

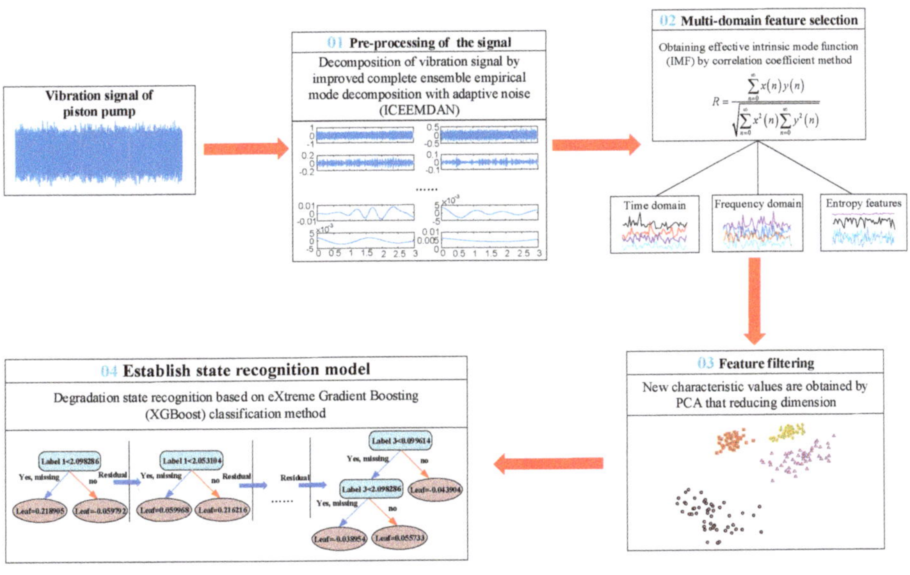

Figure 2. An overview of the proposed methods and analysis steps.

2. Theoretical Background

2.1. Improved Complete Ensemble Empirical Mode Decomposition with the Adaptive Noise (ICEEMDAN) Model

ICEEMDAN adds a special white noise $E_k(w^{(i)})$ that is the k-th IMF component of Gaussian white noise after EMD decomposition [34,37]. Then, by obtaining a unique residual, IMF is defined as the difference between the existing residual signal and its local mean. The results show the residual noise in IMF is greatly reduced, and the problem of illusive components and shortage of modal aliasing in the early stage of decomposition is also solved [38].

Let $E_k(\cdot)$ denote the operator of the kth IMF component obtained by EMD, and $M(\cdot)$ denote the operator for calculating the local mean of the signal. The relationship between the first IMF value $\tilde{c}_1$ and the residual r_1 are as

$$\tilde{c}_1 = \frac{1}{N}\sum_{i=1}^{N} E_1(x) = x - r_1 \tag{1}$$

where $r_1 = \frac{1}{N}\sum_{i=1}^{N} M(x^{(i)})$, x is the original signal, and N is the number of data points of the x. The specific decomposition process of ICEEMDAN can be described as follows.

a. Adding white noise $E_1(w^{(i)})$ to the original signal x, $x^{(i)}$ is obtained as

$$x^{(i)} = x + \beta_0 E_1(w^{(i)}) \tag{2}$$

where $w^{(i)}$ represents the i-th white noise to be added.

b. EMD is used to calculate the local mean of $x^{(i)}$, and the first residual r_1 was obtained by taking the average of them; then, the first IMF value can be calculated by $\tilde{c}_1 = x - r_1$.

c. The second mode component value(IMF2) can be calculated by $\tilde{c}_2 = r_1 - r_2$, where $r_2 = \frac{1}{N}\sum_{i=1}^{N} M(r_1 + \beta_1 E(w^{(i)}))$.

d. Similarly, calculate the k-th IMF value according to $\tilde{c}_k = r_{k-1} - r_k$, where $r_k = \frac{1}{N}\sum_{i=1}^{N} M(r_{k-1} + \beta_{k-1} E_k(w^{(i)}))$.

Until the obtained residual component can be further decomposed by EMD, ICEEMDAN can be used to decompose IMF accurately, which lays a foundation for improving the accuracy of state recognition.

2.2. eXtreme Gradient Boosting (XGBoost) Classifier Design

For the traditional radient Boosting Decision Tree (GBDT) algorithm, only the first order derivative information is used. Moreover, due to the dependency between weak learners, it is difficult for GBDT to train data in parallel [39]. XGBoost takes the Taylor expansion of the loss function up to the second order and adds a regularization term to find the optimal solution, which is used to balance the decline of the objective function and the complexity of the model to avoid overfitting [40]. The XGBoost model is

$$\hat{y}_i = \sum_{k=1}^{K} f_k(x_i), f_k \in F \tag{3}$$

where K is the number of decision-tree, $f_k(x_i)$ is the function of input in the k-th decision-tree, $\hat{y}_i$ is the predicted value, and F is the set of all possible CART. The objective function of XGBoost includes two parts: training error and regularization, which is

$$X_{obj} = \sum_{i=1}^{n} l(y_i, \hat{y}_i) + \sum_{k=1}^{K} \Omega(f_k) \tag{4}$$

where $\sum_{i=1}^{n} l(y, \tilde{y})$ is used to measure the difference between the predicted value and the real value of the loss function. $\sum_{k=1}^{K} \Omega(f_k)$ is the regularization term and $\Omega(f_k) = \gamma T + \frac{1}{2}\lambda\|w\|^2$.

T is the number of leaf node, w is the scores of leaf node, γ is the leaf penalty coefficient, and λ ensures that the scores of leaf node is not too large.

The XGBoost algorithm uses the gradient boosting strategy, adds one new tree at a time instead of getting all the trees at once, and continuously repairs the previous test results by fitting the residuals of the last prediction:

$$\begin{cases} \hat{y}_i^{(0)} = 0 \\ \hat{y}_i^{(1)} = f_1(x_i) = \hat{y}_i^{(0)} + f_1(x_i) \\ \hat{y}_i^{(2)} = f_1(x_i) + f_2(x_i) = \hat{y}_i^{(1)} + f_2(x_i) \\ \cdots\cdots \\ y_i{}^{(K)} = \sum_{k=1}^{K} f_k(x_i) = \hat{y}_i^{(K-1)'} + f_K(x_i) \end{cases} \tag{5}$$

Combined with Formulas (4) and (5), for the t-th decision tree, the objective function can be updated to

$$L^{(K)} = \sum_{i=1}^{n} l(y_i, \hat{y}_{\mathrm{i}}^{(K-1)} + f_K(X_i)) + \Omega(f_k) \tag{6}$$

In order to find the objective function which can be minimized, taking the Taylor expansion of the loss function up to the second order. Then, the objective function is approximately as:

$$L^{(K)} = \sum_{i=1}^{n} [l(y_i, \hat{y}_i^{(K-1)} + f_K(X_i)) + \frac{1}{2} h_i f_K^2(\mathrm{X_i})] + \Omega(f_k) \tag{7}$$

Add the loss function values of each data, and the process is as follows:

$$\begin{aligned} X_{\mathrm{obj}} &= \sum_{i=1}^{n} [g_i f_K(x_i) + \tfrac{1}{2} h_i f_K{}^2(x_i)] + \Omega(f_k) \\ &= \sum_{i=1}^{n} [g_i w_q(x_i) + \tfrac{1}{2} h_i w_q^2(x_i)] + \Omega(f_k) + \lambda T + \tfrac{1}{2}\lambda \sum_{j=1}^{T} w_j^2 \\ &= \sum_{j=1}^{n} [(\textstyle\sum_{i \in I_j} g_i) w_j + \tfrac{1}{2}(\textstyle\sum_{i \in I_j} h_i + \lambda) w_j^2] + \lambda T \end{aligned} \tag{8}$$

where $g_i = \partial \hat{y}^{K-1} l(y_i, \hat{y}^{K-1})$ is the first derivative, $h_i = \partial^2 \hat{y}^{K-1} l(y_i, \hat{y}^{K-1})$ is the second derivative. The optimal w and objective function values obtained by the solution are as

$$w_j = -\frac{G_j}{H_j + \lambda} \tag{9}$$

$$X_{obj} = -\frac{1}{2} \sum_{j=1}^{T} \frac{G_j}{H_j + \lambda} + \lambda T \tag{10}$$

where $G_j = \sum_{i \in I_j} g_i$, $H_j = \sum_{i \in I_j} h_i$.

During the training process, the model continuously calculates the node loss to select the leaf node with the largest gain loss. XGBoost adds new trees by continuously splitting features. Adding a tree each time is actually learning a new function $f_k(X, \theta_k)$ to fit the residual of the last prediction.

When K trees are obtained after training, the features of prediction samples will have a corresponding leaf node in each tree, and each leaf node corresponds to a score. Finally, the corresponding scores of each tree are added up to obtain the recognition prediction value of the sample [30]. The flow chart of XGBoost is shown in Figure 3.

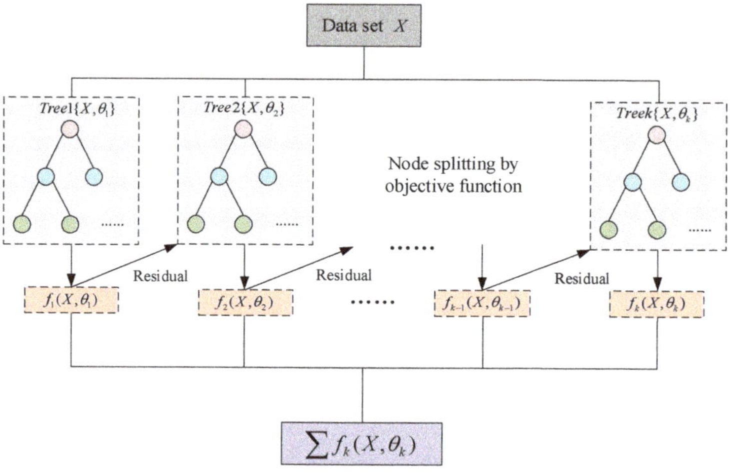

Figure 3. Flow chart of XGBoost.

3. Performance Degradation Test of Piston Pump

In this paper, the piston pump performance degradation test bench could collect the vibration signals of the pump with different wear degrees. During the test, the slipper with different wear degrees was replaced to simulate the phenomenon that the wear of the piston pump gradually increases in actual work. Figure 4 is the physical picture of the test bench. The vibration sensor is installed by magnetic base adsorption, and the vibration signal of pump shell is collected synchronously.

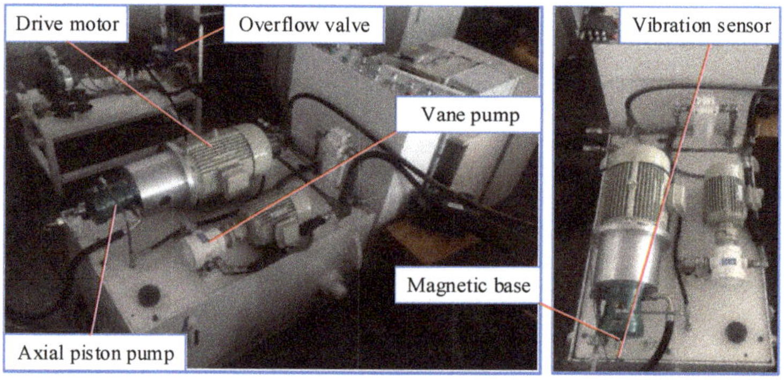

Figure 4. Test bench physical map.

In order to simulate the process of slipper wear degradation that affected the performance of the piston pump, the wear amounts of slipper were 0 mm (no wear), 0.5 mm (slight wear), 1.5 mm (moderate wear) and 2.5 mm (severe wear) in turn. The test adopts 20 kHz sampling frequency to collect the vibration signal. The test hydraulic system was under a pressure of 15 MPa, and the vibration signals of the slipper with different degrees of wear were sequentially collected. In our experiment, the data collected in the last 3 s are taken as effective degradation data, and the data obtained by each degradation degree are divided into 50 groups. Among the 50 sets of data, 30 sets are used for training and 20 sets are used for testing.

4. Test Data Processing of the Vibration Signal

4.1. Pre-Processing of the Vibration Signal by ICEEMDAN

The vibration signals with different degradation degrees are shown in Figure 5, from which it can be seen that there are great differences in the characterization of vibration waveforms under the same working conditions. With the increase of slipper wear, the vibration of piston pump becomes more and more intense, and the amplitude of vibration signal is larger and larger.

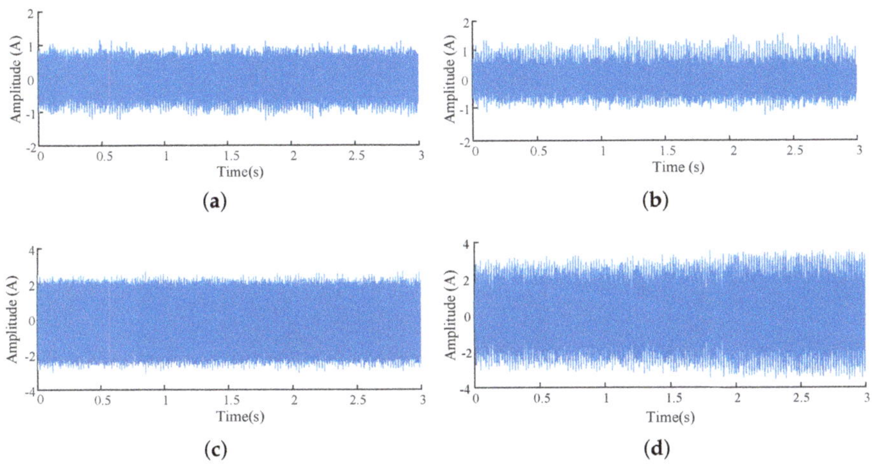

Figure 5. Vibration signals with four wear degrees. (**a**) vibration signal of slipper under normal condition; (**b**) vibration signal of slipper with slight wear; (**c**) vibration signal of slipper under moderate wear; (**d**) vibration signal of slipper under severe wear.

The vibration data of the tested pump under different states are selected and decomposed by ICEEMDAN, and the results are shown in Figure 6. The ICEEMDAN algorithm is used to decompose the vibration signals of four states into respective IMFs and a residual component for model development. The IMFs generated by ICEEMDAN have different sensitivity to the degrees of degradation. If the IMFs are analyzed directly, it is difficult to get the effective features related to the original signal. The appearance of illusive components will cause the extraction of useless features, increase the dimension of the feature set, and bring difficulties to the subsequent analysis [41]. By comparing correlation coefficients, the effective components in IMFs were screened out. The IMF component with the largest correlation coefficient is selected as the research object in this paper.

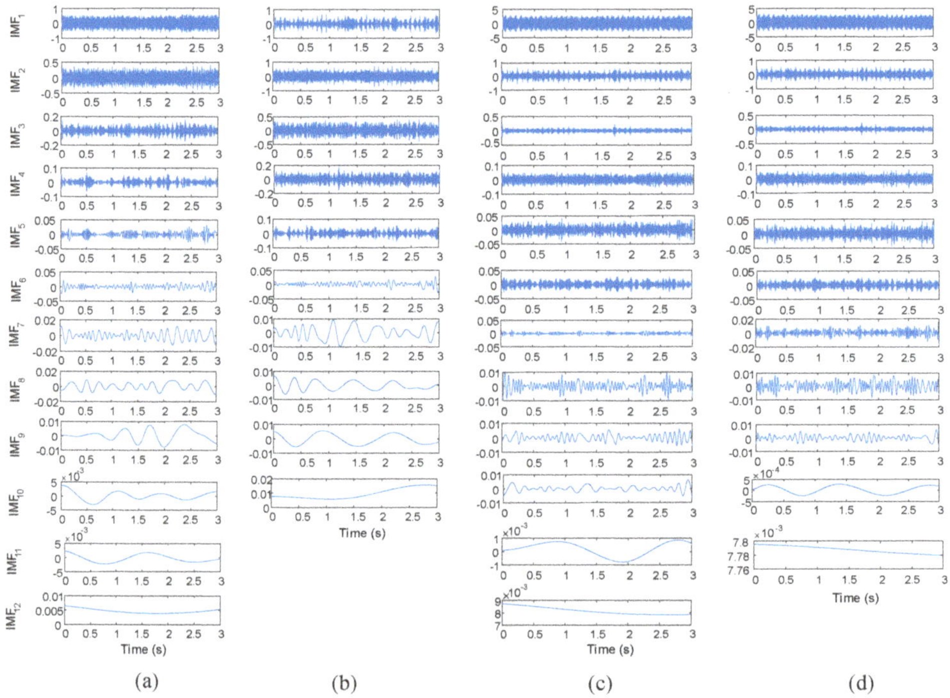

Figure 6. Decomposition results of ICEEMDAN. (**a**) the vibration signal under normal state are decomposed by ICEEMDAN; (**b**) the vibration signal under slight wear are decomposed by ICEEMDAN; (**c**) the vibration signal under moderate wear are decomposed by ICEEMDAN; (**d**) the vibration signal under severe wear are decomposed by ICEEMDAN.

Then, the appropriate IMF component is selected by using the correlation coefficient method [42]. The calculation process of the correlation coefficient is shown in Formula (11):

$$R = \frac{\sum\limits_{n=0}^{\infty} x(n)\, y(n)}{\sqrt{\sum\limits_{n=0}^{\infty} x^2(n) \sum\limits_{n=0}^{\infty} y^2(n)}} \tag{11}$$

where $x(n)$ is the sequence of effective IMF component, $y(n)$ is the sequence of the original signal.

The correlation coefficient method can be used to express the correlation between the selected IMF component and the original signal, the correlation coefficient values of the IMF components of the four states are showed in Table 1. The larger the correlation coefficient is, the greater the correlation between the IMF components and the original signal, the more effective information it contains. On the contrary, the IMF component is more likely to be illusive.

Table 1. The correlation coefficient values of the IMF components of the four states.

Status of the Slippers	No Wear	Mild Wear	Moderate Wear	Severe Wear
IMF_1	0.944	0.891	0.593	0.505
IMF_2	0.762	0.787	0.904	0.994
IMF_3	0.423	0.186	0.415	0.523
IMF_4	0.196	0.119	0.113	0.109
IMF_5	0.135	0.500	0.034	0.011
IMF_6	0.329	0.093	0.008	0.003
IMF_7	0.156	0.028	0.006	0.002
IMF_8	0.068	0.008	0.004	0.000
IMF_9	0.007	0.004	0.007	0.000
IMF_{10}	0.004	0.008	0.008	0.008
IMF_{11}	0.003	–	0.008	0.000
IMF_{12}	0.000	–	0.004	–

From the above table, we can see the correlation coefficient of each IMF component under the four conditions. The IMF_1 component under no wear condition, the IMF_1 component under mild wear condition, the IMF_2 component with moderate wear degree, and the IMF_2 component with severe wear degree are selected respectively, and then the selected IMF is analyzed in time domain, frequency domain, and entropy features.

4.2. Multi-Domain Feature Selection

The selected characteristic parameters of time domain and calculation methods are shown in the Table 2. In the time domain characteristics, the root mean square reflects the irregular continuity of vibration and the energy of signal. The peak–peak value can be used to represent the impact vibration caused by the wear of the the inner edge. The spike degree of vibration signal waveform about piston pump is expressed by peak value. The variance indicates the energy and intensity of vibration signal. The degree to which vibration signals deviates from normal distribution is reflected by skewness and kurtosis.The waveform index is a sensitive and stable parameter, which can well represent the slight damage of slippers with different wear degrees. With the aggravation of wear, the impulsion index and tolerance index increase obviously [43–45].

Table 2. Numerical explanation of time domain.

Time Domain Characteristic	Expression
Root mean square	$F_1 = \sqrt{\frac{1}{N}\sum_{i=1}^{N} x^2(i)}$
Peak index	$F_2 = \frac{\max(x)}{F_1}$
Peak-peak index	$F_3 = \max(x) - \min(x)$
Skewness index	$F_4 = \frac{\sum_{i=1}^{N} (\lvert x(i)\rvert - \bar{x})^3}{N \times F_1^3}$
Variance	$F_5 = \frac{1}{N}\sum_{i=1}^{N} (x(i) - \overline{x})^2$
Tolerance index	$F_6 = \frac{F_1}{\frac{1}{N}\sum_{i=1}^{N} x(i)}$
Waveform index	$F_7 = \frac{\max(x)}{\bar{x}}$
Kurtosis index	$F_8 = \frac{\sum_{i=1}^{N} (\lvert x(i)\rvert - \bar{x})^4}{N \times {F_1}^4}$
Impulsion index	$F_9 = \frac{\max(x)}{\left\lvert \frac{1}{N}\sum_{i=1}^{N} x(i) \right\rvert}$

where x is the selected effective IMF, $\bar{x} = \frac{1}{N}\sum_{i=1}^{N} x(i)$.

In order to obtain degradation information from the frequency domain of vibration signals, we select F_{10}: average frequency, F_{11}: frequency variance, F_{12}: gravity frequency, F_{13}: frequency standard deviation, and statistical indexes about frequency domain in reference [46] are F_{14}, F_{15}, F_{16} in the table. The average frequency reflects the vibration energy in the frequency domain. The frequency variance, F_{14} and F_{15}, indicate the dispersion or concentration of the spectrum. The position change of the main frequency band is obtained according to gravity frequency, frequency standard deviation, and F_{16}. The numerical explanation is shown in Table 3.

Table 3. Numerical explanation of frequency domain.

Frequency Domain Characteristic	Expression	Frequency Domain Characteristic	Expression
F_{10}	$\frac{1}{N}\sum_{i=1}^{N} s(i)$	F_{11}	$\sum_{i=1}^{N} (s(i) - F_{10})^2$
F_{12}	$\frac{\sum_{i=1}^{N} f_i s(i)}{\sum_{i=1}^{N} s(i)}$	F_{13}	$\sqrt{\frac{\sum_{i=1}^{N} (f_i - F_{12})^2 s(i)}{N}}$
F_{14}	$\frac{\sum_{i=1}^{N} (s(i) - F_{10})^3}{N(\sqrt{F_{11}})^3}$	F_{15}	$\frac{\sum_{i=1}^{N} (s(i) - F_{10})^4}{N \times F_{11}^2}$
F_{16}	$\sqrt{\frac{\sum_{i=1}^{N} f_i^2 s(i)}{\sum_{i=1}^{N} s(i)}}$		

$s(i)$ is the spectrum value of effective IMF, $i = 1, 2, 3, \ldots, N$, N is the total number of spectral lines, and f_i is the frequency value of the i-th spectral line.

These characteristic values listed in Table 2 and Table 3 represent the degradation trend of the piston pump from different perspectives, which in time domain and frequency domain vary with the different degradation states. In order to further extract the piston pump characteristic information of different degradation state, the permutation entropy [47], approximate entropy [48], information entropy [49], and energy entropy [50] of effective IMFs are calculated, respectively, forming the feature information set based on IMF frequency band. The four entropy features of each degradation state are shown in Figure 7.

It can be seen from Figure 7 that different degradation states have a good hierarchy. With the degree of wear increasing, the vibration signal over time series becomes relatively chaotic and complex, but its entropy values are more stable.

4.3. Feature Filtering by PCA

Too high feature dimensions lead to over fitting and some of the features are highly correlated, which will increase the unnecessary calculation burden [51]. Therefore, the PCA algorithm is used in this paper to obtain new low-dimensional features, remove redundant features, and retain the main information of original feature vectors while reducing the complexity of classification and recognition model [52]. The results are shown in Figure 8.

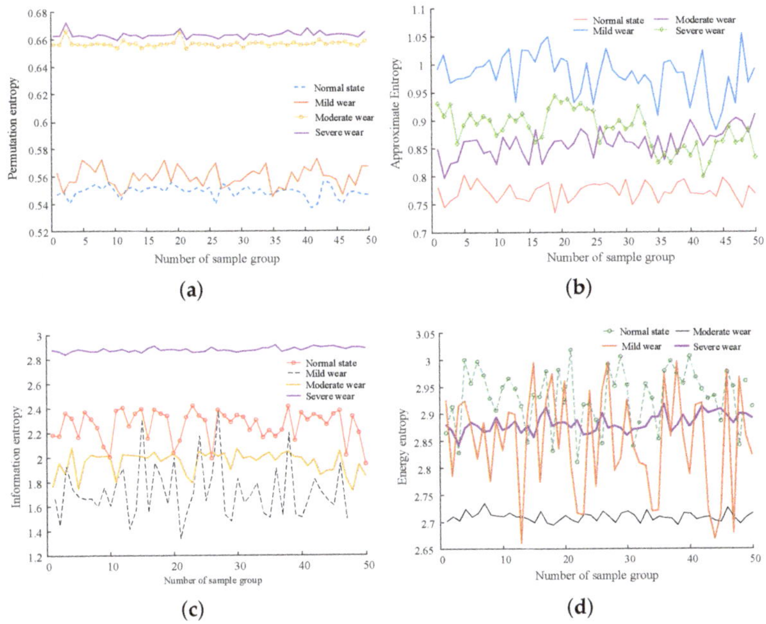

Figure 7. Entropy characteristics of four wear states. (**a**) permutation entropy in different wear states; (**b**) approximate entropy under different wear states; (**c**) information entropy under different wear states; (**d**) energy entropy under different wear conditions.

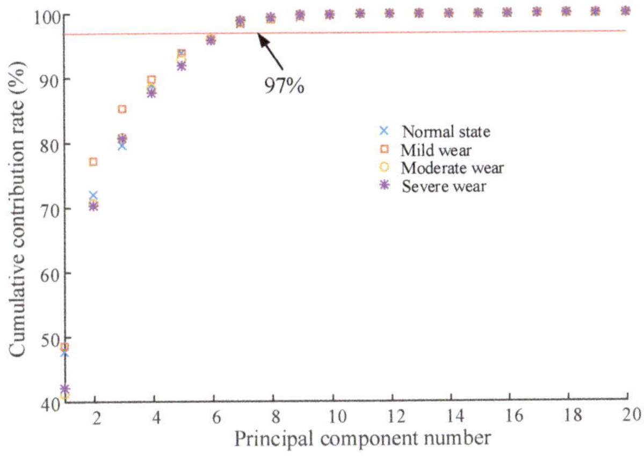

Figure 8. Cumulative contribution rate of different principal components under four wear states.

For the convenience of calculation, the coincidence degree of cumulative contribution rate under different degradation states should be high, and we need the same number of principal components in different degradation states [43]. It can be seen from Figure 8 that, when the number of principal components is 6, the cumulative contribution rate of the four degradation states is about 96.4%, and have a high degree of coincidence and contain most of the degradation information of the feature set. When there are too many principal components, the interpretation degree of redundant principal components is smaller than that of a single variable, which will only increase the calculation time and complexity of the model. On the contrary, when there are too few principal components, the whole feature set can not be expressed, and if different degradation states have different numbers of principal components, the calculation will become more complex and reduce the accuracy of model prediction. Therefore, the first six principal components are used to replace the original 20-dimensional degenerate feature vector, which reduces the data dimension and calculation cost. The first six principal components are used as new feature vectors to construct the training set and test set of XGBoost model respectively.

5. Degradation State Recognition with XGBoost

5.1. Parameter Optimization for the XGBoost Model

In order to improve the performance of the model, it is necessary to adjust the XGBoost parameters. The XGBoost algorithm mainly includes three types of parameters: general parameters, Booster parameters, and task parameters. Each type of parameter includes several specific parameter values, among which the booster parameters have a greater impact on the algorithm performance. In this paper, on the basis of fixing other parameters, changing the booster parameters: max depth (the maximum depth of the tree), the min child weight (minimum sum of instance weight needed in a child) and n estimators (number of base classifiers) to find the optimal parameters, which can avoid the over-fitting of the model and improve the model accuracy.

The maximum depth of the trees and the min child weight are two parameters that affect each other. If the parameters are optimized in turn, they will only become disadvantages of local optimization. Moreover, the amount of data in this paper is small; therefore, we can use the grid search to select the maximum depth of the trees and the min child weight at the same time. The results are shown in Figure 9.

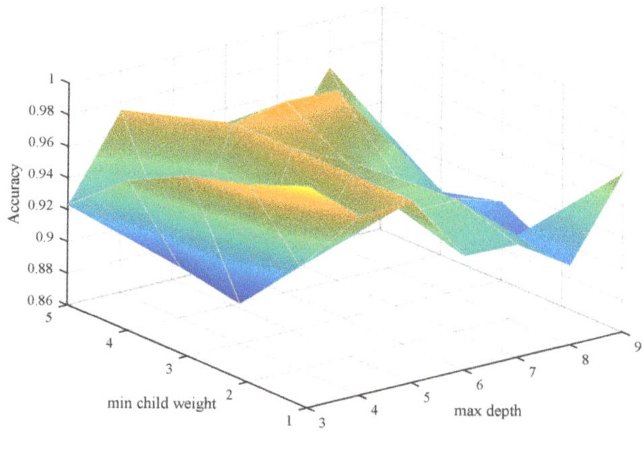

Figure 9. The relationship between recognition accuracy and max depth, min child weight.

When the maximum depth of the tree is 3 and the weight sum of leaf nodes is 1, the average state recognition accuracy is the highest, which is 0.9917. Thus, the max depth = 3 and the min child weight = 1. Fixing the above two parameters, the number of decision trees was selected by logarithmic loss function to evaluate the probability output of XGBoost classifier [53]. The logarithmic loss function is defined as:

$$\sigma = -\frac{1}{N}\sum_{i=1}^{N}\sum_{j=1}^{M} y_{ij}\log(p_{ij}) \tag{12}$$

where N is the input sample size, M is the number of categories, y_{ij} represents the real category of the input data points, and p_{ij} is the probability that the i-th data point predicted by the XGBoost classifier belongs to the j-th class. The results are shown in Figure 10.

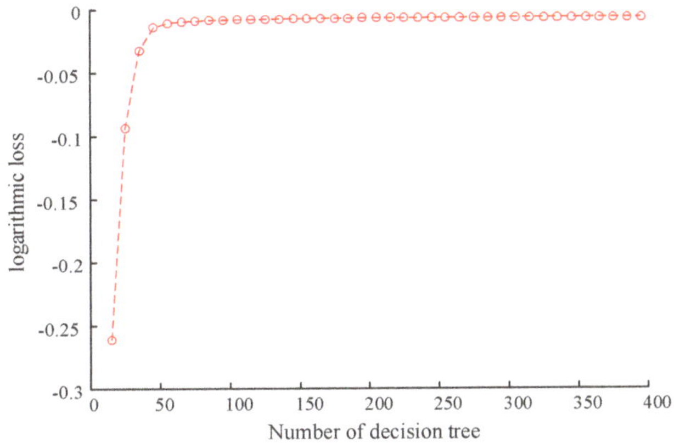

Figure 10. Determination of the number of trees.

The closer the value of logarithmic loss is to 0, the higher the accuracy of the classifier is. In this paper, when the number of trees is 167, the logarithmic loss value is −0.0061, which is the largest, so the number of the trees is equal to 167. The parameters for XGBoost are shown in the Table 4, and the remaining parameters are the default values.

Table 4. Parameters for XGBoost.

Parameter	Numerical Value
Max depth	3
Min child weight	1
N estimators	167
Objective	multi: softmax

5.2. Experimental Results and Analysis

In order to better verify the efficiency and correctness of this method, K-fold cross validation [54] is used to calculate the recognition accuracy of the XGBoost model. The data set is divided into 10 parts, among which six parts were taken as training data, and the remaining four parts were taken as test data in turn for experiments. The average value of recognition accuracy is taken as the final result. Moreover, the method adopted in this paper is also compared with ANN, SVM, and GBDT shown in Table 5. It is found that, when the number of hidden layer neurons of ANN is 12, the effect is better, while the kernel function of SVM is RBF, the penalty coefficient is 4, and gamma is 0.1. By adjusting and optimizing the parameters of GBDT, n estimators = 20, max depth = 3, and min samples split (the minimum number of samples needed when the internal nodes are divided again) = 10.

Table 5. Comparison of degradation state recognition results.

Classification Method	Average Recognition Accuracy	Average Decision Time (s)
ANN	0.991	0.094
SVM	0.997	0.012
GBDT	0.994	0.006
XGBoost	0.997	0.003

The above table shows the average accuracy and average calculation time of multiple status recognition verifications. As you can see from the table, the average recognition accuracy of the four algorithms has little difference, possibly because the feature values extracted by the method proposed in this paper have good discrimination—among which SVM and XGBoost have the highest recognition accuracy of 0.997. SVM takes a long calculation time because of the kernel function. Thus, when the performance of the computer is the same, the average decision time of XGBoost is less than that of SVM, which is 0.03 s. Compared with the traditional GBDT model, XGBoost adds the control of model complexity and pruning processing, which makes the trained model difficult to overfitting, and the calculation time is relatively less. The average recognition rate of GBDT is 0.994, but the average decision time is lower than SVM and ANN. ANN needs to iterate repeatedly to get the ideal classification effect, so it has higher average decision time than three other algorithms. In the complex working environment, the vibration data of piston pump are complex and changeable, so it is necessary to extract a variety of characteristics from it to carry out effective diagnosis. Compared with ANN, the diagnostic accuracy of XGBoost can be improved by 0.001 to 0.013, and there is no need for a tedious parameter optimization process. The XGBoost algorithm can not only ensure high diagnosis accuracy but also reduce calculation time. In order to further prove the effectiveness of the present work in this paper, a variety of methods are introduced for comparison. The comparative results are shown in Table 6.

Table 6. Performance comparison for different methods.

Number	Feature Extraction Method	Feature Reduction Method	Classification Technique	Recognition Rate (%)
1	EEMD, Multi-domain features (time domain, frequency domain, entropy)	PCA	GBDT	98.2
2	EEMD, Multi-domain features	PCA	SVM	98.6
3	EEMD, Multi-domain features	–	XGBoost	92.4
4	EEMD, Multi-domain features	PCA	XGBoost	97.1–100
5	Multi-domain features	PCA	XGBoost	69.5
6	ICEEMDAN, Multi-domain features	–	XGBoost	94.5–99.2
Present work	ICEEMDAN, Multi-domain features	PCA	XGBoost	98.3–100

The results are shown in Table : in the case of no data preprocessing, the recognition rate is low due to the influence of noise. Compared with EEMD, after ICEEMDAN preprocessing, the recognition rate will be relatively high. Therefore, the proposed method effectively identifies degraded states. It can maintain mechanical equipment through timely judgment, which has high practical value in the fault diagnosis of piston pump.

6. Conclusions

In this paper, ICEEMDAN is used to denoise the vibration signal of the piston pump, PCA is used to reduce the feature dimension, and the state recognition based on XGBoost is carried out to identify the wear state and category of the piston pump slipper. The experimental results show that:

a. The ICEEMDAN can decompose the vibration signal of piston pump adaptively, improve the decomposition efficiency, and suppress phenomena of mode mixing. It is feasible to select an effective IMF component by using correlation coefficient.
b. Through time domain, frequency domain, and entropy, the deterioration process of piston pump can be tracked and identified comprehensively. The PCA is used to reduce data dimension and calculation cost, which improves the accuracy and efficiency of state identification.

c. The average recognition accuracy of slipper wear state of piston pump based on ICEEMDAN and XGBoost is 99.7%. Compared with ANN, GBDT, and SVM algorithm, XGBoost identifies four different wear states better and saves the computing time, which highlights the advantages of XGBoost after parameter optimization in pattern recognition.

Author Contributions: All the authors contributed to the original idea and design of the study, to the analysis, to the drafting of the manuscript, and reading and approving the final version. Conceptualization, R.G. and Z.Z.; methodology, R.G.; validation, R.G., T.W., and Z.Z.; formal analysis, G.L. and D.G.; resources, R.G. and J.Z.; data curation, Z.Z.; writing—original draft preparation, Z.Z.; writing—review and editing, R.G. and Z.Z.; visualization, Z.Z.; Software, Z.Z. and T.W.; supervision, T.W.; project administration, R.G. and J.Z.; funding acquisition, R.G., J.Z., and D.G. All authors have read and agreed to the published version of the manuscript.

Funding: This research was funded by the National Key Research and Development Program of China (Grant No. 2019YFB2005204), the National Natural Science Foundation of China (Grant No.52075469, 51675461, 11673040), the Key Research and Development Program of Hebei Province (Grant No. 19273708D), and the Open Foundation of the State Key Laboratory of Fluid Power and Mechatronic Systems (Grant No. GZKF-201922).

Acknowledgments: The authors would like to acknowledge the support provided by Yanshan University.

Conflicts of Interest: The authors declare no conflict of interest.

References

1. Lan, Y.; Hu, J.; Huang, J.; Niu, L.; Zeng, X.; Xiong, X.; Wu, B. Fault Diagnosis on Slipper Abrasion of Axial Piston Pump based on Extreme Learning Machin. *Measurement* **2010**, *124*, 378–385. [CrossRef]
2. Guo, S.; Chen, J.; Lu, Y.; Yan, W.A.N.G.; Hongkang, D.O.N.G. Hydraulic piston pump in civil aircraft: Current status, future directions and critical technologies. *Chin. J. Aeronaut.* **2020**, *33*, 16–30. [CrossRef]
3. Son, Y.K.; Savage, G.J. A New Sample-Based Approach to Predict System Performance Reliability. *IEEE Trans. Reliab.* **2008**, *52*, 322–330.
4. Zamanian, A.H.; Ohadi, A. Wear prediction for metals. *Tribol. Int.* **1997**, *30*, 377–383.
5. Lin, J.S.; Chen, Q. Fault diagnosis of rolling bearings based on multifractal detrended fluctuation analysis and Mahalanobis distance criterion. *Mech. Syst. Signal Process.* **2013**, *38*, 515–533. [CrossRef]
6. Xu, B.; Zhang, J.; Yang, H. Investigation on structural optimization of anti-overturning slipper of axial piston pump. *Sci. China Technol. Sci.* **2012**, *55*, 3010–3018. [CrossRef]
7. Liu, H.; Yuan, S.H.; Jing, C.B.; Zhao, Y.M. Effects of Wear Profile and Elastic Deformation on the Slipper's Dynamic Characteristics. *J. Mech. Eng.* **2012**, *28*, 608–621. [CrossRef]
8. Tian, Z.K.; Li, H.R.; Sun, J.; Li, B.C. Degradation state identification method of hydraulic pump based on improved MF-DFA and SSM-FM. *Chin. J. Sci. Instrum.* **2016**, *37*, 1851–1860.
9. Xiao, W.B.; Chen, J.; Zhou, Y.; Wang, Z.Y.; Zhao, F.G. Wavelet packet transform and hidden Markov model based bearing performance degradation assessment. *J. Mech. Eng.* **2011**, *30*, 32–35.
10. Singh, D.S.; Zhao, Q. Pseudo-fault signal assisted EMD for fault detection and isolation in rotating machines. *Mech. Syst. Signal Process.* **2016**, *81*, 202–218. [CrossRef]
11. Lei, Y.G. Machinery Fault Diagnosis Based on Improved Hilbert-Huang Transform. *J. Mech. Eng.* **2011**, *47*, 71–77. [CrossRef]
12. Lei, Y.G.; He, Z.J.; Zi, Y.Y. Application of the EEMD method to rotor fault diagnosis of rotating machinery. *Mech. Syst. Signal Process.* **2009**, *23*, 1327–1338. [CrossRef]
13. Lee, D.-H.; Ahn, J.-H.; Koh, B.-H. Fault Detection of Bearing Systems through EEMD and Optimization Algorithm. *Sensors* **2017**, *17*, 2477. [CrossRef]
14. Wu, Z.H.; Huang, N.E. Ensemble empirical mode decomposition: A noise-assisted data analysis method. *Adv. Adapt. Data Anal.* **2009**, *1*, 1–41. [CrossRef]
15. Torres, M.E.; Colominas, M.A.; Schlotthauer, G.; Flandrin, P. A complete ensemble empirical mode decomposition with adaptive noise. In Proceedings of the 2011 IEEE International Conference on Acoustics, Speech and Signal Processing (ICASSP), Prague, Czech Republic, 22–27 May 2011; pp. 4144–4147.
16. Tian, Z.K.; Li, H.R.;Gu, H.Q.; Xu, B.H. Degradation status identification of a hydraulic pump based on local characteristic-scale decomposition and JRD. *J. Vib. Shock* **2016**, *35*, 54–59.

17. Wang, B.; Li, H.R.; Chen, Q.H.; Xu, B.H. Rolling Bearing Performance Degradative State Recognition Based on Mathematical Morphological Fractal Dimension and Fuzzy Center Means. *Acta Armamentarii* **2015**, *36*, 1982–1990.
18. Zhang, S.; Li, W. Bearing Condition Recognition and Degradation Assessment under Varying Running Conditions Using NPE and SOM. *Math. Probl. Eng.* **2014**, *9*, 781583. [CrossRef]
19. Si, L.; Wang, Z.; Liu, X.; Tan, C. A sensing identification method for shearer cutting state based on modified multi-scale fuzzy entropy and support vector machine. *Eng. Appl. Artif. Intell.* **2019**, *78*, 86–101. [CrossRef]
20. Ji, Y.; Sun, S. Multitask multiclass support vector machines: Model and experiments. *Pattern Recognit.* **2013**, *46*, 914–924. [CrossRef]
21. Azadeh, A.; Saberi, M.; Kazem, A.; Ebrahimipour, V.; Nourmohammadzadeh, A.; Saberi, Z. A flexible algorithm for fault diagnosis in a centrifugal pump with corrupted data and noise based on ANN and support vector machine with hyper-parameters optimization. *Appl. Soft Comput.* **2013**, *13*, 1478–1485. [CrossRef]
22. Siddiqui, S.A.; Verma, K.; Niazi, K.R.; Fozdar, M. Real-Time Monitoring of Post-Fault Scenario for Determining Generator Coherency and Transient Stability through ANN. *IEEE Trans. Ind. Appl.* **2017**, *54*, 685–692. [CrossRef]
23. Wang, Y.S.; Liu, N.N.; Guo, H.; Wang, X.L. An engine-fault-diagnosis system based on sound intensity analysis and wavelet packet pre-processing neural network. *Eng. Appl. Artif. Intell.* **2020**, *94*, 103765. [CrossRef]
24. Yan, X.; Jia, M. A novel optimized SVM classification algorithm with multi-domain feature and its application to fault diagnosis of rolling bearing. *Neurocomputing* **2018**, *313*, 47–64. [CrossRef]
25. Wang, S.; Xiang, J.; Zhong, Y.; Tang, H. A data indicator-based deep belief networks to detect multiple faults in axial piston pumps. *Mech. Syst. Signal Process.* **2018**, *112*, 154–170. [CrossRef]
26. Chen, C.-H.; Zhao, W.D.; Pang, T.; Lin, Y.-Z. Virtual metrology of semiconductor PVD process based on combination of tree-based ensemble model. *ISA Trans.* **2020**, *103*, 192–202. [CrossRef]
27. Ampomah, E.K.; Qin, Z.; Nyame, G. Evaluation of Tree-Based Ensemble Machine Learning Models in Predicting Stock Price Direction of Movement. *Information* **2020**, *11*, 332. [CrossRef]
28. Adler, W.; Potapov, S.; Lausen, B. Classification of repeated measurements data using tree-based ensemble methods. *Comput. Stat.* **2011**, *26*, 355–369. [CrossRef]
29. Kotsiantis, S. Combining bagging, boosting, rotation forest and random subspace methods. *Artif. Intell. Rev.* **2011**, *35*, 223–240. [CrossRef]
30. Chen, T.; Guestrin, C. Xgboost: A scalable tree boosting system. In Proceedings of the 22nd ACM SIGKDD International Conference on Knowledge Discovery and Data Mining, San Francisco, CA, USA, 13–17 August 2016; pp. 785–794.
31. Friedman, J.H. Greedy Function Approximation: A Gradient Boosting Machine. *Ann. Stat.* **2001**, *29*, 1189–1232. [CrossRef]
32. Wu, Z.; Wang, X.; Jiang, B. Fault Diagnosis for Wind Turbines Based on ReliefF and eXtreme Gradient Boosting. *Appl. Sci* **2020**, *10* , 3258. [CrossRef]
33. Debaditya, C.; Hazem, E. Advanced machine learning techniques for building performance simulation: A comparative analysis. *J. Build. Perform. Simul.* **2019**, *12*, 193–207.
34. Colominas, M.A.; Schlotthauer, G.; Torres. M.E. Improved complete ensemble EMD: A suitable tool for biomedical signal processing. *Biomed. Signal Process. Control.* **2014**, *14*, 19–29. [CrossRef]
35. Ji, N.; Ma, L.; Dong, H.; Zhang, X. Selection of Empirical Mode Decomposition Techniques for Extracting Breathing Rate From PPG. *IEEE Signal Process. Lett.* **2019**, *26*, 592–596.
36. Ali, M.; Prasad, R. Significant wave height forecasting via an extreme learning machine model integrated with improved complete ensemble empirical mode decomposition. *Renew. Sustain. Energy Rev.* **2019**, *104*, 281–295. [CrossRef]
37. Huang, N.E.; Shen, Z.; Long, S.R.; Wu, M.C.; Shih, H.H.; Zheng, Q.; Yen, N.C.; Tung, C.C.; Liu, H.H. The empirical mode decomposition and the Hilbert spectrum for nonlinear and non-stationary time series analysis. *Proc. Math. Phys. Eng. Sci.* **1998**, *454*, 903–995. [CrossRef]
38. Li, Y.; Xu, F. Signal processing and identification on the surface of Pinus massoniana Lamb. glulam using acoustic emission and improvement complete ensemble empirical mode decomposition with adaptive noise. *Measurement* **2019**, *148*, 106978. [CrossRef]

39. Wang, L.; Zhou, D.; Zhang, H.; Zhang, W.; Chen, J. Application of Relative Entropy and Gradient Boosting Decision Tree to Fault Prognosis in Electronic Circuits. *Symmetry* **2018**, *10*, 495. [CrossRef]
40. Chen, Y.; Wang, X.; Jung, Y.H.; Abedi, V.; Zand, R.; Bikak, M.; Adibuzzaman, M. Classification of short single lead electrocardiograms (ECGs) for atrial fibrillation detection using piecewise linear spline and XGBoost. *Physiol. Meas.* **2018**, *39*, 104006. [CrossRef]
41. Zheng, H.M.; Dang, C.L.; Gu, S.M.; Peng, D.; Chen, K. A quantified self-adaptive filtering method: Effective IMFs selection based on CEEMD. *Meas. Sci. Technol.* **2018**, *29*, 085701. [CrossRef]
42. Jiang, L.; Li, B.; Li, X.J. An Improved HHT Method and its Application in Fault Diagnosis of Roller Bearing. *Appl. Mech. Mater.* **2013**, *273*, 264–268. [CrossRef]
43. Lu, C.; Wang, S.; Maids, V. Fault severity recognition of aviation piston pump based on feature extraction of EEMD paving and optimized support vector regression model. *Aerosp. Sci. Technol.* **2017**, *67*, 105–117. [CrossRef]
44. Eriştia, H.; Uçarb, A.; Demir, Y. Wavelet-based feature extraction and selection for classification of power system disturbances using support vector machines. *Electr. Power Syst. Res.* **2010**, *80*, 743–752. [CrossRef]
45. Hasan, M.J.; Kim, J.; Kim, C.H.; Kim, J.-M. Health State Classification of a Spherical Tank Using a Hybrid Bag of Features and K-Nearest Neighbor. *Appl. Sci.* **2020**, *10*, 2525. [CrossRef]
46. Lei, Y.G.; He, Z.J.; Zi, Y.Y.; Hu, Q. Fault diagnosis of rotating machinery based on multiple ANFIS combination with Gas. *Mech. Syst. Signal Process.* **2007**, *21*, 2280–2294. [CrossRef]
47. Bandt, C.; Pompe, B. Permutation Entropy: A Natural Complexity Measure for Time Series. *Phys. Rev. Lett.* **2002**, *88*, 174102. [CrossRef]
48. Jiang, Z.; Deng, A. Identification of Rotating Machinery Rub and Impact Fault Using Hurst Exponent and Approximate Entropy as Characteristic Parameters of Acoustic Emission. *Proc. CSEE*, **2010**, *30*, 96–102.
49. Li, J.L.; Huang, S.H. Fault Diagnosis of a Rotary Machine Based on Information Entropy and Rough Set. *Int. J. Plant Eng. Manag.* **2007**, *12*, 199–206.
50. Zhu, L.D.; Liu, C.F.; Ju, C.Y.; Guo, M. Vibration recognition for peripheral milling thin-walled workpieces using sample entropy and energy entropy. *Int. J. Adv. Manuf. Technol.* **2020**, *108*, 3251–3266. [CrossRef]
51. Bect, P.; Simeu-Abazi, Z.; Maisonneuve, P.L. Diagnostic and decision support systems by identification of abnormal events: Application to helicopters. *Aerosp. Sci. Technol.* **2015**, *46*, 339–350. [CrossRef]
52. Wan, X.; Wang, D.; Tse, P.W.; Xu, G.; Zhang, Q. A critical study of different dimensionality reduction methods for gear crack degradation assessment under different operating conditions. *Measurement* **2016**, *78*, 138–150. [CrossRef]
53. Hitchcock, J.M. Fractal Dimension and Logarithmic Loss Unpredictability. *Theor. Comput.* **2003**, *304*, 431–441. [CrossRef]
54. Rodriguez, J.D.; Perez, A.; Lozano, J.A. Sensitivity Analysis of k-Fold Cross Validation in Prediction Error Estimation. *IEEE Trans. Pattern Anal. Mach. Intell.* **2010**, *32*, 569–575. [CrossRef] [PubMed]

applied sciences

MDPI

Article

Intelligent Fault Diagnosis of Rotating Machinery Using Hierarchical Lempel-Ziv Complexity

Bing Han [1], Shun Wang [2], Qingqi Zhu [1], Xiaohui Yang [1] and Yongbo Li [2,*]

1 Shaanxi Engineering Laboratory for Transmissions and Controls, Northwestern Polytechnical University, Xi'an 710072, China; npuhanbing@nwpu.edu.cn (B.H.); 1115012259@mail.nwpu.edu.cn (Q.Z.); yangxh@nwpu.edu.cn (X.Y.)

2 MIIT Key Laboratory of Dynamics and Control of Complex System, School of Aeronautics, Northwestern Polytechnical University, Xi'an 710072, China; wangshun@mail.nwpu.edu.cn

* Correspondence: yongbo@nwpu.edu.cn; Tel.: +86-1880-918-3733

Received: 11 May 2020; Accepted: 17 June 2020; Published: 19 June 2020

Abstract: The health condition monitoring of rotating machinery can avoid the disastrous failure and guarantee the safe operation. The vibration-based fault diagnosis shows the most attractive character for fault diagnosis of rotating machinery (FDRM). Recently, Lempel-Ziv complexity (LZC) has been investigated as an effective tool for FDRM. However, the LZC only performs single-scale analysis, which is not suitable to extract the fault features embedded in vibrational signal over multiple scales. In this paper, a novel complexity analysis algorithm, called hierarchical Lempel-Ziv complexity (HLZC), was developed to extract the fault characteristics of rotating machinery. The proposed HLZC method considers the fault information hidden in both low-frequency and high-frequency components, resulting in a more accurate fault feature extraction. The superiority of the proposed HLZC method in detecting the periodical impulses was validated by using simulated signals. Meanwhile, two experimental signals were utilized to prove the effectiveness of the proposed HLZC method in extracting fault information. Results show that the proposed HLZC method had the best diagnosing performance compared with LZC and multi-scale Lempel-Ziv complexity methods.

Keywords: feature extraction; fault diagnosis; Lempel-Ziv complexity; rotating machinery

---

1. Introduction

Rotating machinery is commonly used in modern industries, such as the aero-engine, vehicle, ship, and railway industries [1]. In industrial applications, the strict working environment may result in localized damage on rotating machinery. If the damage cannot be timely diagnosed, it may cause disastrous failure and serious economic loss. Therefore, it is crucial to conduct the fault diagnosis of rotating machinery (FDRM) so as to ensure its safety operation [2].

Until now, many advanced techniques have been developed to accomplish the FDRM, such as the vibration-based fault diagnosis method [3,4], acoustic emission-based fault diagnosis method [5,6], and rotary encoder-based fault diagnosis method [7]. Among these techniques, the vibration signal method is most widely applied in industrial applications due to the its advantage of easy measurement and high scalability [8]. Generally, three main stages are included in the vibration-based method: data collection, fault feature extraction, and pattern identification. During the three stages, the fault feature extraction lays a good foundation for FDRM. Some advanced fault feature approaches have been proposed for the FDRM, such as blind source separation [9], wavelet-based method [10], and adaptive decomposition methods [11].

Unfortunately, the complex dynamical structure and complex operating conditions of rotating machinery often generate nonlinear and non-stationary characteristics in the measured vibration signals, resulting in the difficulty in extracting the weak fault characteristics from the vibration signals [12]. Thanks to the rapid development of complexity theory, some complexity indexes are used to extract fault features of rotating machinery, such as Lempel-Ziv complexity (LZC) [13], approximate entropy [14], fuzzy entropy [12,15], permutation entropy [16,17], symbolic dynamic entropy [18], and multi-scale entropy [19]. Related research have indicated that LZC is powerful in analyzing vibration signals of mechanical systems. Yan et al. [13] used LZC to distinguish different bearing failure severity. Hong et al. [20] calculated LZC after wavelet transformation to recognize the fault severity of bearing. Cui et al. [21] proposed a signal decomposition and reconstruction method based on LZC and the double-dictionary matching pursuit. Meanwhile, Cui et al. [22] proposed a fault diagnosis method that was based on Sparsogram and LZC. Moreover, Bai et al. [23] utilized the permutation of the LZC to quantify the complexity of the signal, whereas Yin et al. [24] proposed a novel symbolic aggregate approximation and LZC method for fault diagnosis of rolling bearings.

However, the existing LZC methods have one common problem when analyzing the vibration signals. Since the fault information is usually embedded in vibration signals over different scale domains [25], the LZC-based methods only perform single-scale analysis, and thus the fault characteristics cannot be comprehensively described. In order to match the fault characteristics comprehensively, we proposed multi-scale Lempel-Ziv complexity (MLZC) to extract the fault features over multiple scales by multi-scale analysis [26]. However, the coarse-grained process is actually a linear smoothing process, which only considers the low-frequency components through the averaging process, and thus the fault information of the high-frequency component is discarded. In actual application, the fault information is embedded in both the low–high and high-frequency components of measured vibration signal. For example, a real vibration signal and its corresponding frequency spectrum are shown in Figure 1a,b, respectively. As can be seen, both the low-frequency (0–500 Hz) and high-frequency (2500–3500 Hz) components contain the main fault information of rotating machinery. Therefore, the fault extraction performance of MLZC still needs to be improved.

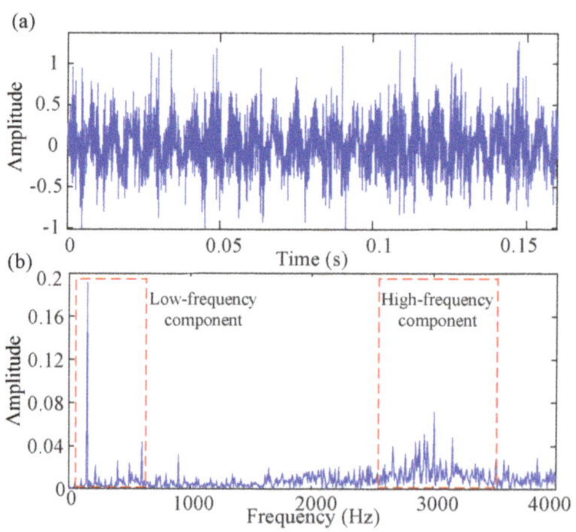

Figure 1. (**a**) The waveform of measured rating machinery vibration signals, and (**b**) its corresponding frequency spectrum.

To overcome such defects, this paper developed a novel approach called hierarchical Lempel-Ziv complexity (HLZC) to quantify the complexity from the measured time series. Compared with the MLZC method, HLZC utilizes both the low-frequency components generated by averaging the components and the high-frequency components generated by taking the difference of components to produce the sub time series in each layer. The merits of our proposed HLZC method in fault feature extraction are verified using both synthetic signals and experimental signals. Results demonstrated that our proposed HLZC method is superior to LZC and MLZC in extracting fault characteristics with high stability. After the HLZC-based feature extraction, we combined the HLZC with support vector machine (SVM) classifier [27] to accomplish the intelligent fault diagnosis of rotating machinery.

The remainder of this paper is organized as follows. Section 2 introduces the fundamentals of our proposed HLZC method. Moreover, the superiority of HLZC is validated using simulated impulsive signal through comparing with LZC and MLZC methods. Section 3 describes the framework of HLZC-based intelligent fault diagnosis method. Section 4 provides the experimental variation using two case studies. Finally, Section 5 draws the final conclusion of this paper.

2. Proposed Hierarchical Lempel-Ziv Complexity

2.1. Lempel-Ziv Complexity

LZC, as a nonlinear method, has been proven to be an efficient tool to measure the complexity for a given time series. LZC consists of two basic operations: copy and insert [20]. The LZC algorithm can be detailed as

(1) Cover the finite sequence $x(t)$ into 0–1 sequence by comparing with the median value T_d using Equation (1). Then, we can obtain the symbol series $S_N = \{s_1 s_2 \ldots s_N\}$.

$$s_i = \begin{cases} 0, \text{ if } x(i) < T_d \\ 1, \text{ otherwise} \end{cases} \tag{1}$$

(2) Set the initial value $S_{v,0} = \{\}$, $Q_0 = \{\}$, $C_N(0) = 0$, and $i = 1$. Note that S_v and Q represent the substrings of the symbol series S_N, and C_N represents complexity counter.

(3) Let $Q_i = \{Q_{i-1} s_i\}$ and judge whether Q_i belongs to $S_{v,i-1} = \left\{S_{v,i-2} s_{i-1}\right\}$. If so, set $C_N(i) = C_N(i-1)$ and $i = i+1$. Otherwise, set $Q_i = \{\}$, $C_N(i) = C_N(i-1)+1$, and update $i = i+1$.

(4) Repeat Step (3) until the end of symbol series $S_N = \{s_1 s_2 \ldots s_N\}$, and then the $C_N(N)$ can be obtained. $C_N(N)$ is the last complexity counter, which reflects upon the number of all different subsequences contained in the original data sequence.

(5) Normalize the $C_N(N)$ to obtain relatively independent indicator $C_{n,N}$ using Equations (2) and (3).

$$C_{n,N} = \frac{C_N(N)}{C_{UL}} \tag{2}$$

$$C_{UL} = \lim_{N \to \infty} C_N(N) \approx \frac{N}{\log_2 N} \tag{3}$$

2.2. Multi-Scale Lempel-Ziv Complexity

On the basis of LZC and coarse-grained procedure [26], MLZC can be summarized into two steps. (1) Conduct the multiple series by the coarse-graining analysis; (2) calculate the LZC values of each coarse-grained time series.

(1) Given an arbitrary time series $X\{x(k), k = 1, 2, \cdots, N\}$, construct consecutive coarse-grained time series $\left\{y^{(\tau)}\right\}$ according to Equation (4).

$$y_j^\tau = \frac{1}{\tau} \sum_{i=(j-1)\tau+1}^{j\tau} x_i \quad 1 \le j \le \frac{N}{\tau} \tag{4}$$

where τ is a positive integer. The obtained coarse-grained time series $\left\{y^{(1)}\right\}$ is equal to the original time series when $\tau = 1$.

(2) Calculate the LZC for each coarse-grained time series according to the definition of LZC as written in Equation (5).

$$MLZC(x, \tau) = LZC(y^\tau) \tag{5}$$

The flowchart of MLZC is drawn in Figure 2. Additionally, we set the parameter scale τ = 20 in the paper. The parameter τ = 20 has been proven to be effective using experimental tests for multi-scale analysis in [12,15,18,28].

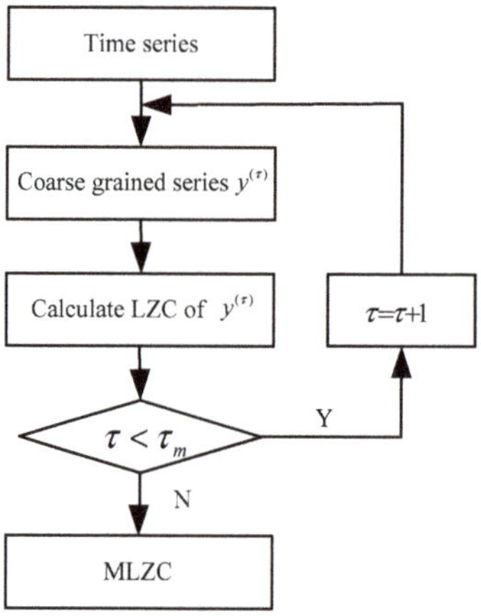

Figure 2. The flowchart of multi-scale Lempel-Ziv complexity (MLZC).

However, the MLZC algorithm only uses the low-frequency components generated by the multi-scale procedure for feature extraction, which unavoidably discards some useful fault information hidden in the high-frequency components. To address these shortcomings of MLZC, this paper developed a novel method called HLZC. First, the hierarchical decomposition is adopted to generate sub time series called hierarchical series. Second, the LZC values of all hierarchical nodes are all computed for comprehensive complexity estimation.

2.3. Hierarchical Lempel-Ziv Complexity

In this subsection, a novel method called HLZC is proposed by combining the hierarchical decomposition and Lempel-Ziv complexity. The hierarchical decomposition can decompose an arbitrary time series into high-frequency components and low-frequency components [29]. Figure 3 gives an example of the structure of hierarchical components decomposed by hierarchical decomposition. The detailed calculation procedures of our proposed HLZC method can be summarized into four steps as follows.

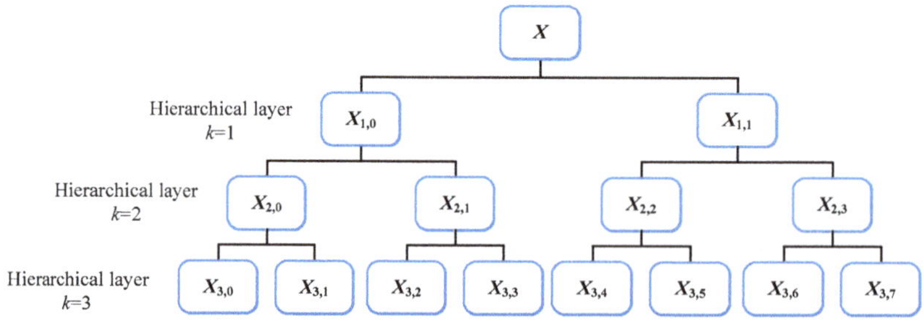

Figure 3. The structure of hierarchical components with three hierarchical layers.

(1) For an arbitrary time series $X\{x(i), i = 1, 2, \cdots, N\}$, the averaging operator Q_0 and differential operator Q_1 can be expressed as follows:

$$Q_0(x) = \frac{x(i) + x(i+1)}{2} \quad i = 1, 2, \cdots, N-1 \tag{6}$$

$$Q_1(x) = \frac{x(i) - x(i+1)}{2} \quad i = 1, 2, \cdots, N-1 \tag{7}$$

where $Q_0(x)$ and $Q_1(x)$ denote the low-frequency component and high-frequency component for a given time series, respectively.

(2) Conducting of the operators Q_j matrix (j = 0 or 1) can be adaptively generated according to the length of the time series N as follows:

$$Q_j = \begin{bmatrix} \frac{1}{2} & \frac{(-1)^j}{2} & 0 & 0 & \cdots & 0 & 0 \\ 0 & 0 & \frac{1}{2} & \frac{(-1)^j}{2} & \cdots & 0 & 0 \\ \vdots & \vdots & \vdots & \vdots & \vdots & \vdots & \vdots \\ 0 & 0 & 0 & 0 & \cdots & \frac{1}{2} & \frac{(-1)^j}{2} \end{bmatrix}_{N/2 \times N} \tag{8}$$

(3) Repeat step (2) to obtain the hierarchical components $X_{k,e}$ as Equation (9).

$$X_{k,e} = Q_{r_k}^{k} \cdot Q_{r_{k-1}}^{k-1} \cdot \cdots \cdot Q_{r_1}^{1} \cdot x \tag{9}$$

where e is the number of hierarchical nodes. For hierarchical layer k, e can be obtained as follows:

$$e = \sum_{m=1}^{k} 2^{k-m} r_m \tag{10}$$

where $[r_1, r_2 \cdots, r_k]$ represents the unique vector corresponding to the integer e, and $\{r_m, m = 1, \ldots, k\} \in \{0, 1\}$ indicates the averaging or differential operator at the kth layer.

(4) Compute all the hierarchical components by repeatedly using steps (1)–(3). Then, calculate LZC values of all the nodes. The HLZC is the set of all LZC values expressed as

$$HLZC(X, k) = LZC\left(X_{k,e}\right) \tag{11}$$

The calculation process of HLZC is shown in Figure 4.

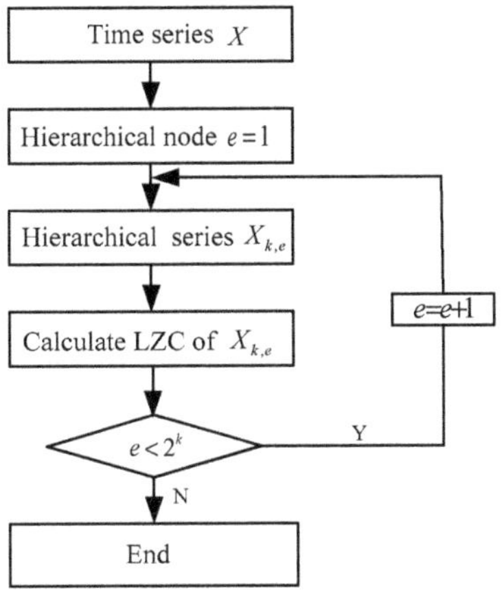

Figure 4. Flowchart of the hierarchical Lempel-Ziv complexity (HLZC) method.

2.4. Simulated Impulsive Signal

In this subsection, one impulsive signal was adopted to verify the advantage of proposed HLZC in detecting various fault types. Three different bearing fault types were utilized in the simulated signal: bearing with ball fault, bearing with inner race fault, and bearing with outer race fault. The number of sample points of the synthetic signal was 24,600, which was cut out by a sliding window of 2048 points with a step length of 256 points. Three simulated faulty bearing signals in time domains are shown in Figure 5a. For comparison purposes, LZC, MLZC, and HLZC were all utilized to process the impulsive signals. In this paper, three commonly used distance measures—Euclidean distance (ED), Chebychev distance (CD), and Minkowski distance (MD)—were all applied to verify the advantage of our proposed method in tracking the impulses. Note that we averaged the first 10 samples as normal samples. The distance value between each sample and normal samples was computed for comparisons. Here, we set the scale $\tau = 1{:}20$ for MLZC and the number of hierarchical layers $k = 4$ for HLZC methods.

The obtained results are shown in Figures 5–7. Two conclusions can be drawn from Figures 5–7 as follows. First, it can be observed from (b) and (c) of three figures that the original LZC and MLZC could not detect the impulses derived from three different bearing fault types, resulting in failure of bearing feature extraction. In contrast, our proposed HLZC method not only identified three different bearing types by tracking the impulse, but also generated less fluctuation and high accuracy, as shown in (d) of the three figures. Second, the results using the three distance measures had a high consistency, and thus the fault detection ability of three methods can be listed as HLZC > MLZC > LZC. This further demonstrated that our proposed HLZC method has a significant advantage in fault feature extraction.

To explain the advantage of our proposed HLZC in detecting the impulses, we enlarged the bearing with outer race fault signal and conducted the fast Fourier transform (FFT) analysis. The time domain waveforms and its corresponding frequency spectra are shown in Figure 8a,b, respectively. As can be seen, the fault information was mainly located in the high-frequency component (1000–1500 Hz). The fault information embedded in the high-frequency component was ignored using the traditional MLZC method, resulting in worse impulsive detection performance. Compared with the MLZC method, HLZC utilized both averaging and differencing process to extract the fault information hidden in both low-frequency and the high-frequency components. Therefore, the proposed HLZC method generates the best performance in fault feature extraction.

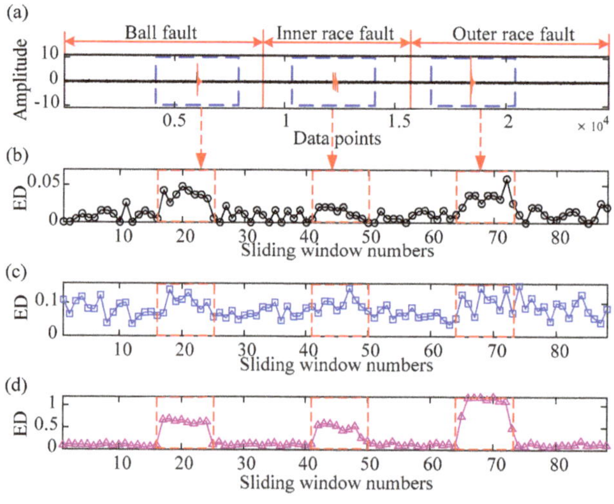

Figure 5. Results of sliding window analysis of Lempel-Ziv complexity (LZC), MLZC, and HLZC methods: (**a**) the simulated impulsive signal, (**b**) the Euclidean distance (ED) value of LZC, (**c**) the ED value of MLZC, and (**d**) the ED value of HLZC.

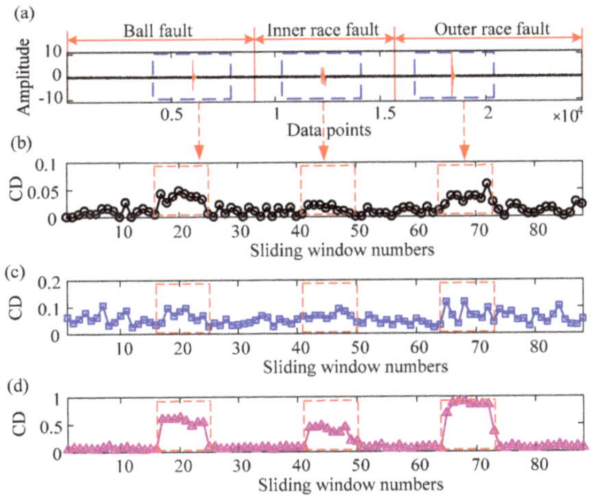

Figure 6. Results of sliding window analysis of LZC, MLZC, and HLZC methods with Chebychev distance (CD): (**a**) the simulated impulsive signal, (**b**) the CD value of LZC, (**c**) the CD value of MLZC, and (**d**) the CD value of HLZC.

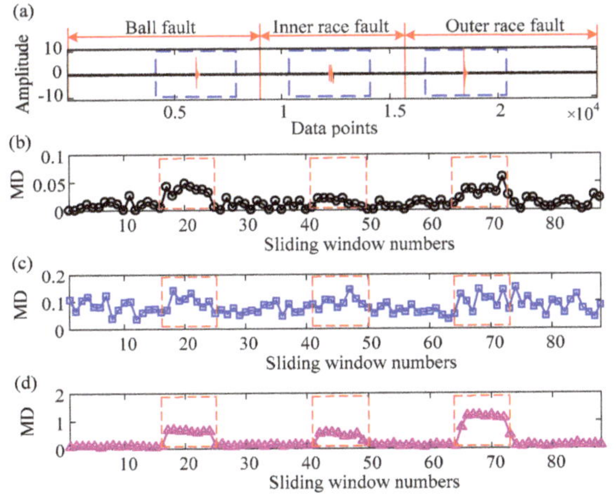

Figure 7. Results of sliding window analysis of LZC, MLZC, and HLZC methods with Minkowski distance (MD): (**a**) the simulated impulsive signal, (**b**) the MD value of MLZC, (**c**) the MD value of LZC, and (**d**) the MD value of HLZC.

Note that the fault frequency (BPFO) obtained using the envelope spectrum analysis was equal to the side band interval frequency of bearing intrinsic frequency in the conventional FFT method. First, the frequency spectrum using FFT could not detect the bearing fault frequency (BPFO) directly. It can be seen in Figure 8b that the bearing faults could be diagnosed by observing the intrinsic frequency and its harmonics. Second, we also conducted the envelope spectrum analysis. The fault frequency (BPFO) and its harmonics can be clearly observed in Figure 8c. Because the bearing fault can generate repetitive impulses, the measured vibration signal is thus a typical amplitude- and frequency-modulated (AM-FM) signal. The amplitude-modulated (AM) signal can be obtained by using envelope demodulated analysis, and then the fault frequency (BPFO) can be obtained by conducting the FFT on the AM signal.

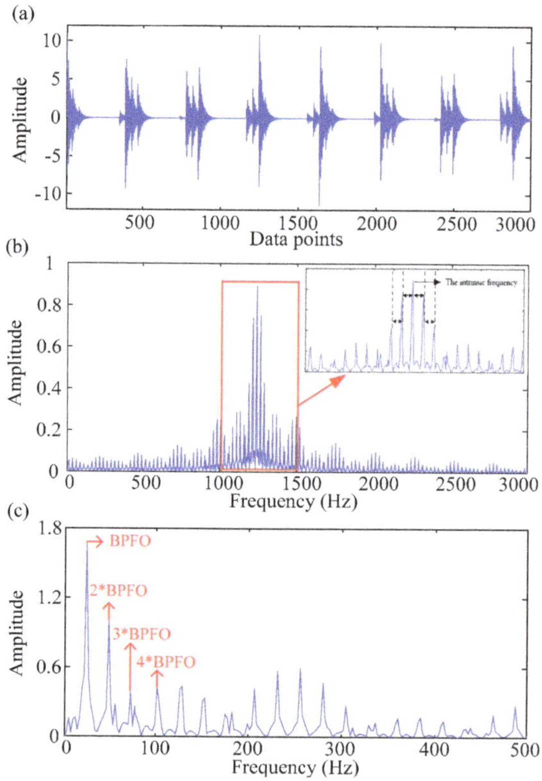

Figure 8. The simulated bearing signal with outer race fault: (**a**) the waveform, (**b**) its corresponding fast Fourier transform (FFT) spectrum, (**c**) the envelope spectrum.

3. Proposed Fault Diagnosis Framework

3.1. Support Vector Machine

In this paper, support vector machine (SVM) [30] was taken as the classifier to accomplish the pattern identification. SVM is a typical supervised learning method for recognition and regression analysis. The kernel function plays a significant role in the SVM classifier, which is not only important to reduce the computation cost but also useful in transforming the features into high dimension so as to construct the hyper-plane [28,31].

Three different used kernel functions of SVM consist of linear kennel function, polynomial kernel functions, and radial basis function (RBF) kernel function, which can be expressed as follows:

(1) Linear kernel

$$K(x, x_i) = < x \cdot x_i > \tag{12}$$

(2) Polynomial kernel

$$K(x, x_i) = (< x \cdot x_i > +c)^d \tag{13}$$

(3) Radial basis function (RBF) kernel

$$K(x, x_i) = \exp\left\{-\gamma \|x - x_i\|^2\right\} \tag{14}$$

where $\gamma > 0$, and γ is the kernel parameter.

Among these functions, the RBF function is most widely used due to its good performance, in which there are two parameters—penalty parameter C and kernel parameter γ—which require optimization. Here, the grid search method was utilized to optimize the two parameters [28]. It is worth mentioning that the dataset was randomly split into training and testing subsets through a k-fold cross-validation (CV). Every k subset takes turns to perform as an independent test set for the rest of the $(k-1)$ training subsets. The test sets being independent provides the necessary compensation for CV so as to enhance the consistency in the output. In this paper, the fivefold CV was adopted to adjust the model parameters. The coarse grid points were firstly selected through the exponential growing sequence 2^{-I} to 2^{+I}, where I is an integer. The optimal parameters for C and γ are assumed as (j,k), and the values are further optimized by using finer grid, in which the respective search area is $2^{j\pm f}$ and $2^{k\pm f}$. In this paper, we set the range of f as $-0.75 \leq f \leq 0.75$, with an interval of 0.25.

It is worth noticing that the one-against-one approach is applied to solve the multi-class classification problem in this paper [32]. To deal with the k number classes, the $k(k-1)/2$ SVM models are required for classification. For the present work, LIBSVM software package was used to deal with the multiple-fault diagnosis of rotating machinery.

3.2. Proposed Method

With the help of the advantages of HLZC and SVM, a novel intelligent fault diagnosis scheme called HLZC-SVM is proposed in this paper. There are two stages in the proposed intelligent fault diagnosis framework. In the first stage, HLZC is employed to extract the fault features from the vibration signals of rotating machinery. In the second stage, SVM is adopted to identify different fault types. Five steps are included in the FDRM method as follows:

(1) Measure the vibration data for various conditions of rotating machinery;
(2) Partition the measured vibration data into training datasets and testing datasets;
(3) Utilize HLZC to extract fault features from the vibration signals. Note that the hierarchical decomposition layers of HLZC is set as $k = 4$, and thus 31 features will be obtained;
(4) Train SVM classifier using the training features;
(5) Test the trained SVM classifier, wherein the output of SVM can be used to recognize the different fault types of rotating machinery.

The flowchart of the proposed fault diagnosis method is shown in Figure 9.

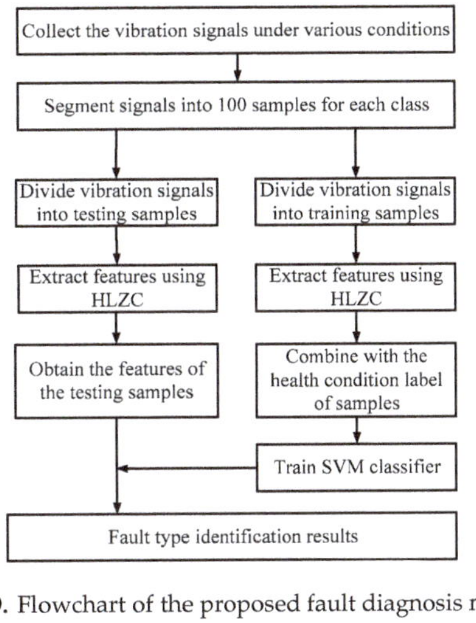

Figure 9. Flowchart of the proposed fault diagnosis method.

4. Experimental Validations

In order to validate the effectiveness of the proposed HLZC method in extracting the fault features, we designed two experiments in this paper. The two experiments were conducted on a fault simulator made by SpectraQuest called machinery fault simulator (MFS), which is drawn in Figure 10. The MFS consists of rolling bearings, a driven motor, and a three-way gearbox. At the rear of the gearbox, a magnetic clutch was used to generate the radial load. To collect the vibration data, we installed an acceleration transducer on the top of the gearbox. The rotating speed of the motor was kept at constant vale of 3000 rpm. Note that a 5 in-lbs of torque was added to simulate the machine load environment. The sampling frequency was 12,800 Hz. To simulate different faults, we replaced the test bearing and test gear by artificial damaged gear and artificial local damaged bearing, as shown in Figure 11.

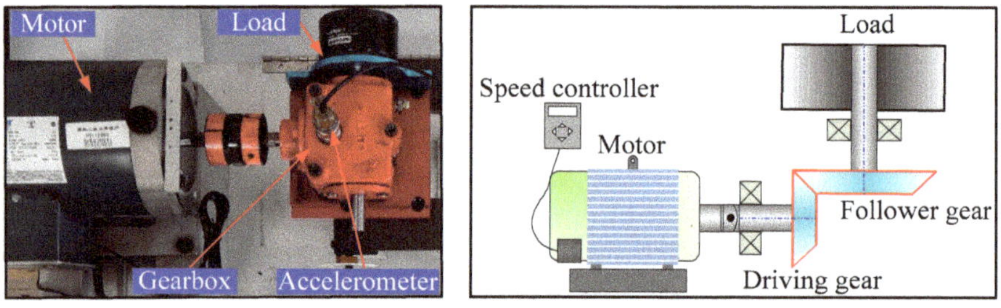

Figure 10. The experimental platform and its schematic diagram.

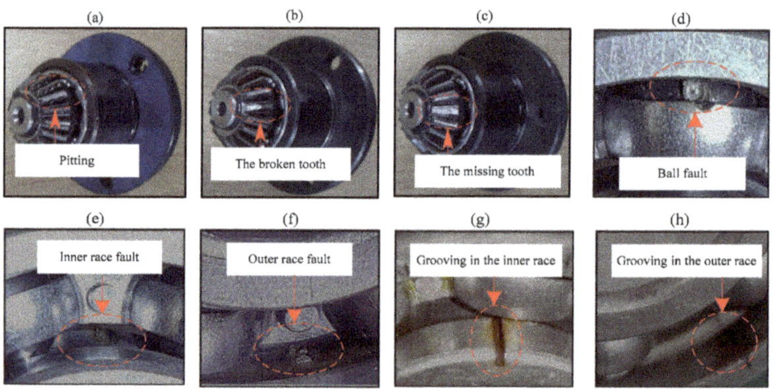

Figure 11. Faulty gears and bearings: (**a**) pitting in the driving tooth, (**b**) broken tooth in the driving tooth, (**c**) missing tooth in the driving tooth, (**d**) ball fault, (**e**) inner race fault, (**f**) outer race fault, (**g**) grooving in the inner race, (**h**) grooving in the outer race.

Experiment 1 aimed to show the superiority of the proposed HLZC method for single FDRM, which only covers single fault types of rotating machinery. Experiment 2 aimed to simulate the compound fault of rotating machinery including the bearing and gear fault, which was used to validate the superiority of the proposed HLZC method for compound FDRM. In this experiment, 50% of samples were randomly chosen as training samples, and the remainder of samples were used as testing samples. For comparison purposes, LZC and MLZC were all applied to process the data collected from the two experiments.

4.1. Experiment 1

Experiment 1 consisted of one healthy condition and five single fault conditions, including inner race fault (IRF), ball fault (BF), grooving in the inner race (GIR), grooving in the outer race (GOR), and outer race fault (ORF). There were 100 samples in each class and 600 samples in total. Meanwhile, the data length was 2048 points. The waveforms under six healthy conditions are shown in Figure 12. Table 1 gives the detailed information of six healthy conditions, including class label, damage diameter, and the numbers of training and testing data.

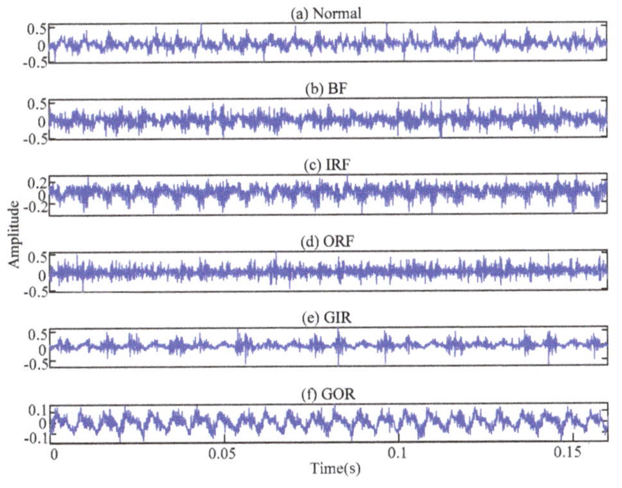

Figure 12. The time domain waveforms of rotating machinery under six healthy conditions in Experiment 1: (**a**) health condition (Normal), (**b**) ball fault (BF), (**c**) inner race fault (IRF), (**d**) outer race fault (ORF), (**e**) grooving in the inner race (GIR), (**f**) grooving in the outer race (GOR).

Table 1. Detailed information of six conditions in Experiment 1.

Fault Class	Class Label	Damage Diameter (mm)	Number of Training Samples	Number of Testing Samples
Normal	1	0	50	50
Ball fault	2	0.01	50	50
Inner race fault	3	0.01	50	50
Outer race fault	4	0.01	50	50
Grooving in the inner race	5	0.2	50	50
Grooving in the outer race	6	0.2	50	50

Following the steps in Section 3, we firstly utilized the proposed HLZC method to extract the fault features. Then, the obtained features were fed into SVM for classification. The obtained results are shown in Figure 13. It can be seen from 0 that 16 samples were misclassified and the final accuracy was 94.67%. For comparison, the MLZC and LZC were also tested. To avoid randomness, we carried out 20 trials. Figure 14 and Table 2 illustrate the detailed recognition results using three methods. First, the proposed HLZC method achieved the highest average classification accuracy of 94.3% (ranging from 91.33% to 95.67%). This can be attributed to the high frequency components considered in the HLZC method, which can contribute more information in the high frequency to generate a more accurate estimation of complexity. Second, the MLZC method obtained the second-highest average classification accuracy of 91.72% (ranging from 89.33% to 93.67%). Third, the LZC method had the lowest classification accuracy of 63.47% (ranging from 61.67% to 66.33%) due to the ineffectiveness of single analysis. It is indicated that HLZC had the best performance in extracting fault features among the three methods.

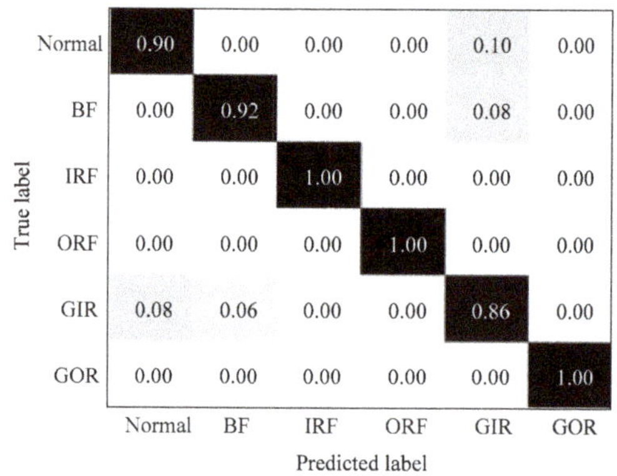

Figure 13. Confusion matrix using the proposed HLZC method for Experiment 1.

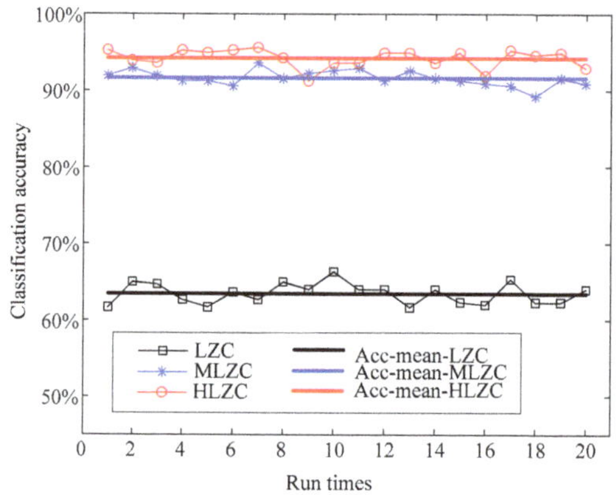

Figure 14. Diagnosis results of 20 trials using three methods in Experiment 1.

Table 2. Detailed classification accuracy of the experimental datasets in Experiment 1 and Experiment 2.

Experiments	HLZC Accuracy (%)			MLZC Accuracy (%)			LZC Accuracy (%)		
	Max	Min	Mean	Max	Min	Mean	Max	Min	Mean
1	95.67	91.33	94.30	93.67	89.33	91.72	66.33	61.67	63.47
2	97.20	92	94.72	90	84.80	87.82	45.20	36	41.22

4.2. Experiment 2

Experiment 2 aimed to investigate the performance of HLZC in compound fault diagnosis of rotating machinery. Experiment 2 was composed of five compound fault types: health condition (Normal), broken tooth in the driving tooth with inner race fault (BI), missing tooth in the driving tooth with inner race fault (MI), health tooth in the driving tooth with inner race fault (NI), and pitting in the driving tooth with inner race fault (PI). There were 100 samples in each class and 500 samples in total. In addition, the data length was 2048 points. The waveforms under five working types are shown in Figure 15. The detailed information of five compound fault conditions is shown in Table 3.

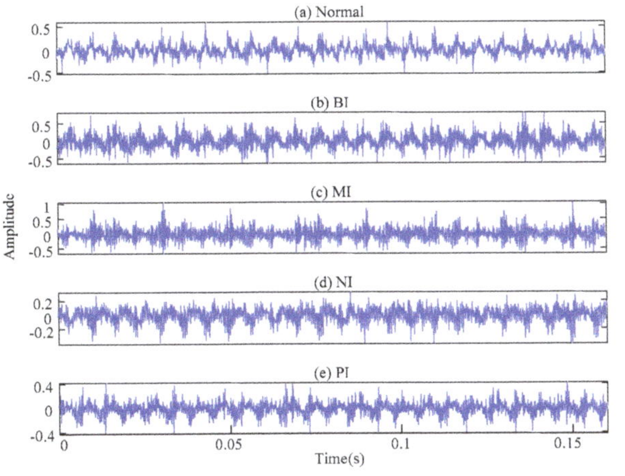

Figure 15. The time domain waveforms of rotating machinery in Experiment 2: (**a**) health condition (Normal), (**b**) broken tooth in the driving tooth with inner race fault (BI), (**c**) missing tooth in the driving tooth with inner race fault (MI), (**d**) health tooth in the driving tooth with inner race fault (NI), (**e**) pitting in the driving tooth with inner race fault (PI).

Table 3. Detailed information of five conditions in Experiment 2.

Fault Class	Class Label	Damage Diameter (mm)	Number of Training Samples	Number of Testing Samples
Normal	1	0	50	50
BI	2	0.01	50	50
MI	3	0.01	50	50
NI	4	0.01	50	50
PI	5	0.01	50	50

Like Experiment 1, the proposed HLZC-SVM method was also utilized for fault type identification of rotating machinery. Figure 16 shows the classification results. From Figure 16, we can see that there were a total of 10 samples misclassified with an accuracy of 96%. For comparison, the LZC and MLZC methods were also tested. The testing accuracies of the four methods are listed in Table 2 and Figure 17. As can be seen, the LZC-SVM was not effective, with an average classification accuracy of 41.22%. Second, combined with multi-scale analysis, the diagnosing performance of the MLZC-SVM method was enhanced with the average classification accuracy of 87.82%. Lastly, the proposed HLZC-SVM method had the highest average classification accuracy of 94.72%.

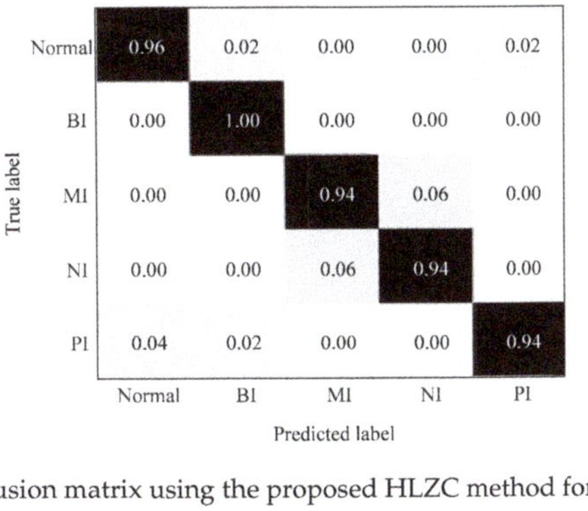

Figure 16. Confusion matrix using the proposed HLZC method for Experiment 2.

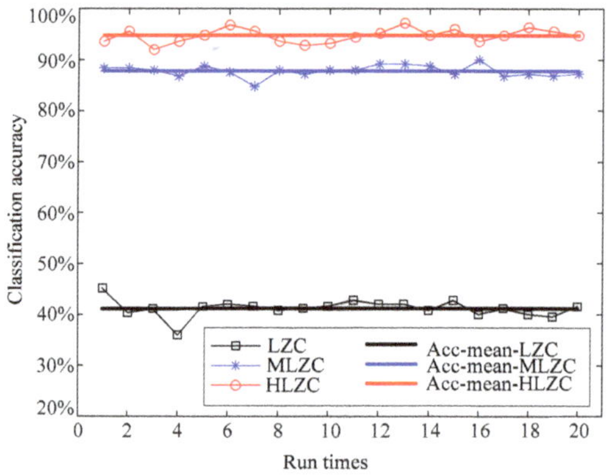

Figure 17. Diagnosis results of 20 trials using three methods in Experiment 2.

To better evaluate the classifier performance, we calculated the receiver operating characteristic (ROC) curves of HLZC-SVM, MLZC-SVM, and LZC-SVM methods using the experimental data. Here, we ran each method 20 times. The average ROC curve chart and AUC mean value of five health conditions of rotating machinery using three methods are shown in Figure 18. It can be observed from Figure 18a that our proposed HLZC-SVM method had the best performance with the highest AUC value for five health conditions (Normal with 1, BI with 1, MI with 0.99, NI with 0.99, and PI with 0.99). In contrast, the AUC values of MLZC-SVM and LZC-SVM for five health conditions showed a decreasing trend, as shown in Figure 18b,c, respectively. Among the three methods, the LZC-SVM method performed worst, in which the AUC values of five health conditions were only 0.5, 0.93, 0.53, 0.50, and 0.86. The comparison results further demonstrate that our proposed HLZC-SVM method had the best classification ability compared with the other two methods.

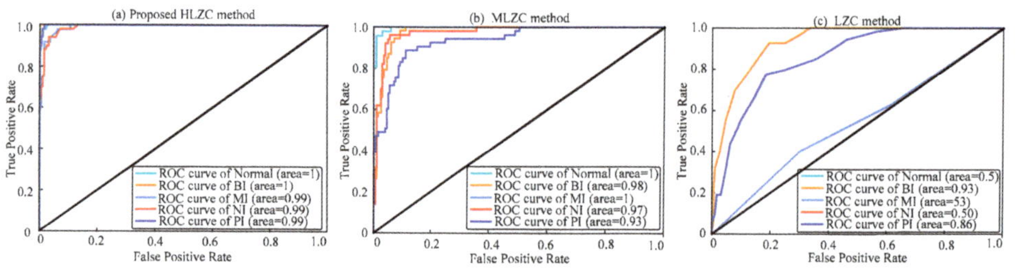

Figure 18. Performance comparison between three methods: (**a**) HLZC-support vector machine (SVM), (**b**) MLZC-SVM, and (**c**) LZC-SVM.

Moreover, random forest (RF) classifier was also applied for pattern identification for comparison. The ROC curve chart and AUC value were used to evaluate classification performance. Figure 19 shows average ROC curve chart and AUC mean value after running 20 times. As can be seen, the SVM classifier obtained a larger AUC value of 0.99 compared with RF classifier (AUC with 0.97), which meant the SVM had a better classification performance in recognizing various fault types of rotating machinery.

To gain a clear sense of the cluster ability of the extracted features using HLZC and MLZC methods, we used two-dimensional projection for visualization with PCA, as drawn in Figure 20. In 0a, it can be observed that the HLZC features of the five health conditions had a clear boundary and each cluster was individually separate. However, for the MLZC method, a few features were mixed, resulting in difficulty for classification. This phenomenon indicated that the fault features extracted using HLZC had more cluster ability than the MLZC method.

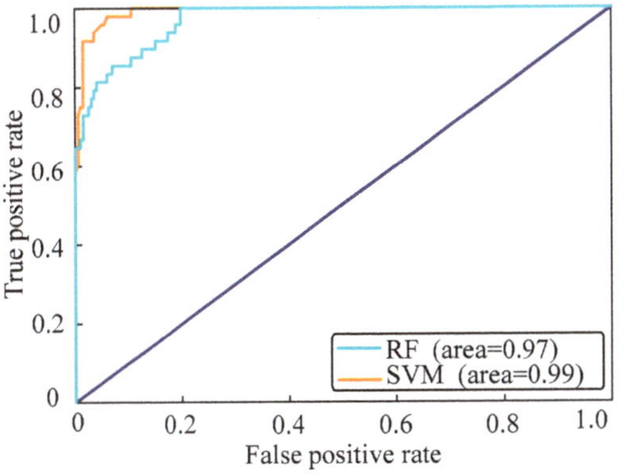

Figure 19. Performance comparison for different random forest (RF) and SVM.

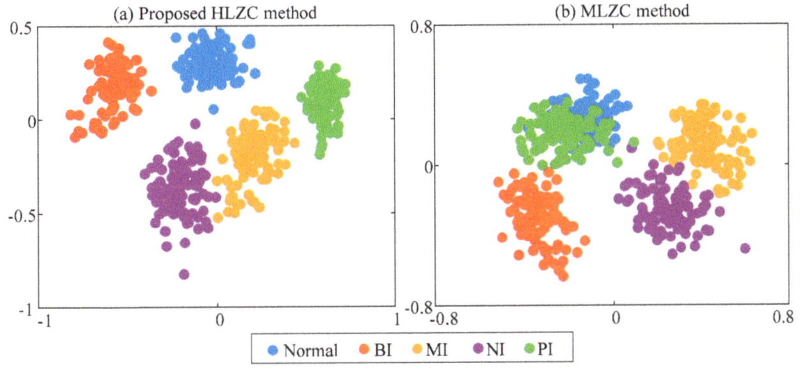

Figure 20. Projections of two-dimensional visualization of the obtained features in Experiment 2: (**a**) our proposed HLZC method, (**b**) traditional MLZC method.

We also tested the performance of our proposed HLZC method using different percentages of samples for training (the remaining samples will be considered as testing samples). Eight percentages were tested: 10% to 80%. To reduce randomness, 20 trials were conducted for each percentage. The averaging of training and testing accuracies were calculated and their corresponding standard deviations are illustrated in Figure 21. When the percentage increased to 50%, it achieved the highest classification accuracy at 94.72%. Therefore, we selected 50% of samples for training to demonstrate the advantage of our proposed HLZC method.

In order to discuss the influence of layer k, we applied the data of Experiment 2 for validation. Figure 22 shows the obtained classification accuracies using different layer k. As can be seen, when the layer $k < 4$, the classification accuracy will be significantly enhanced as the layer k rises. However, a larger layer k will greatly enhance the central processing unit (CPU) time. We observed that when the layer $k = 4$ increased to 5, the obtained classification accuracy was only improved by 0.08% from 94.72% (layer $k = 4$) to 94.80% (layer $k = 5$). However, the CPU time for layer $k = 5$ was 2116 s, which was almost double that of layer $k = 4$ at 1095 s.

Moreover, there was one extremely useful byproduct of RF—variable importance measures [33]—which was calculated to show the contribution of different components for the final classification accuracy, as shown in Figure 23. Note that a lager importance value indicated that the component had a greater influence on final predictions. It can be seen that X_{50} generated by layer $k = 5$ had a lower feature importance compared with that of component X_{40} generated by layer $k = 4$. This phenomenon further indicated that the additional features generated by HLZC with $k = 5$ had a small contribution on the final classification accuracy, which was well consistent with the classification results in Figure 22. Considering both the classification accuracy and CPU time, we selected the layer $k = 4$ for HLZC in this study.

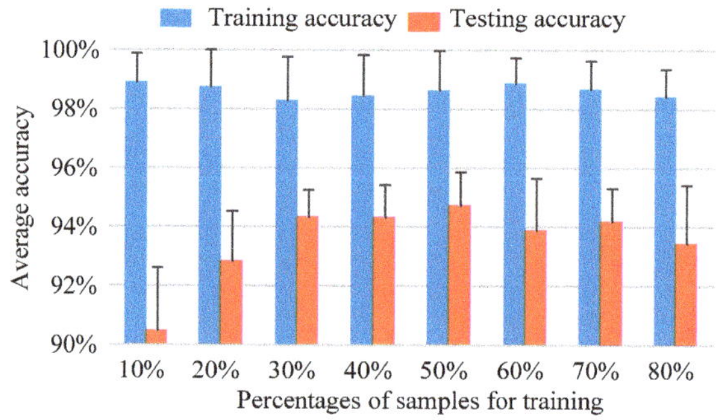

Figure 21. The classification results of the proposed method using training by different percentages of samples.

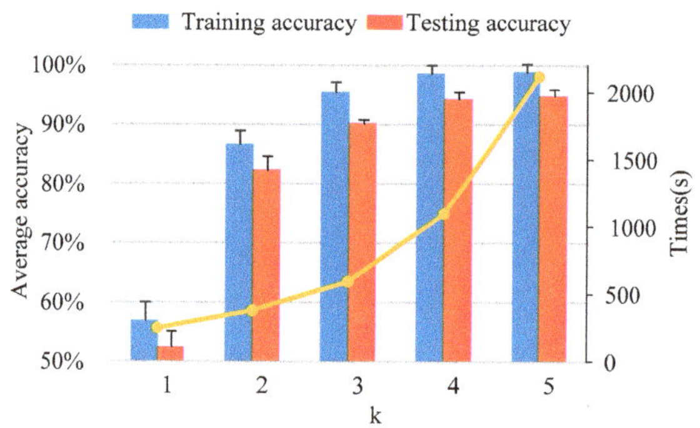

Figure 22. The mean classification accuracy and computation time using HLZC with different *k*.

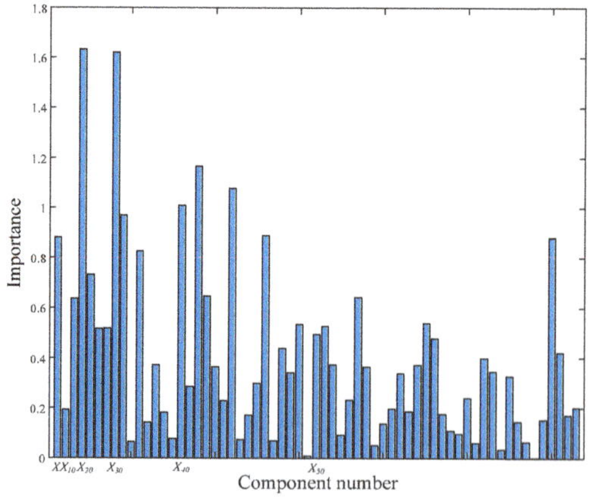

Figure 23. The importance of each component for the HLZC method.

For the MLZC method, we calculated the importance of each scale to show the contribution of different scales using data collected from Experiment 2. The obtained results are shown in Figure 24. It is worth mentioning that a lager importance value indicates that the component has a greater influence on final predictions. It can be seen that when the scale factor $\tau > 20$, the feature importance values of components showed a decreasing trend. This phenomenon indicated that the larger scale components ($\tau > 20$) had a small contribution on the final classification accuracy.

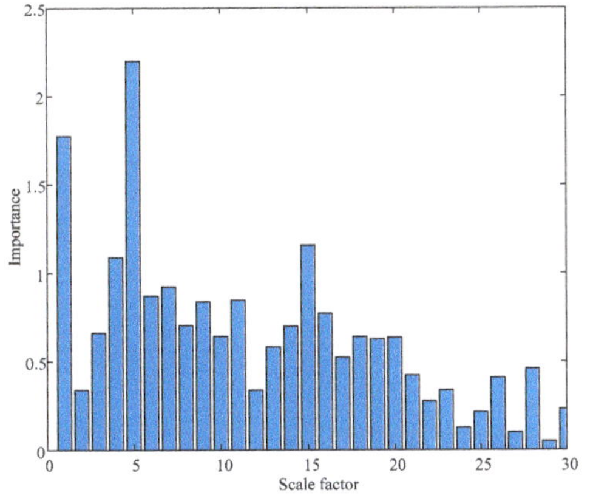

Figure 24. The importance of each scale factor for the MLZC method.

Additionally, the data of Experiment 2 were applied for validation to discuss the influence of scale factor τ. To reduce randomness, we conducted 20 trials conducted for each percentage. The testing accuracies were calculated and their corresponding CPU time is illustrated in Figure 25 and Table 4. As can be seen, when $\tau < 21$, the classification accuracy will be significantly enhanced as τ rises. When scale factor $\tau = 20$, it achieves the highest classification accuracy with 87.82%. Moreover, the larger scale means low calculation efficiency; thereby, we set scale factor $\tau = 1{:}20$ in this paper.

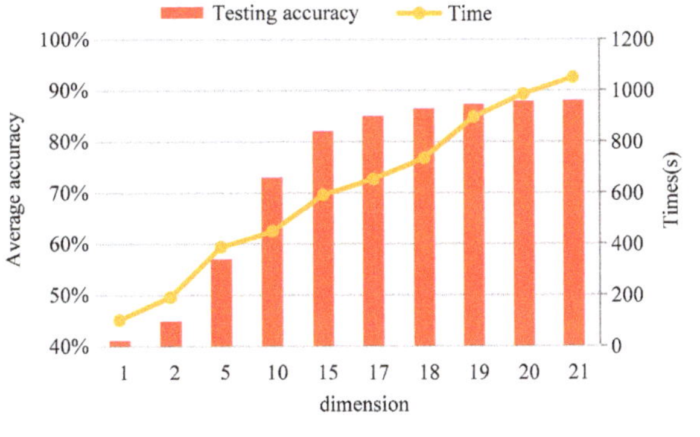

Figure 25. The mean classification accuracy and computation time using MLZC with different scale factor τ.

Table 4. Detailed classification accuracy of the experimental data using different scale factor τ.

Scale Factor τ	1	2	5	10	15	17	18	19	20	21
Max (%)	45.2	52	60	75	85	88	89	90	90	90
Min (%)	36	43	52	70	78	82	84	83	85	85
Mean (%)	41.22	45	57	73	82	85	86.4	87.2	87.82	87.2

5. Conclusions

A novel complexity analysis algorithm called HLZC was proposed for fault diagnosis of rotating machinery. The proposed HLZC can extract the fault information hidden in both low and high frequency components through the hierarchical decomposition. After the fault extraction, we utilized the SVM classifier to recognize different fault types of rotating machinery. To evaluate the performance of the proposed HLZC-SVM method, we used one simulated signal and two experimental signals with different fault types to verify the effectiveness of the HLZC-SVM method in FDRM. The comparison results demonstrated that the proposed HLZC-SVM method yielded the highest average classification accuracy of 94.3% and 94.72% for two cases, which was significantly higher than that of the LZ-SVM method (63.47% and 41.22%) and the MLZC-SVM method (91.72% and 87.82%). This further reinforces the fact that HLZC has certain advantages in fault feature extraction of rotating machinery. The main contributions of this paper include:

(1) LZC was extended to hierarchical decomposition analysis, namely, HLZC;
(2) HLZC considered the fault information hidden in both low-frequency and high-frequency components through conducting the averaging and differencing operations;
(3) A novel fault diagnosis scheme was proposed by combining HLZC and SVM;
(4) The proposed method was verified using both simulated and experimental signals.

There were some limitations for the proposed HLZC in fault diagnosis applications. First, although the proposed HLZC was demonstrated to be effective for the fault diagnosis of rotating machinery, it requires a large amount of labeled data for feature extraction and training the intelligent model for classification, which is difficult to meet sufficient labeled requirement in real industrial application scenarios. This limitation can be overcome by combing the feature knowledge transfer strategy with HLZC in future work. Second, the proposed HLZC lacks the denoising process to remove strong background noises so that it is difficult to extract the weak fault features from the strong noisy signal, especially at the early fault stage. This issue can hopefully be solved by applying the symbolic dynamic filtering to remove the noise-related fluctuations and reverse the fault-related information in future work.

Author Contributions: B.H. conceived the paper. B.H. and S.W. processed the data and wrote the paper. Q.Z. and X.Y. edited the manuscript. Y.L. helped to refined the whole paper and complete the final submission. All authors have read and agree to the published version of the manuscript.

Funding: This research was funded by the Key Project of National Natural Science Foundation of China (grant no. 51535009), the 111 Project (grant no. B13044), and the National Natural Science Foundation of China under grant 51805434, and in part by the Key Research Program, Shaanxi Province under grant 2019KW-017.

Conflicts of Interest: The authors declare no conflict of interest.All authors have read and agreed to the published version of the manuscript.

References

1. Liu, R.; Yang, B.; Zio, E.; Chen, X. Artificial intelligence for fault diagnosis of rotating machinery: A review. *Mech. Syst. Signal Process.* **2018**, *108*, 33–47. [CrossRef]
2. Li, Y.; Liu, F.; Wang, S.; Yin, J. Multi-scale Symbolic Lempel-Ziv: An Effective Feature Extraction Approach for Fault Diagnosis of Railway Vehicle Systems. *IEEE Trans. Ind. Inform.* **2020**, 1. [CrossRef]
3. Feng, Z.; Chen, X.; Liang, M. Iterative generalized synchrosqueezing transform for fault diagnosis of wind turbine planetary gearbox under nonstationary conditions. *Mech. Syst. Signal Process.* **2015**, *52*, 360–375. [CrossRef]
4. Abdeljaber, O.; Avci, O.; Kiranyaz, S.; Gabbouj, M.; Inman, D.J. Real-time vibration-based structural damage detection using one-dimensional convolutional neural networks. *J. Sound Vib.* **2017**, *388*, 154–170. [CrossRef]
5. Elasha, F.; Greaves, M.; Mba, D.; Fang, D. A comparative study of the effectiveness of vibration and acoustic emission in diagnosing a defective bearing in a planetry gearbox. *Appl. Acoust.* **2017**, *115*, 181–195. [CrossRef]

6. Sharma, R.B.; Parey, A. Modelling of acoustic emission generated due to pitting on spur gear. *Eng. Fail. Anal.* **2018**, *86*, 1–20. [CrossRef]
7. Zhao, M.; Lin, J. Health Assessment of Rotating Machinery Using a Rotary Encoder. *IEEE Trans. Ind. Electron.* **2018**, *65*, 2548–2556. [CrossRef]
8. Peng, Z.; Tse, P.W.; Chu, F. An improved Hilbert–Huang transform and its application in vibration signal analysis. *J. Sound Vib.* **2005**, *286*, 187–205. [CrossRef]
9. Li, G.; Tang, G.; Wang, H.; Wang, Y. Blind source separation of composite bearing vibration signals with low-rank and sparse decomposition. *Measurement* **2019**, *145*, 323–334. [CrossRef]
10. Qin, Y. A New Family of Model-Based Impulsive Wavelets and Their Sparse Representation for Rolling Bearing Fault Diagnosis. *IEEE Trans. Ind. Electron.* **2018**, *65*, 2716–2726. [CrossRef]
11. Li, J.; Yao, X.; Wang, H.; Zhang, J. Periodic impulses extraction based on improved adaptive VMD and sparse code shrinkage denoising and its application in rotating machinery fault diagnosis. *Mech. Syst. Signal Process.* **2019**, *126*, 568–589. [CrossRef]
12. Zheng, J.; Pan, H.; Cheng, J. Rolling bearing fault detection and diagnosis based on composite multiscale fuzzy entropy and ensemble support vector machines. *Mech. Syst. Signal Process.* **2017**, *85*, 746–759. [CrossRef]
13. Yan, R.; Gao, R. Complexity as a Measure for Machine Health Evaluation. *IEEE Trans. Instrum. Meas.* **2004**, *53*, 1327–1334. [CrossRef]
14. Yan, R.; Gao, R.X. Approximate Entropy as a diagnostic tool for machine health monitoring. *Mech. Syst. Signal Process.* **2007**, *21*, 824–839. [CrossRef]
15. Zheng, J.; Cheng, J.; Yang, Y.; Luo, S. A rolling bearing fault diagnosis method based on multi-scale fuzzy entropy and variable predictive model-based class discrimination. *Mech. Mach. Theory* **2014**, *78*, 187–200. [CrossRef]
16. Yan, R.; Liu, Y.; Gao, R.X. Permutation entropy: A nonlinear statistical measure for status characterization of rotary machines. *Mech. Syst. Signal Process.* **2012**, *29*, 474–484. [CrossRef]
17. Zheng, J.; Pan, H.; Yang, S.; Cheng, J. Generalized composite multiscale permutation entropy and Laplacian score based rolling bearing fault diagnosis. *Mech. Syst. Signal Process.* **2018**, *99*, 229–243. [CrossRef]
18. Li, Y.; Yang, Y.; Li, G.; Xu, M.; Huang, W. A fault diagnosis scheme for planetary gearboxes using modified multi-scale symbolic dynamic entropy and mRMR feature selection. *Mech. Syst. Signal Process.* **2017**, *91*, 295–312. [CrossRef]
19. Wu, S.-D.; Wu, C.-W.; Lin, S.-G.; Lee, K.-Y.; Peng, C.-K. Analysis of complex time series using refined composite multiscale entropy. *Phys. Lett. A* **2014**, *378*, 1369–1374. [CrossRef]
20. Hong, H.; Liang, M. Fault severity assessment for rolling element bearings using the Lempel–Ziv complexity and continuous wavelet transform. *J. Sound Vib.* **2009**, *320*, 452–468. [CrossRef]
21. Cui, L.; Gong, X.; Zhang, J.; Wang, H. Double-dictionary matching pursuit for fault extent evaluation of rolling bearing based on the Lempel-Ziv complexity. *J. Sound Vib.* **2016**, *385*, 372–388. [CrossRef]
22. Cui, L.; Li, B.; Ma, J.; Jin, Z. Quantitative trend fault diagnosis of a rolling bearing based on Sparsogram and Lempel-Ziv. *Measurement* **2018**, *128*, 410–418. [CrossRef]
23. Bai, Y.; Liang, Z.; Li, X. A permutation Lempel-Ziv complexity measure for EEG analysis. *Biomed. Signal Process. Control.* **2015**, *19*, 102–114. [CrossRef]
24. Yin, J.; Xu, M.; Zheng, H. Fault diagnosis of bearing based on Symbolic Aggregate approXimation and Lempel-Ziv. *Measurement* **2019**, *138*, 206–216. [CrossRef]
25. Li, Y.; Xu, M.; Wei, Y.; Huang, W. A new rolling bearing fault diagnosis method based on multiscale permutation entropy and improved support vector machine based binary tree. *Measurement* **2016**, *77*, 80–94. [CrossRef]
26. Costa, M.; Goldberger, A.L.; Peng, C.-K. Multiscale Entropy Analysis of Complex Physiologic Time Series. *Phys. Rev. Lett.* **2002**, *89*, 068102. [CrossRef]
27. Li, Y.; Wang, S.; Deng, Z. Intelligent fault identification of rotary machinery using refined composite multi-scale Lempel–Ziv complexity. *J. Manuf. Syst.* **2020**. [CrossRef]
28. Li, Y.; Xu, M.; Wang, R.; Huang, W. A fault diagnosis scheme for rolling bearing based on local mean decomposition and improved multiscale fuzzy entropy. *J. Sound Vib.* **2016**, *360*, 277–299. [CrossRef]
29. Jiang, Y.; Peng, C.-K.; Xu, Y. Hierarchical entropy analysis for biological signals. *J. Comput. Appl. Math.* **2011**, *236*, 728–742. [CrossRef]

30. Elton, R.J.; Vasuki, P.; Mohanalin, J. Voice Activity Detection Using Fuzzy Entropy and Support Vector Machine. *Entropy* **2016**, *18*, 298. [CrossRef]
31. Li, Y.; Wang, X.; Si, S.; Huang, S. Entropy Based Fault Classification Using the Case Western Reserve University Data: A Benchmark Study. *IEEE Trans. Reliab.* **2020**, *69*, 754–767. [CrossRef]
32. Martínez-Morales, J.D.; Palacios-Hernández, E.R.; Campos-Delgado, D.U. Multiple-fault diagnosis in induction motors through support vector machine classification at variable operating conditions. *Electr. Eng.* **2016**, *100*, 59–73. [CrossRef]
33. Archer, K.J.; Kimes, R.V. Empirical characterization of random forest variable importance measures. *Comput. Stat. Data Anal.* **2008**, *52*, 2249–2260. [CrossRef]

applied sciences

MDPI

Article

Exploiting Generative Adversarial Networks as an Oversampling Method for Fault Diagnosis of an Industrial Robotic Manipulator

Ziqiang Pu [1], Diego Cabrera [2], René-Vinicio Sánchez [2], Mariela Cerrada [2], Chuan Li [3] and José Valente de Oliveira [4,*]

1 National Research Base of Intelligent Manufacturing Service, Chongqing Technology and Business University, China and Universidade do Algarve, 8005-139 Faro, Portugal; a66835@ualg.pt

2 GIDTEC Research Group, Universidad Politécnica Salesiana, Cuenca 010105, Ecuador; dcabrera@ups.edu.ec (D.C.); rsanchezl@ups.edu.ec (R.-V.S.); mcerrada@ups.edu.ec (M.C.)

3 National Research Base of Intelligent Manufacturing Service, Chongqing Technology and Business University, Chongqing 400067, China; chuanli@ctbu.edu.cn

4 Universidade do Algarve and with the Center of Intelligent Systems, IDMEC, Instituto Superior Técnico, Universidade de Lisboa, 1049-001 Lisboa, Portugal

* Correspondence: jvo@ualg.pt; Tel.: +351-289-800-10

Received: 30 September 2020; Accepted: 28 October 2020; Published: 31 October 2020

Abstract: Data-driven machine learning techniques play an important role in fault diagnosis, safety, and maintenance of the industrial robotic manipulator. However, these methods require data that, more often that not, are hard to obtain, especially data collected from fault condition states and, without enough and appropriated (balanced) data, no acceptable performance should be expected. Generative adversarial networks (GAN) are receiving a significant interest, especially in the image analysis field due to their outstanding generative capabilities. This paper investigates whether or not GAN can be used as an oversampling tool to compensate for an unbalanced data set in an industrial manipulator fault diagnosis task. A comprehensive empirical analysis is performed taking into account six different scenarios for mitigating the unbalanced data, including classical under and oversampling (SMOTE) methods. In all of these, a wavelet packet transform is used for feature generation while a random forest is used for fault classification. Aspects such as loss functions, learning curves, random input distributions, data shuffling, and initial conditions were also considered. A non-parametric statistical test of hypotheses reveals that all GAN based fault-diagnosis outperforms both under and oversampling classical methods while, within GAN based methods, an average accuracy difference as high as 1.68% can be achieved.

Keywords: feature extraction; generative adversarial network; random forest; unbalance data; fault diagnosis

1. Introduction

For data-driven machine learning techniques [1], it plays an important role in the machinery fault diagnosis and prognostics [2,3]. Recently, deep learning [4] emerges as one that has been progressively adopted to develop health monitoring systems for the machinery. One way of looking at deep learning is as a feature engineering method [5] that automatically extracts features from the collected signals. Propagating these signals from the input layer to layers with less and fewer neurons, the neural network is forced to represent the input data space into a lower dimensional feature space, which, in general, reduces overfitting and increases the accuracy.

Deep learning models such as convolutional neural networks [6], deep autoencoders [7], deep belief networks [8], and deep Boltzmann machines [9] have achieved outstanding results in fault diagnosis, and other fields essentially due to the above-described dimensionality reduction effect. In the industrial manipulator field, Nho Cho et al. [10] proposed an algorithm based on the multilayer perceptrons for the manipulator actuator fault detection. Wang et al. [11] proposed a multi-data fusion with an optimized convolutional neural network (CNN) for fault detection of rotating machinery. Ma et al. [12] proposed the convolutional multi-time scale echo state network (ESN) to efficient classification. Three different ESNs with different time scales were used to allow the recurrent neural network to successively refine the features in a similar way of a kernel in a CNN. Due to the fast training of ESN, Long et al. [13] proposed a deep echo state network optimized by particle swarm optimization to fault diagnosis of a wind turbine gearbox. Hu et al. [14] present an approach with deep Boltzmann machine and multi-grained scanning forest to effectively deal with industrial fault diagnosis. Wang et al. [15] proposed a new deep neural network model based on a deep Boltzmann machine for condition prognosis. Lee et al. [16] proposed a real-time fault diagnosis model using a deep neural network. Shao et al. [17] presented a continuous deep belief network (CDBN) for bearings fault detection. Shen et al. [8] used a deep belief network with an optimized function of Nesterov momentum (NM) for bearing fault diagnosis. Polic et al. [18] presented a new method of a convolutional neural network encoder for feature extraction in tactile robotics. D'Elia et al. [19] based on the study of how the power flows inside the time synchronous average of the ring gear and a modified statistical parameter for planet gears fault diagnosis. Zaidan et al. [20] use a Bayesian hierarchical model with utilizing fleet data from multiple assets to perform probabilistic estimation of remaining useful life for civil aerospace gas turbine engines.

However, all the above models are critically dependent on a representative, balanced, large enough data set that require data that, more often that not, are hard to obtain [21]. While data from a healthy state are abundant, data from faulty states are rare, sparse, and hardly representative of all possible faults. This could lead to low diagnosis precision in these intelligent fault diagnosis techniques. Without appropriated data, these machine learning methods simply do not have acceptable performance [22], and it is important to consider a method that mitigates such a data shortage.

As Robotic manipulators, they are wildly used in the industry as they can be used in a range of tasks such as assembly, painting, or welding [23]. However, the transmission system of a manipulator is prone to faults due to prolonged working periods [24]. Typically, these faults manifest in the connection parts, bearings, gears, or gear shafts. A faulty manipulator will be less precise, less efficient, less productive, and less secure—even though it could be a tricky task to obtain fault data for such precision machinery. Therefore, the monitoring of the manipulator health condition with limited data sources is of paramount interest.

For mitigating this, some fault diagnosis works resorted to the SMOTE. SMOTE stands for Synthetic Minority Oversampling Technique and aims at compensating for the data unbalance of a given class by increasing the number of samples in that class. Roughly, it creates a new synthetic sample by interpolating two existing similar samples of the same class, the new sample having the same class of the two originating samples [25]. SMOTE is a popular method but sometimes, by increasing the number of samples, it will also increase the overlapping between classes. An alternative approach to cope with the unbalanced data set problem is to resort to deep learning and in particular to generative adversarial networks (GAN). The idea is to use the generative model of a GAN to generate enough samples for effective training of the diagnoser. GAN was first proposed by Goodfellow [26] and consists of two adversarial models (neural networks): a generator and a discriminator.

The learning process can be described as a min-max game. The generator produces synthetic examples while the discriminator tries to decide whether the current input is either a real or a synthetic example. Both models improve their performance simultaneously up to a Nash equilibrium using a gradient-based optimization technique. The number of GAN applications has been steadily increasing. A sample of recent applications is offered next. In the image processing field, Ledig et al. [27] presented SRGAN, a generative adversarial network for image super-resolution. Isola et al. [28] demonstrated that the conditional adversarial network is a promising approach for image-to-image translation tasks. Based on the above comment, Zhu et al. [29] presented an approach for learning to translate an image from a source domain to a target domain in the absence of paired examples. Shao et al. [30] develop an auxiliary classifier GAN (ACCGAN) to learn from mechanical sensor signals and generate realistic one-dimensional raw data. Li et al. [31] proposed a novel fault detection method for 3D printers using GAN which consider only normal condition signals with outstanding performance. Mao et al. [32] used a GAN for an unbalanced data driven fault diagnosis of rolling bearings. In addition, for rolling bearings, Jiang et al. [33] proposed a novel anomaly detection approach based on GAN with only health data. Li et al. [34] applied a GAN for the feature space learning in fault diagnosis of 3D printers using only one sample of each faulty state. Wang et al. [35] proposed a method based on a conditional variational auto-encoder and a generative adversarial network for unbalanced fault diagnosis of a planetary gearbox.

To the best of our knowledge, no work reported model based generation of synthetic samples for fault diagnosis of robotic manipulators. Hence, the main contributions of our work are: (1) The application of GAN to generate synthetic examples (signals) representing fault states for mitigating the presence of an unbalanced data set in a fault diagnosis task of an industrial robotic manipulator. More concretely, GAN generates a synthetic wavelet packet transform based feature of a vibrational signal as acquired by an accelerometer; (2) A comprehensive study taking into account six different scenarios for mitigating the unbalanced data, including classical under and oversampling (SMOTE) methods as well as for assessing the effect of factors such as generator selection, the number of training examples in each class, data shuffling in training data, the distribution used for sampling input random data and initial conditions.

The rest of this paper is organized as follows. The proposed GAN based fault diagnosis scheme is specified in Section 2. The manipulator experiment was presented in Section 3. The fault diagnosis of the manipulator was analyzed in Section 4 to validate the experiment result. Finally, the conclusions and the future work were detailed in Section 5.

2. Methodology

2.1. Feature Extraction

Wavelet packet transform (WPT) can be viewed as a time frequency conversion technique of a non-stationary signal [36]. It complements the shortage of wavelet transforms that only decomposes the low frequency components but cannot extract high-resolution on high frequency components. The discrete wavelet transform of the discrete signal $f(t)$ is given by [37]:

$$w_{m,n}(t) =< f(t), \Psi_{m,n}(t) >= \frac{1}{\sqrt{m}} \int_{-\infty}^{+\infty} \Psi_{m,n}(\frac{t-n}{m}) f(t) dt, \tag{1}$$

where m is the scaling factor and n is the sifting factor, which are given, respectively, by:

$$m = a_0^m, \tag{2}$$

$$n = b_0 a_0^m. \tag{3}$$

When $a_0 = 2, b_0 = 1$, (1) can be re-written as

$$\Psi_{m,n}(t) = 2^{-\frac{m}{2}} \Psi(2^{-m} t - n) \tag{4}$$

In the wavelet transform, the signal $u(t)$ can be separated in the Hilbert space by a scaling and by a wavelet function. The scaling function $\Phi(t)$ corresponds to the low frequency part of the original signal while the wavelet function $\varphi(k)$ corresponds to the high frequency part of the original signal with initial conditions:

$$\Psi_{m,n}^0(t) = \Phi(t) \tag{5}$$

$$\Psi_{m,n}^1(t) = \varphi(t) \tag{6}$$

Figure 1 shows a 3-level decomposition of a WPT of a signal $u(t)$. This is decomposed in a high-frequency part $h_k(t)$ and in a low-frequency part $g_k(t)$. Each part is computed by a filter, i.e.,

$$h_k(t) = \frac{t_k + t_{k+1}}{2}, \tag{7}$$

$$g_k(t) = \frac{t_k - t_{k+1}}{2}, \tag{8}$$

The function using the above filters can be given by:

$$u_{2n}(t) = \sum_k h_k u_n(2t-k), \tag{9}$$

$$u_{2n+1}(t) = \sum_k g_k u_n(2t-k) \tag{10}$$

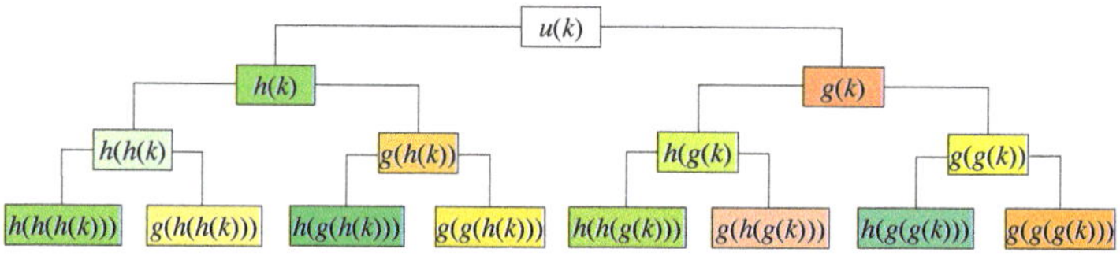

Figure 1. The decomposition levels of a wavelet packet transform of the signal $u(t)$.

As illustrated in the figure, this procedure can be recursively applied to both low and high-frequency parts. However, the number of decomposition levels will be limited by the actual application. Due to its smoothness and nonlinear characteristics, in this paper, we applied the Daubechies WPT with seven levels (Db7).

We further compute an informative statistics from the WPT as follows [38]:

$$p(m) = \sqrt{\sum_{n=1}^{N} (w_{m,n}(t))^2} \tag{11}$$

where N denotes the number of data in each node of the 7th decomposition level and $w_{m,n}(t)$ and is given by (1). Hence, a feature vector $\mathbf{p}$ can be defined for the signal $u(t)$ as follows:

$$u(t) \leftrightarrow \mathbf{p} = [p(1), p(2), \ldots, p(d)]^{\mathrm{T}} \tag{12}$$

where d is the number of features, i.e., with seven levels $d = 2^7 = 128$.

2.2. Generative Adversarial Network

A Generative adversarial network consists of two models: the generative model G(z) and the discriminative model D(x). The goal of the generative model is to produce synthetic samples such that the discriminate model could not distinguish them from the real samples. At the same time, the objective of the discriminative model is to accept real samples and reject synthetic ones with the highest possible accuracy. In equilibrium, the discriminative model cannot identify the source of the data, meaning that synthetic data are indistinguishable from real data. Figure 2 shows a block diagram of a GAN as used in this work

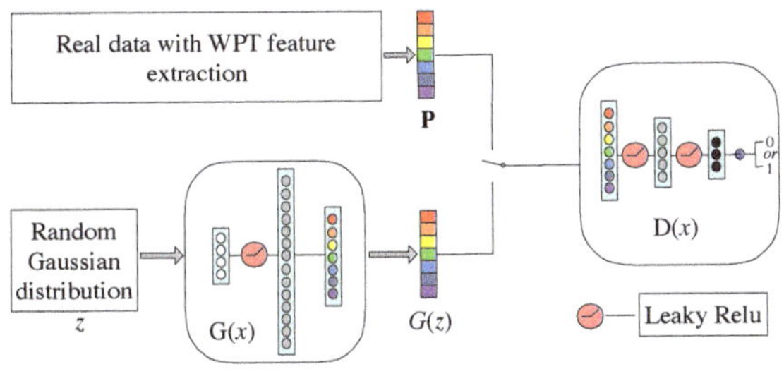

Figure 2. A block diagram of a GAN as used in this work.

For learning, a GAN implements an adversarial competition between the generator G(z) and the discriminator D(x). Initially, a real sample $u(t)$ is processed by the above described WPT feature extraction technique, and (12) is set as the real input of D(x). A random signal $z(n)$ with a given distribution is input into the generator which in turn produces a synthetic feature vector G(z). The discriminator is trained with a target value of 1 when a real sample is presented at its input, and with 0 for a synthetic example. This process repeats until a Nash equilibrium is reached. In general terms, the GAN optimization problem can be given by: In general terms, the GAN optimization problem can be given by:

$$\mathcal{G}^* = \arg\min\max V(\mathrm{G}, \mathrm{D}) \tag{13}$$

where max stands for the maximization of the probability of the generator G while min refers to the minimization of the probability in the discriminator D; $V(\mathrm{G}, \mathrm{D})$ is the GAN objective function that can be given by:

$$V(\mathrm{G}, \mathrm{D}) = \underset{x \sim P_{data}}{\mathbb{E}}[\log(\mathrm{D}(x))] + \underset{x \sim P_{\mathrm{G}}}{\mathbb{E}}[\log(1 - \mathrm{D}(\mathrm{G}(z)))] \tag{14}$$

where $E_{x \sim P_{data}}$ represents the expectation of real probability distribution, whereas $E_{x \sim P_G}$ represents the expectation of the random distribution. Since x and z are real-value random variables on the probability space, the expect values of x and z can defined as the integral of x and z, respectively. Therefore, (14) can be re-written as

$$V(\mathrm{G}, \mathrm{D}) = \int_x [P_{data}(x) \log \mathrm{D}(x) + P_{\mathrm{G}}(x) \log(1 - \mathrm{D}(x))] dx \tag{15}$$

where $\int_x P_{data}(x)dx$ stands for the expectation of $E_{x \sim P_{data}}(.)$ and $\int_x P_G(x)dx$ denote the expectation of $E_{x \sim P_G}(.)$. Let $f(x) = P_{data}(x) \log \mathrm{D}(x) + P_G(x) log(1 - \mathrm{D}(x))$; then, the derivative of $f(x)$ is given by

$$\dot{f}(x) = \frac{P_{data}(x)}{\log \mathrm{D}(x)} + \frac{P_{\mathrm{G}}(x)}{\log(1 - \mathrm{D}(x))} \tag{16}$$

where $P_{data}(x)$ is the distribution of the real data while $P_{\mathrm{G}}(x)$ is the distribution of generated data. When $\dot{f} \to 0$, D tends to the maximum value, which is given by

$$\mathrm{D}^* = \frac{P_{data}(x)}{P_{data}(x) + P_{\mathrm{G}}(x)} \tag{17}$$

When $P_G(x) = 0$, D^* becomes 1 meaning that the discriminator can effectively recognize synthetic data, when $P_G(x)$ is close to $P_{data}(x)$, D^* tends to the optimal value of 0.5, which means that synthetic data are indistinguishable from real data. Plugging in (17) into (15), one has:

$$V(\mathrm{G}, \mathrm{D}^*) \quad = \quad \int_x [P_{data}(x) \log \frac{P_{data}(x)}{P_{data}(x) + P_G(x)} + P_G(x) \log(1 - \frac{P_{data}(x)}{P_{data}(x) + P_G(x)})] dx \tag{18}$$

As the objective of generative part is to shrink the distance between real and generated data, the loss function of generative model can be defined as

$$V(G, D^*) = \int_x [P_G(x) log(1 - \frac{P_{data}(x)}{P_{data}(x) + P_G(x)})]dx \quad (19)$$

Under the above loss and, in general, the training process of a GAN is not stable and gives rise to model collapse. To mitigate this problem, the Wasserstein GAN was proposed where the KL divergence of the classic GAN was replaced by the 1-Wasserstein distance [39]. The Wasserstein GAN loss function is therefore given by:

$$V(G, D) = \mathop{\mathbb{E}}_{x \sim P_{data}} [D(x)] + \mathop{\mathbb{E}}_{x \sim P_G} [(D(G(z))] \quad (20)$$

This change in the loss function can make the convergence of the generator faster, but it can be further improved. In [40], an improved Wasserstein GAN was proposed in which an additional gradient penalty was added to (20), i.e.,

$$\begin{aligned} V(G, D) &= E_{x \sim P_{data}} [D(x)] + \mathop{\mathbb{E}}_{x \sim P_G} [(D(G(z))] \\ &\quad + \lambda \, \mathbb{E}[(|\delta D(\alpha x - (1 - \alpha G(z)))| - 1)^2] \end{aligned} \quad (21)$$

where α is a user-defined scaling factor and λ stands for the gradient penalty coefficient.

In another approach, Cabrera et al. [41] proposed a metric aiming at keeping track of the best current generator while training progresses. The metric is given by:

$$||D_r - D_g|| = (R_{mean} - G_{mean})^2 + (R_{std} - G_{std})^2 \quad (22)$$

where R_{mean} and G_{mean} are the centroids of the real and generated data clusters, respectively, while R_{std} and G_{std} are the real and generated data dispersion (standard deviation), respectively. The smaller (22) the closer the generated data are from real data. In each training step, (22) is computed for the current generator, and the generator exhibiting the lower current distance is viewed as the best current generator. Hereafter, we refer to this process as (model) generator selection.

A Vapnik Loss Inspired GAN

The loss function is a key issue for model selection. Therefore, we are adopting a recently proposed loss function within the GAN and more concretely within the Wasserstein GAN framework. This loss function, first proposed by Vapnik et al. [42], considers the geometry distance between the predict and original data. In brief, the classical loss function used in regression, given a set of N examples, can be given by:

$$L_c = \frac{1}{N} \sum_{i=1}^{N} (h(x_i) - y_i)^2 + \frac{1}{N} \sum_{i=1}^{N} \sum_{j=1}^{N} (h(x_i) - y_i) I (h(x_i) - y_i) \quad (23)$$

where x_i and y_i are the i-th independent and dependent observation, respectively, $h(x)$ is the regressor hypothesis, and I is the identity matrix. Based on the VC theory, in [42], the identity matrix is replaced by the so-called V-matrix, i.e., (23) becomes:

$$L_v = \frac{1}{N} \sum_{i=1}^{N} \sum_{j=1}^{N} (h(x_i) - y) V (h(x_i) - y) \quad (24)$$

where V is the V-matrix. For data in $\mathbb{R}^d$, the V-matrix can be computed for all $i, j = 1, \ldots, d$ as

$$V(i,j) = \sum_{k=1}^{d} (c_k - max(x_i^k - x_j^k)) \tag{25}$$

where d denotes the number of data dimensions, $0 \leq xi \leq c_i$ and $c_1, \ldots, c_d$ are non-negative constants. This approach has shown good results in the regression problems. Motivated by both the theoretical background and the experimental results obtained in regression problems, including in the framework of SVR [43], we are proposing to apply this loss function in the framework of GAN as a regularization term in (22), which becomes:

$$Rf_v(G,D) = E_{x\sim P_{data}}[D(x)] + E_{x\sim P_G}[(1D(x)] + \lambda E[(|\delta D(\alpha x - (1-\alpha G(z)))| - 1)^2] + L_v \tag{26}$$

where L_v is of the form (24).

2.3. Random Forests for Fault Classification

Ensemble learning uses a group of algorithms to get a better prediction than any of its base algorithms. A random forest (RF) is a homogeneous ensemble classifier that uses a set of decision trees (DT) [44,45]. Each DT is grown independently using the Bagging technique. In addition, and to increase diversity (reducing the correlation between trees), an RF grows each tree from a random selection of data features. Once trained, an RF uses a majority voting mechanism for making its classification (or regression) prediction.

The CART algorithm is frequently used to grow a decision tree. In the CART algorithm, the Gini index is the metric used for selecting the data set feature to be used in a given node of the tree. Given a node m and the estimated probability $p(c|m)(c = 1, 2, 3, \ldots, C)$, the Gini impurity index is defined as:

$$G(m) = \sum_{c_1 \neq c_2} p(c_1|m)p(c_2|m) = 1 - \sum_{c=1}^{C} p^2(c|m) \tag{27}$$

Let n be the splitting point of node m that separates the node into two portions in in which a proportion P_a of the samples in m is assigned to m_a and a proportion P_b is assigned to m_b, i.e., $P_a + P_b = 1$. Thus, the decrease in the Gini impurity index is defined as follows:

$$\delta G(m,n) = G(m) - P_a G(m_a) - P_b G(m_b) \tag{28}$$

The optimal feature j^* and the optimal splitting point n^* that produce the largest decrease in the Gini impurity corresponds to

$$n^*, j^* = \arg\max \delta G(m,n) \tag{29}$$

The flowchart for building an RF is shown in Figure 3.

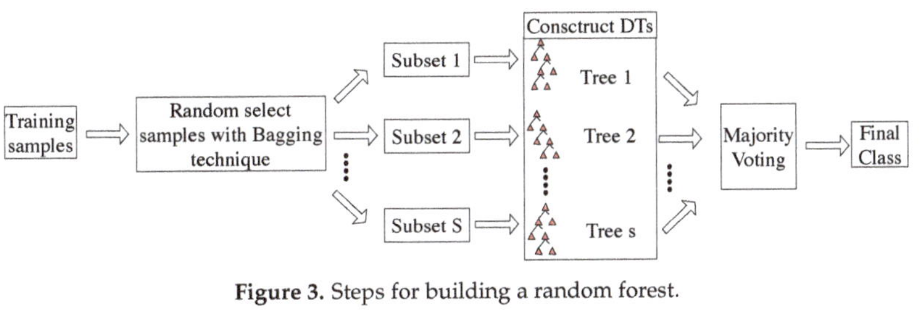

Figure 3. Steps for building a random forest.

2.4. Data Generation for Fault Classification

Based on the feature extraction process that uses Wavelet Packet Transform, illustrated in Section 2.1, the GAN network described in Section 2.2 and a random forest classifier detailed in the previous section, one can setup the learning scheme for the manipulator fault diagnoser. As shown in Figure 4, a real data observation of from each class is sent to the WPT to extract the vector of features (12). Meanwhile, a random signal z is input into the generator that will produce a vector of synthetic features $G(z)$.

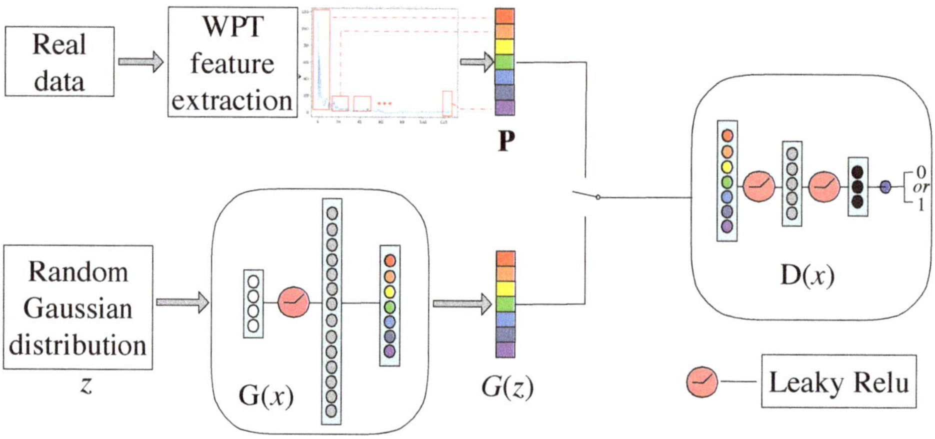

Figure 4. The learning scheme of the fault diagnoser.

The goal of the discriminator $D(x)$ is to distinguish between the real vector of features, outputting a 1, and the synthetic vector of features, outputting a 0. The learning process described in Section 2.2 is applied. Once both the generator and discriminator are trained, the generator can be used to generate as many synthetic data as required. Notice that the learning process and subsequent synthetic generation are carried out for each faulty class, i.e., for each class, we need to increase the number of observations. Finally, (health and faulty) real data together with faulty generated data are used in the random forest classifier for fault classification. Based on the above description, the procedure can be outlined in the following flow chart (Figure 5).

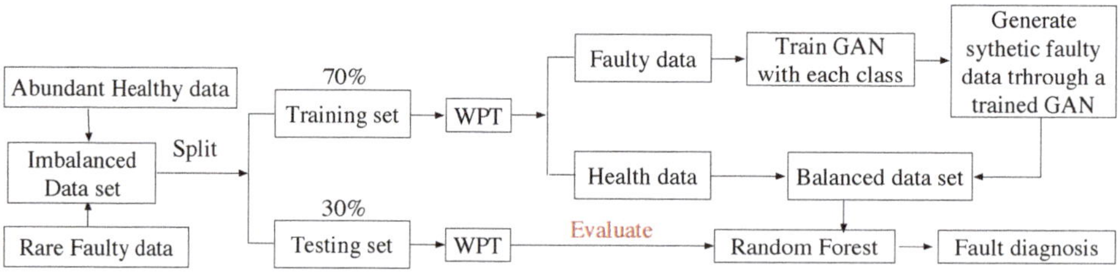

Figure 5. The flow chart of the procedure of the proposed approach.

In addition, this whole process is illustrated in Figure 6.

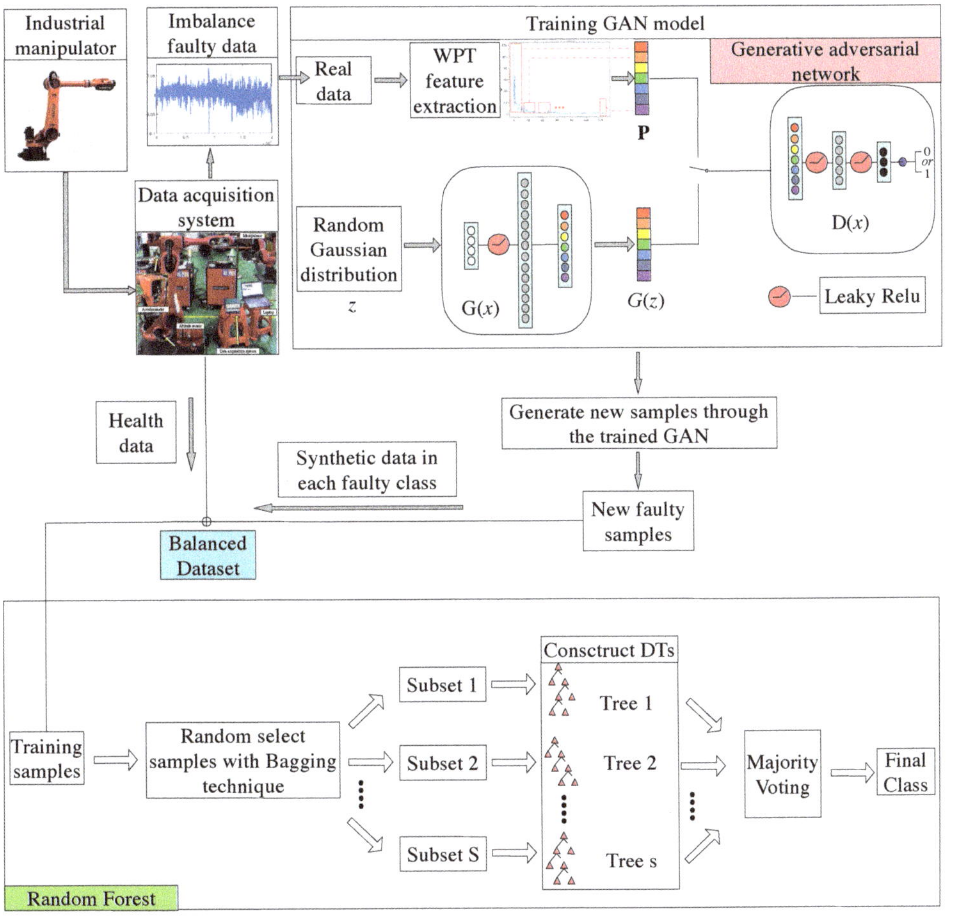

Figure 6. The complete data pipeline for fault diagnosis of the manipulator.

3. Experiment

3.1. Experimental Test-Rig

Experiments were carried out in the gear shaft system of a Brtirus1510A 6 degrees of freedom industrial manipulator. The gear system, which is the main driver component of the manipulator, consists of two planetary gears and two sun gears. The objective is to monitor the health state of gears by measuring the vibrational signals using an PCB 622B01 accelerometer. The accelerometer was installed on the basis of the sixth axis of the manipulator. See Figure 7 for its exact location. Cracking, pitting, and broken tooth are the main gear fault types in manipulators. Table 1 shows the fault type we are interested in. Figure 8 shows an example of each one of the four types of faults.

The robot is moved by the motors, and the teaching box gives the instructions to the robot to start its next movement. At the beginning of the process, the robot is in its original position of 0 degree. Firstly, it will start back and forth movement from −115 degrees to 140 degrees of the limit range point in the first axis. Secondly, the same movement and the same limit range which is from −50 degrees to 35 degrees. Thirdly, the robot will move from −60 degrees to 90 degrees. Fourthly, the same configuration of movement is from −180 degrees to 180 degrees. Fifthly, the movement range will be decreased such that the range is from −90 degrees to 90 degrees. Finally, the robot will move from −180 degrees to 180 degrees and stop in the original place. This series of dynamic movement is only one experiment process. In the next step, we replace the faulty part in Table 1 to restart the above movement for the next experiment. Finally, the signal in each channel is collected by the NI acquisition system which is an analog-to-digital conversion system that the digital samples are collected with an interface on the laptop.

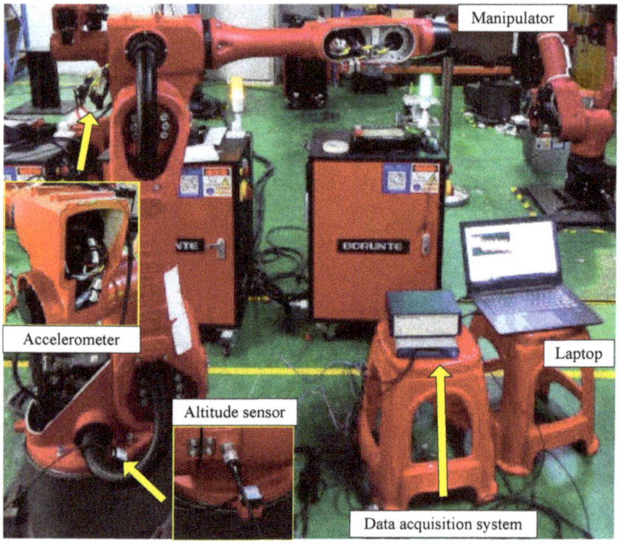

Figure 7. Monitoring the health state of the gears using vibrational signals acquired by an accelerometer located in the basis of the sixth axis of the manipulator.

Table 1. Different fault patterns in the industrial manipulator.

Fault Id	Part	Fault Type
A	None	Healthy
B	Sun gear 1	Pitting
C	Sun gear 1	Broken tooth
D	Planetary gear 1	Cracking

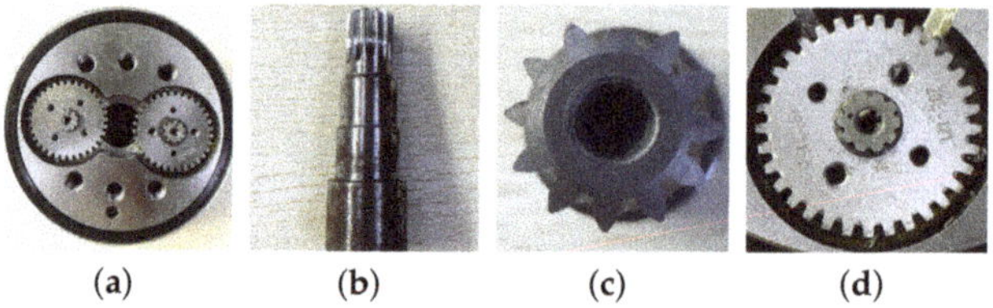

Figure 8. Examples of each one of the 4 monitoring conditions: (**a**) Healthy state; (**b**) Pitting in Sun gear 1; (**c**) Broken tooth in Sun gear 1; and (**d**) Cracking in Planetary gear 2.

3.2. Collected Vibration Signals

As mentioned above, the experimental measurements include four fault types shown in Table 1. The sampling rate is 100 kHz. The duration of each measurement is 20 s. The sampling interval was set to 0.2 s. Thus, 20,000 observations were obtained in each fault type, and 20 k points are chosen for each observations. Therefore, a dataset of 80,000 $\times$ 20 k can be acquired during the experiment. Figure 9 shows an example of a vibration signal acquired in each one of the fault types.

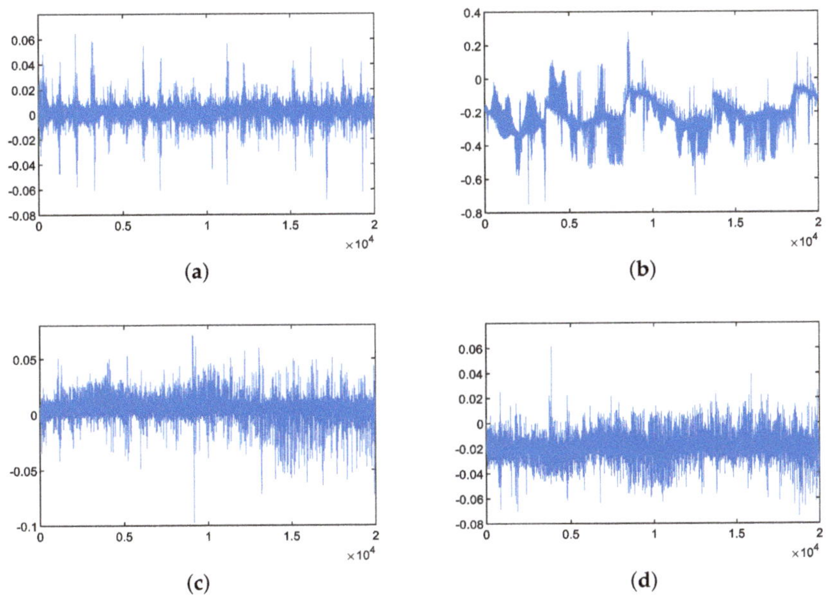

Figure 9. Vibration signals acquired in the following fault conditions: (**a**) Healthy state; (**b**) Pitting in Sun gear 1; (**c**) Broken tooth in Sun gear 1; and (**d**) Cracking in Planetary gear 2.

For hold-out validation, the data set was divided into two disjoint subsets, the training and the test (sub)sets. The training set has 70% of data while the test set has the remaining 30%. Table 2 shows the number of observations for each one of these sets.

Table 2. Number of observations in the training and test sets used for hold out validation. The training set is clearly unbalanced.

Fault Id	Training Set	Test Set
A	14,000	6000
B	140	6000
C	140	6000
D	140	6000

4. Results and Discussion

4.1. On Different Scenarios for Dealing with the Unbalanced Data Set

Several scenarios were considered to assess the effectiveness of the proposed model. In all these scenarios, we all used an RF with 1000 trees for classification. The different scenarios identified as follows: RF-i denotes a random forest trained with the unbalanced dataset described by Table 2; RF-b2 denotes a random forest trained with a subsampled balanced dataset with only 140 observations per condition; RF-GAN and RF-GAN2 stands rand forests trained with data sets that have the real 14,000 healthy observations and 14,000 faulty observations, the only difference between these scenarios is that RF-GAN uses the technique described in Section 2.2 to select the best current model for generating samples while RF-GAN2 uses the model obtained in the last iteration (i.e., iteration 20,000) of training process to generate the samples. RF-GAN1 is similar to RF-GAN with the difference that only 13,860 synthetic faulty states were generated while the remaining 140 are the original faulty states presented in the training data set. For comparison purposes, we also considered a random forest trained with a data set previously processed by SMOTE, a popular oversampling method for dealing with unbalanced data sets.

In all GAN based models, the generators are multi-layer-perceptrons with a 64:1014:128 fully connected topology while discriminators are also multi-layer-perceptrons with 128:1024:2048:1 fully connected topology. These were selected empirically after some preliminary tests. The Adam optimizer is used with its key parameter settings of $\beta_1 = 0.9$, $\beta_2 = 0.999$ and $\epsilon = 1 \times 10^{-8}$. The learning rate for the generators is set to 1×10^{-5}, while, for the discriminators, is set to 1×10^{-4}. The α and λ are set to 1×10^{-4} and 1.0, respectively. A maximum number of 10,000 iterations was set for training.

Figure 10 shows the distribution of the obtained accuracy for each scenario using boxplots. A boxplot summarizes a data distribution stressing five of its characteristic values: minimum, lower quartile, median, upper quartile, and maximum value. The red line denotes the median value. The distribution pertains to 20 independent repetitions.

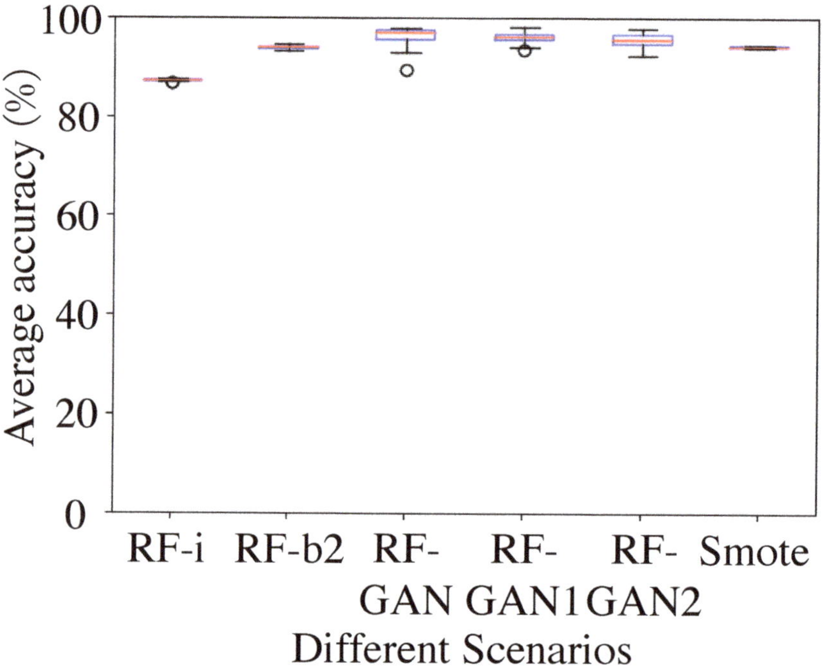

Figure 10. Boxplots exhibiting the relative distributions of accuracy obtained with the different scenarios considered for fault classification. See text for details.

The results presented in Figure 10 were analyzed by the Friedman test, a non parametric statistic test of hypotheses to evaluate whether or not there is a statistically significant difference between the results (boxplots) of the different scenarios. The Friedman null hypothesis is that there is no statistically significant difference between the results of the different scenarios. Given a significant level, α, this hypothesis cannot be rejected whenever the p_{Friedman}, the p-value generated by the test, satisfies $p_{\text{Friedman}} > \alpha$. The null hypothesis is rejected otherwise, meaning that there is a statistically significant difference between the analyzed scenarios. In such a case, we can detect which of the scenario is responsible for such difference resorting to a pairwise *posthoc* test. A ranking can be obtained by counting the number of times that a method was a winner in the pairwise comparison. See [46,47] for further details. Here, we are using the usual $\alpha = 0.05$ and the Wilcoxon test as *posthoc*.

Applying Friedman to the results in Figure 10 yielded $p_{\text{Friedman}} = 1.865 \times 10^{-19} < 0.05$, meaning there is a statistically significant difference between the six scenarios. Table 3 shows the subsequent *posthoc* results. From these, one should conclude that there is no statistically significant difference between scenarios RF-GAN and RF-GAN1 and that these outperform all the others. This is an interesting observation as both RF-GAN and RF-GAN1 use the technique described in Section 2.2 to select the best current generator and that the only difference between these scenarios is that in RF-GAN all the faulty state data are synthetic while in RF-GAN1 only 13,860 faulty examples are synthetic while the remaining 140 are the original faulty states presented in the training data set. This further endorses the quality of the obtained generators.

Table 3. Wilcoxon *posthoc* pairwise tests for the different scenarios.

Pair	*p*-Value	Winner
RF-i vs. RF-b2	8.857×10^{-5}	RF-b2
RF-i vs. RF-GAN	8.857×10^{-5}	RF-GAN
RF-i vs. RF-GAN1	8.857×10^{-5}	RF-GAN1
RF-i vs. RF-GAN2	8.857×10^{-5}	RF-GAN2
RF-i vs. SMOTE	8.857×10^{-5}	SMOTE
RF-b2 vs. RF-GAN	8.857×10^{-5}	RF-GAN
RF-b2 vs. RF-GAN1	8.857×10^{-5}	RF-GAN1
RF-b2 vs. RF-GAN2	8.857×10^{-5}	RF-GAN2
RF-b2 vs. SMOTE	1.204×10^{-5}	SMOTE
RF-GAN vs. RF-GAN1	0.079	–
RF-GAN vs. RF-GAN2	8.857×10^{-5}	RF-GAN
RF-GAN vs. SMOTE	8.857×10^{-5}	RF-GAN
RF-GAN1 vs. RF-GAN2	8.845×10^{-5}	RF-GAN1
RF-GAN1 vs. SMOTE	8.844×10^{-5}	RF-GAN1
		RF-GAN/RF-GAN1

The average accuracy of RF-i was 87.88% while, with an undersampling balanced dataset, RF-b2 reached 94.5%. RF-GAN had an average accuracy of 97.06%, RF-GAN1 97.75%, and RF-GAN2 95.38%. SMOTE had an averaged accuracy of 95.17% . We observe that the GAN based average accuracies are all higher than the unbalanced (RF-i), undersampling (RF-b2), and oversampling (SMOTE) scenarios. Within the GAN based scenarios, RF-GAN has shown a difference of 1.68% relatively to RF-GAN2 in terms of average accuracy.

Curiously enough, we observed no advantage on the application of (24). For the moment, we keep in mind that (26) is not the only possibility to used the Vapnik-loss in a GAN and that other forms are currently being studied.

4.2. On of the Performance in Each Class

To further analyze the above results, the recall indicator of each fault class is studied. As it is shown in Figure 11, the recall indicator in the RF-i model of the health state reaches 100% while, for the other three faulty classes, comes down to 63.88% in gear pitting; 84.3% in gear broken tooth, and 63.35% in gear cracking for an average of 77.88%. This clearly shows the effect of the unbalanced data set. For RF-GAN, the recall indicator in each class is 99.53%, 99.73%, 99.63%, and 99.87%, respectively. In RF-GAN1, the recall indicator in each class is 98.63%, 98.67%, 98.28%, and 95.40%, respectively. For the RF-GAN2 model, the recall indicator in each class is 99.30%, 93.08%, 98.15%, and 90.08%, respectively. The high recall indicators are due to the existence of sufficient examples in each class. This is also visible in the recall of SMOTE.

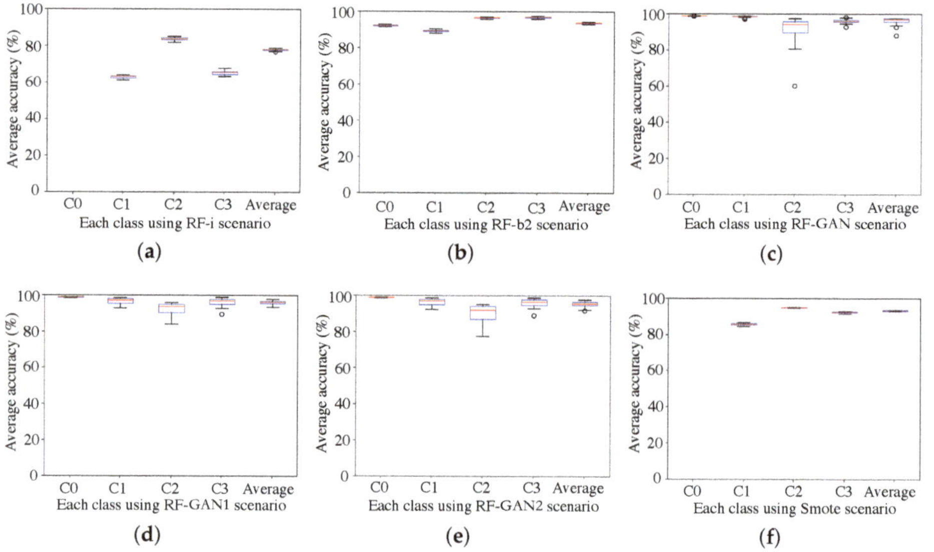

Figure 11. Recall indicators for the different scenarios: (**a**) RF-i; and (**b**) RF-b2; (**c**) RF-GAN; (**d**) RF-GAN1; (**e**) RF-GAN2; and (**f**) SMOTE.

The fit score is another metric that can be used to compare the relative performance of the different scenarios in each class. From Figure 12, one can see that the performance of RF-i as measured by the F1-score is 70.38% in fault A, 77.54% in fault B, 90.03% in fault C, and 76.49% in fault D. The performance of RF-b2 in each faulty condition is 91.05% in fault A, 92.25% in fault B, 96.50% in fault C and 94.3% in fault D, respectively. The performance of RF-GAN is 98.52% in fault A, 98.83% in fault B, 95.66% in fault C and 95.23 in fault D. The performance of RF-GAN1 is 98.55% in fault A, 98.95% in fault B, 96.87% in fault C and 96.62 in fault D. The performance of RF-GAN2 is 95.70% in fault A, 96.54% in fault B, 90.01% in fault C, and 94.57% in fault D. For SMOTE, the performance is 89.72% in fault A, 93.08% in fault B, 97.56% in fault C, and 95.28% in fault D.

For completeness, the confusion matrices are presented in Figure 13. All these matrices consider the 6000 test observations as per Table 2. For RF-i (Figure 13a), one can see a high number of misclassifications in non-healthy states due to the unbalanced training set. These misclassifications are strongly reduced (especially in the GAN-based scenarios) when enough data are generated and used for training.

4.3. *On the Training Set*

4.3.1. Learning Curves

The performance of the proposed model under different data set sizes is now considered. Figure 14 shows the average performance over 20 independent runs in the testing set for the scenario RF-GAN when only a given percentage of faulty observations are available for training. More concretely, the following percentages were considered i =1, 2, 4, 6, 8, 10, 20, 60, 80, and 100 %. For instance, when $i = 4\%$, the number of training examples in each faulty state is $14,000 \times 0.04 = 560$.

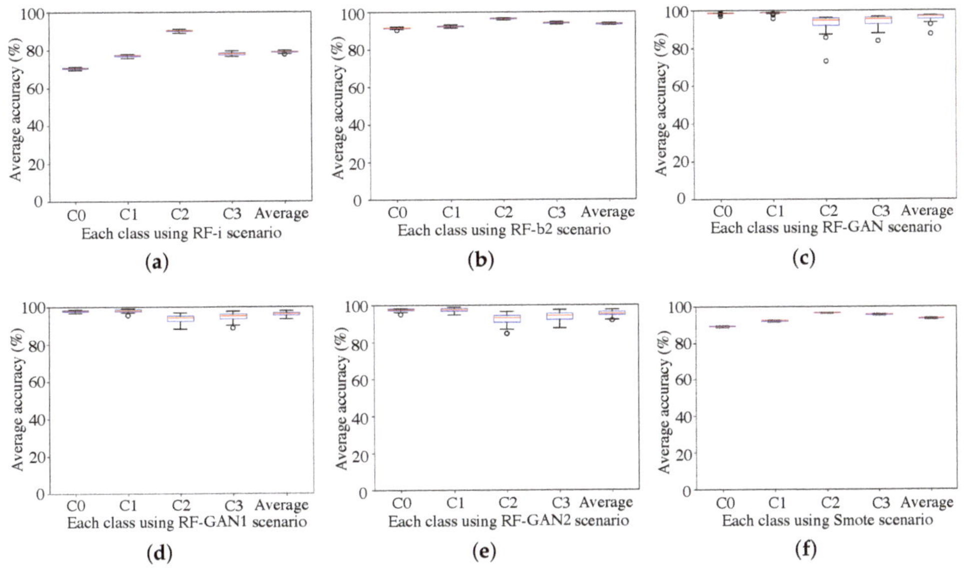

Figure 12. F1-score for scenario: (**a**) RF-i; and (**b**) RF-b2; (**c**) RF-GAN; (**d**) RF-GAN1; (**e**) RF-GAN2; and (**f**) SMOTE.

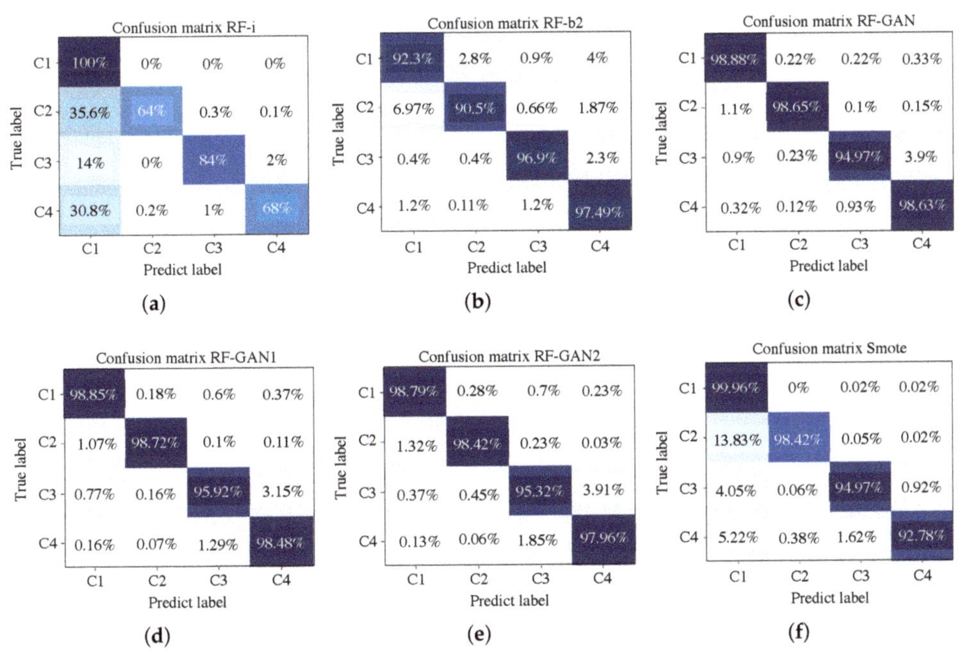

Figure 13. The confusion matrix for: scenario: (**a**) RF-i; and (**b**) RF-b2; (**c**) RF-GAN; (**d**) RF-GAN1; (**e**) RF-GAN2; and (**f**) SMOTE.

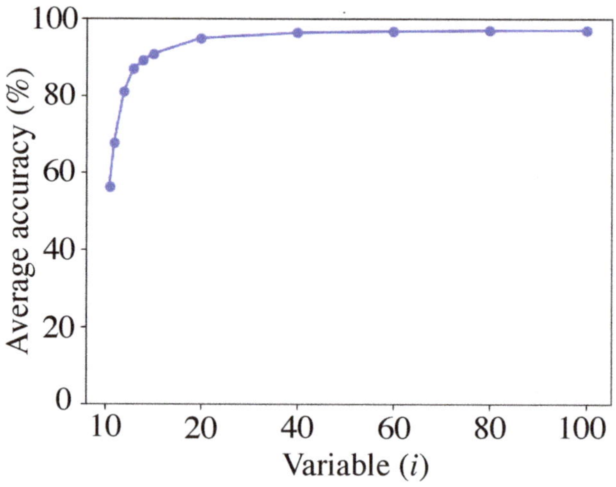

Figure 14. Learning curve of scenario RF-GAN for i =1, 2, 4, 6, 8, 10, 20, 60, 80, and 100% of the training set.

It can be seen that the average accuracy increased from 56.25% to 97.05% by increasing the availability of faulty data from 1% (140 examples) to 100% (14,000 examples) in each fault type. There was a strong increased in performance up to 20%; after that point, the improvement in accuracy was slower and slower until about 80%. After this value, the improvement was neglectable. That is, adding more data after a certain point hardly improves the performance.

4.3.2. Shuffling Data

When generating training examples from a GAN based model, shuffling the training data are a key important factor for obtaining an acceptable performance. Figure 15 illustrates the importance of data shuffling. The results presented in this figure were obtained with exactly the same RF-GAN configuration, the only difference being the way data are presented to the GAN for training. For non shuffled data, the model is simply not able to generate no matter the other initial conditions (weights).

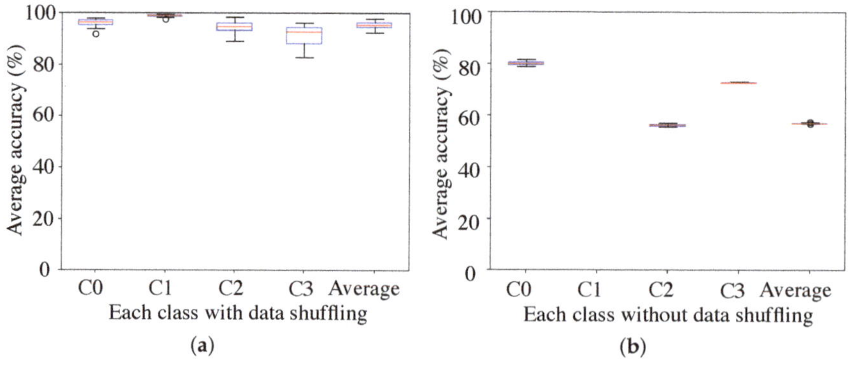

Figure 15. The effect of shuffling input data for training a GAN based model: (**a**) with shuffling and (**b**) without shuffling.

4.4. On the Distribution Used for Sampling Random Inputs

In a GAN based model, the z signal presented to the generator (recall Figure 4) can be drawn from any distribution. However, for this particular application, some distributions are better than others for the training process. Figure 16 shows the classification results for RF-GAN when (a) z is sampled from the standard normal distribution (0 mean and variance 1) and (b) a uniform distribution in $[-1,1]$. Undoubtedly, the former outperforms the latter.

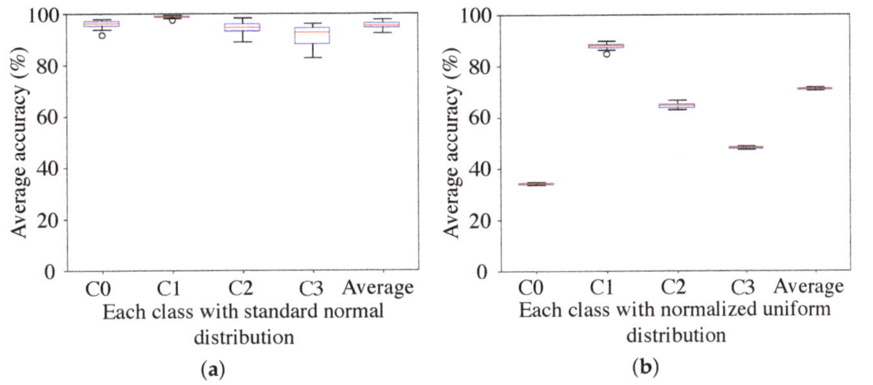

Figure 16. The effect of the distribution used for sampling the input z of the GAN generator: (**a**) standard normal distribution and (**b**) normalized uniform distribution.

4.5. On the Initial Conditions

GAN is trained using a gradient based method that is sensitive to initial conditions (weights). Figure 17 illustrates the impact of the initial conditions on the fault classification results. As shown in that figure, RF-GAN was able to produce an acceptable accuracy for any of the initial set of weights used. However, and as expected for local optimization methods, in 30 independent runs (initializations), it was possible to identify a particular set of initial random weights that outperformed all the others.

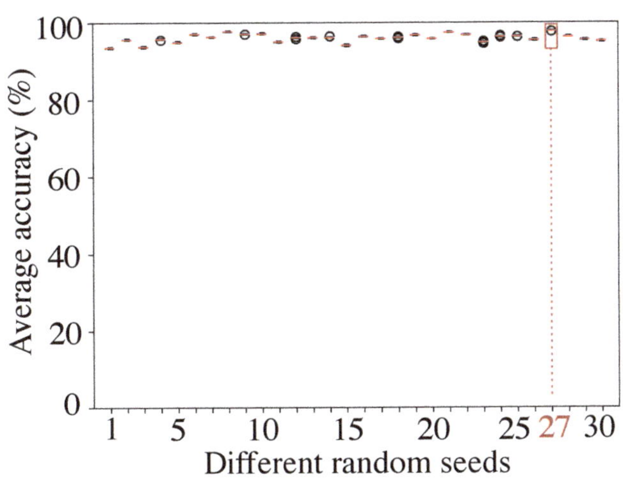

Figure 17. The effect of initial weights (generated from different random generator seeds) on the classification accuracy of RF-GAN.

5. Conclusions

Robotic manipulators are wildly used in the industry and their maintenance and monitoring systems are resorting more and more to data intensive machine learning methods. Methods such as multilayer perceptrons, convolutional neural network, echo state networks, or deep Boltzmann machines have all been used for such endeavors. However, all of these methods rely on a representative, balanced, and large enough training set, which, due to the very nature of (some) faults, is very hard to collect from the equipment. Motivated by the recent success of generative adversarial network (GAN), in this work, we have exploited, for the first time, this type of generative model as an oversampling method for fault classification in an industrial robotic manipulator.

A comprehensive empirical analysis was performed taking into account six different scenarios for mitigating the unbalanced data, including classical under and oversampling (SMOTE) methods. In all of these, a wavelet packet transform combined with GAN is used for feature generation while a random forest is used for fault classification. Studies were also conducted for assessing the sensibility of aspects such as generator selection, the number of training examples in each class, training data shuffling, the distribution used for sampling input random data, and initial conditions.

The main conclusion is that it was possible to increase the performance of the fault diagnoser for an industrial robotic manipulator for any of GAN based models over classical undersampling and oversampling (SMOTE) methods. This is accomplished at the expense of a much higher design and computational effort. Training a GAN is not an easy task due to the model collapse and other factors, and it is certainly a quite time-consuming process. After training a GAN for each fault, one will have a set of generators able to produce as much synthetic data as required in an efficient way though.

Within the GAN based models, those that keep track of the best current generator during training yielded the best results. No statistically significant difference was observed between the scenario that uses exclusively synthetic data for the faulty states and the scenario that uses the available real data for such states. This is yet another piece of evidence on the quality of the obtained GAN generators.

In many cases like prognostics and health management (PHM), enough data can effectively improve the fault monitoring capability of the industrial system. However, it can not completely be executed due to a lack of faulty data. GAN is an efficient tool to get rid of the limitation of data imbalance state, which can enhance the monitoring capability in PHM. Therefore, this approach can provide a data background for PHM. However, this approach still has limitations, one is that this model can only learn the data distribution from a limited faulty data source while there are many kinds of faulty data in the industrial system. For the new faulty data, they need to be sent to this model to learn the new distribution again to generate enough data that will bring time cost for training. Another one is that this approach is trained by sending one single faulty class as an input to GAN in order to obtain a generator. This means we need to train several GAN models for several faulty classes, which is computationally demanding and time consuming.

Author Contributions: Writing—original draft preparation, and investigation, Z.P.; methodology, D.C.; writing—review and supervision, J.V.d.O.; funding acquisition, C.L.; investigation, R.-V.S.; investigation, M.C. All authors have read and agreed to the published version of the manuscript.

Funding: This work is financed by national funds through FCT—Foundation for Science and Technology, I.P., through IDMEC, under LAETA, project UIDB/50022/2020. This work was supported in part by the National Natural Science Foundation of China under Grant 51775112, the Natural Science Foundation of Chongqing cstc2019jcyj-zdxmX003, the CTBU Project under Grant KFJJ2019060.

Acknowledgments: The authors are thankful for the finance support from FCT—Foundation for Science and Technology, I.P., through IDMEC, under LAETA and in part to the National Natural Science Foundation of China.

Conflicts of Interest: The authors declare no conflict of interest.

References

1. Jin, X.; Que, Z.; Sun, Y.; Guo, Y.; Qiao, W. A data-driven approach for bearing fault prognostics. *IEEE Trans. Ind. Appl.* **2019**, *55*, 3394–3401. [CrossRef]
2. Schmidt, S.; Heyns, P.S.; Gryllias, K.C. A discrepancy analysis methodology for rolling element bearing diagnostics under variable speed conditions. *Mech. Syst. Signal Process.* **2019**, *116*, 40–61. [CrossRef]
3. Zaidan, M.A.; Harrison, R.F.; Mills, A.R.; Fleming, P.J. Bayesian hierarchical models for aerospace gas turbine engine prognostics. *Expert Syst. Appl.* **2015**, *42*, 539–553. [CrossRef]
4. He, M.; He, D. Deep learning based approach for bearing fault diagnosis. *IEEE Trans. Ind. Appl.* **2017**, *53*, 3057–3065. [CrossRef]
5. Liu, H.; Liu, C.; Huang, Y. Adaptive feature extraction using sparse coding for machinery fault diagnosis. *Mech. Syst. Signal Process.* **2011**, *25*, 558–574. [CrossRef]
6. Wu, C.; Jiang, P.; Ding, C.; Feng, F.; Chen, T. Intelligent fault diagnosis of rotating machinery based on one-dimensional convolutional neural network. *Comput. Ind.* **2019**, *108*, 53–61. [CrossRef]
7. Zhang, Y.; Li, X.; Gao, L.; Chen, W.; Li, P. Intelligent fault diagnosis of rotating machinery using a new ensemble deep auto-encoder method. *Measurement* **2020**, *151*, 107232. [CrossRef]
8. Shen, C.; Xie, J.; Wang, D.; Jiang, X.; Shi, J.; Zhu, Z. Improved hierarchical adaptive deep belief network for bearing fault diagnosis. *Appl. Sci.* **2019**, *9*, 3374. [CrossRef]
9. Deng, S.; Cheng, Z.; Li, C.; Yao, X.; Chen, Z.; Sanchez, R.V. Rolling bearing fault diagnosis based on Deep Boltzmann machines. In Proceedings of the 2016 Prognostics and System Health Management Conference (PHM-Chengdu), Chengdu, China, 19–21 October 2016; pp. 1–6.
10. Cho, C.N.; Hong, J.T.; Kim, H.J. Neural network based adaptive actuator fault detection algorithm for robot manipulators. *J. Intell. Robot. Syst.* **2019**, *95*, 137–147. [CrossRef]
11. Wang, H.; Li, S.; Song, L.; Cui, L. A novel convolutional neural network based fault recognition method via image fusion of multi-vibration-signals. *Comput. Ind.* **2019**, *105*, 182–190. [CrossRef]
12. Ma, Q.; Chen, E.; Lin, Z.; Yan, J.; Yu, Z.; Ng, W.W. Convolutional Multitimescale Echo State Network. *IEEE Trans. Cybern.* **2019**. [CrossRef] [PubMed]
13. Long, J.; Zhang, S.; Li, C. Evolving deep echo state networks for intelligent fault diagnosis. *IEEE Trans. Ind. Inform.* **2019**, *16*, 4928–4937. [CrossRef]
14. Hu, G.; Li, H.; Xia, Y.; Luo, L. A deep Boltzmann machine and multi-grained scanning forest ensemble collaborative method and its application to industrial fault diagnosis. *Comput. Ind.* **2018**, *100*, 287–296. [CrossRef]
15. Wang, J.; Wang, K.; Wang, Y.; Huang, Z.; Xue, R. Deep Boltzmann machine based condition prediction for smart manufacturing. *J. Ambient Intell. Humaniz. Comput.* **2019**, *10*, 851–861. [CrossRef]
16. Lee, K.P.; Wu, B.H.; Peng, S.L. Deep-learning-based fault detection and diagnosis of air-handling units. *Build. Environ.* **2019**, *157*, 24–33. [CrossRef]
17. Shao, H.; Jiang, H.; Li, X.; Liang, T. Rolling bearing fault detection using continuous deep belief network with locally linear embedding. *Comput. Ind.* **2018**, *96*, 27–39. [CrossRef]
18. Polic, M.; Krajacic, I.; Lepora, N.; Orsag, M. Convolutional autoencoder for feature extraction in tactile sensing. *IEEE Robot. Autom. Lett.* **2019**, *4*, 3671–3678. [CrossRef]
19. D'Elia, G.; Mucchi, E.; Cocconcelli, M. On the identification of the angular position of gears for the diagnostics of planetary gearboxes. *Mech. Syst. Signal Process.* **2017**, *83*, 305–320. [CrossRef]
20. Zaidan, M.A.; Mills, A.R.; Harrison, R.F.; Fleming, P.J. Gas turbine engine prognostics using Bayesian hierarchical models: A variational approach. *Mech. Syst. Signal Process.* **2016**, *70*, 120–140. [CrossRef]
21. Ali, J.B.; Fnaiech, N.; Saidi, L.; Chebel-Morello, B.; Fnaiech, F. Application of empirical mode decomposition and artificial neural network for automatic bearing fault diagnosis based on vibration signals. *Appl. Acoust.* **2015**, *89*, 16–27.
22. Yan, K.; Ji, Z.; Lu, H.; Huang, J.; Shen, W.; Xue, Y. Fast and accurate classification of time series data using extended ELM: Application in fault diagnosis of air handling units. *IEEE Trans. Syst. Man, Cybern. Syst.* **2017**, *49*, 1349–1356. [CrossRef]
23. Iqbal, J.; Islam, R.U.; Abbas, S.Z.; Khan, A.A.; Ajwad, S.A. Automating industrial tasks through mechatronic systems—A review of robotics in industrial perspective. *Teh. Vjesn.* **2016**, *23*, 917–924.

24. Caccavale, F.; Cilibrizzi, P.; Pierri, F.; Villani, L. Actuators fault diagnosis for robot manipulators with uncertain model. *Control Eng. Pract.* **2009**, *17*, 146–157. [CrossRef]
25. Chawla, N.V.; Bowyer, K.W.; Hall, L.O.; Kegelmeyer, W.P. SMOTE: Synthetic minority over-sampling technique. *J. Artif. Intell. Res.* **2002**, *16*, 321–357. [CrossRef]
26. Goodfellow, I.; Pouget-Abadie, J.; Mirza, M.; Xu, B.; Warde-Farley, D.; Ozair, S.; Courville, A.; Bengio, Y. Generative adversarial nets. In Proceedings of the Advances in Neural Information Processing Systems 27, Montreal, QC, Canada, 8-13 December 2014; pp. 2672–2680.
27. Ledig, C.; Theis, L.; Huszár, F.; Caballero, J.; Cunningham, A.; Acosta, A.; Aitken, A.; Tejani, A.; Totz, J.; Wang, Z.; et al. Photo-realistic single image super-resolution using a generative adversarial network. In Proceedings of the IEEE Conference on Computer Vision and Pattern Recognition, Honolulu, HI, USA, 21–26 July 2017; pp. 4681–4690.
28. Isola, P.; Zhu, J.Y.; Zhou, T.; Efros, A.A. Image-to-image translation with conditional adversarial networks. In Proceedings of the IEEE Conference on Computer Vision and Pattern Recognition, Honolulu, HI, USA, 21–26 July 2017; pp. 1125–1134.
29. Zhu, J.Y.; Park, T.; Isola, P.; Efros, A.A. Unpaired image-to-image translation using cycle-consistent adversarial networks. In Proceedings of the IEEE International Conference on Computer Vision, Venice, Italy, 22–29 October 2017; pp. 2223–2232.
30. Shao, S.; Wang, P.; Yan, R. Generative adversarial networks for data augmentation in machine fault diagnosis. *Comput. Ind.* **2019**, *106*, 85–93. [CrossRef]
31. Li, C.; Cabrera, D.; Sancho, F.; Sánchez, R.V.; Cerrada, M.; Long, J.; de Oliveira, J.V. Fusing convolutional generative adversarial encoders for 3D printer fault detection with only normal condition signals. *Mech. Syst. Signal Process.* **2020**, *147*, 107108. [CrossRef]
32. Mao, W.; Liu, Y.; Ding, L.; Li, Y. Imbalanced fault diagnosis of rolling bearing based on generative adversarial network: A comparative study. *IEEE Access* **2019**, *7*, 9515–9530. [CrossRef]
33. Jiang, W.; Hong, Y.; Zhou, B.; He, X.; Cheng, C. A GAN-based anomaly detection approach for imbalanced industrial time series. *IEEE Access* **2019**, *7*, 143608–143619. [CrossRef]
34. Li, C.; Cabrera, D.; Sancho, F.; Sánchez, R.V.; Cerrada, M.; de Oliveira, J.V. One-shot fault diagnosis of 3D printers through improved feature space learning. *IEEE Trans. Ind. Electron.* **2020**, *147*, 107108.
35. Wang, Y.R.; Sun, G.D.; Jin, Q. Imbalanced sample fault diagnosis of rotating machinery using conditional variational auto-encoder generative adversarial network. *Appl. Soft Comput.* **2020**, *92*, 106333. [CrossRef]
36. Gokhale, M.; Khanduja, D.K. Time domain signal analysis using wavelet packet decomposition approach. *Int. J. Commun. Netw. Syst. Sci.* **2010**, *3*, 321. [CrossRef]
37. Bruce, L.M.; Koger, C.H.; Li, J. Dimensionality reduction of hyperspectral data using discrete wavelet transform feature extraction. *IEEE Trans. Geosci. Remote Sens.* **2002**, *40*, 2331–2338. [CrossRef]
38. Li, C.; Sanchez, R.V.; Zurita, G.; Cerrada, M.; Cabrera, D.; Vásquez, R.E. Gearbox fault diagnosis based on deep random forest fusion of acoustic and vibratory signals. *Mech. Syst. Signal Process.* **2016**, *76*, 283–293. [CrossRef]
39. Arjovsky, M.; Chintala, S.; Bottou, L. Wasserstein gan. *arXiv* **2017**, arXiv:1701.07875.
40. Gulrajani, I.; Ahmed, F.; Arjovsky, M.; Dumoulin, V.; Courville, A.C. Improved training of wasserstein gans. In Proceedings of the Advances in Neural Information Processing Systems 30, Long Beach, CA, USA, 4–9 December 2017; pp. 5767–5777.
41. Cabrera, D.; Sancho, F.; Long, J.; Sánchez, R.V.; Zhang, S.; Cerrada, M.; Li, C. Generative adversarial networks selection approach for extremely imbalanced fault diagnosis of reciprocating machinery. *IEEE Access* **2019**, *7*, 70643–70653. [CrossRef]
42. Vapnik, V.; Izmailov, R. Rethinking statistical learning theory: Learning using statistical invariants. *Mach. Learn.* **2019**, *108*, 381–423. [CrossRef]
43. Smola, A.J.; Schölkopf, B. A tutorial on support vector regression. *Stat. Comput.* **2004**, *14*, 199–222. [CrossRef]
44. Friedl, M.A.; Brodley, C.E. Decision tree classification of land cover from remotely sensed data. *Remote Sens. Environ.* **1997**, *61*, 399–409. [CrossRef]

45. Liaw, A.; Wiener, M. Classification and regression by randomForest. *R News* **2002**, *2*, 18–22.
46. Demsar, J. Statistical Comparisons of Classifiers over Multiple Data Sets. *J. Mach. Learn. Res.* **2006**, *7*, 1–30.
47. Pacheco, F.; Valente de Oliveira, J.; Sánchez, R.V.; Cerrada, M.; Cabrera, D.; Li, C.; Zurita, G.; Artés, M. A statistical comparison of neuroclassifiers and feature selection methods for gearbox fault diagnosis under realistic conditions. *Neurocomputing* **2016**, *194*, 192–206. [CrossRef]

MDPI
St. Alban-Anlage 66
4052 Basel
Switzerland
Tel. +41 61 683 77 34
Fax +41 61 302 89 18
www.mdpi.com

Applied Sciences Editorial Office
E-mail: applsci@mdpi.com
www.mdpi.com/journal/applsci

www.ingramcontent.com/pod-product-compliance
Lightning Source LLC
Chambersburg PA
CBHW061450210926
51850CB00008B/1668
* 9 7 8 3 0 3 6 5 5 1 0 9 8 *